Numerical Solution
of Ordinary Differential Equations

This is Volume 74 in
MATHEMATICS IN SCIENCE AND ENGINEERING
A series of monographs and textbooks
Edited by RICHARD BELLMAN, *University of Southern California*

A complete list of the books in this series appears at the end of this volume.

Numerical Solution of Ordinary Differential Equations

LEON LAPIDUS
DEPARTMENT OF CHEMICAL ENGINEERING
PRINCETON UNIVERSITY
PRINCETON, NEW JERSEY

and

JOHN H. SEINFELD
DEPARTMENT OF CHEMICAL ENGINEERING
CALIFORNIA INSTITUTE OF TECHNOLOGY
PASADENA, CALIFORNIA

ACADEMIC PRESS   New York and London   1971

ACADEMIC PRESS, INC.
111 Fifth Avenue, New York, New York 10003

United Kingdom Edition published by
ACADEMIC PRESS, INC. (LONDON) LTD.
Berkeley Square House, London W1X 6BA

LIBRARY OF CONGRESS CATALOG CARD NUMBER: 73-127689
AMS (MOS) 1970 Subject Classification: 65L05

PRINTED IN THE UNITED STATES OF AMERICA

TO
Mary and Jay

Contents

3.   Stability of Multistep and Runge–Kutta Methods

4.   Predictor–Corrector Methods

5. Extrapolation Methods

6. Numerical Integration of Stiff Ordinary Differential Equations

Preface

Most scientists and engineers who have faced the problem of solving numerically a set of nonlinear ordinary differential equations are probably aware of the multitude of techniques available for such a problem. Further pursuit would undoubtedly reveal the large amount of published material on this subject in the pure and applied mathematics—as well as the engineering —literature. The abundant literature on the subject of numerical solution of ordinary differential equations is, on the one hand, a result of the tremendous variety of actual systems in the physical and biological sciences and engineering disciplines that are described by nonlinear ordinary differential equations, and, on the other hand, a result of the fact that the subject is currently quite active.

The existence of a large number of methods, each having special advantages, has been a source of confusion as to what methods are best for certain classes of problems. It is this observation which particularly motivated the writing of this book; thus we have attempted to bring together all the well-known classical methods as well as the most recent techniques in the literature in a consistent and comprehensive framework. This book is intended to contain the most complete treatment of numerical methods for the integration of ordinary differential equations yet published.

In this book we pursue three objectives. We have avoided, on one hand, a completely theoretical treatment and, on the other hand, a collection of computer programs which can be applied in a cookbook manner. Our first objective, then, is to bridge theory and practice, providing sufficient theory to motivate the various methods and algorithms, yet devoting considerable attention to the practical capabilities of the methods. Thus, we hope to show the student why a method is derived in the first place and then how the method performs in actual applications with its advantages and disadvantages.

Our second objective is to give advice. Many people reading this book will be doing so purely because they have a specific problem to solve. At appropriate places in the book we have attempted to synthesize the extensive computational experience gained by the authors and others in order to provide the reader with definite recommendations on which methods to use in which circumstances. Each of Chapters 2–6 contains a short review of key published numerical results together with detailed numerical examples which we have solved. These computations, carried out on a wide variety of problems, provide concrete results upon which to construct useful and important recommendations.

Our third objective is to prepare the reader for the published literature on numerical integration of ordinary differential equations. Problems of current interest in the literature are considered in detail. These include, for example, hybrid methods, extrapolation methods, and the generation of highly stable algorithms.

Chapter 1 provides an introduction to the problem of numerical integration of differential equations and the techniques available for derivation of numerical algorithms. Chapter 2 presents an extensive study of Runge–Kutta and allied single-step methods. The problem of numerical instability is treated in Chapter 3. Chapter 4 includes a comprehensive treatment of multistep and predictor–corrector methods. Chapters 5 and 6 treat extrapolation methods and the numerical integration of stiff ordinary differential equations, respectively, both topics not having appeared in book form previously.

The book is designed for use in a junior, senior, or first-year graduate course in numerical analysis. Little formal background beyond a course in differential equations is required. Every effort has been made to substitute learning by means of specific examples for rigorous generality. The worked examples will provide convenient homework problems to test the students' ability to transform theory into practice.

We are pleased to acknowledge in particular the following people who are among those who have contributed to the numerical results presented: Messrs. M. Hwang, J. A. Bertucci, V. J. Corbo, J. H. Ellsworth, H. S. Kao, R. H. Rossen, E. J. Schlossmacher, and S. D. Weinrich.

**Numerical Solution
of Ordinary Differential Equations**

1

Fundamental Definitions and Equations

It seems clear that many areas of engineering and scientific analysis require methods for solving sets of initial value ordinary differential equations (ODEs). The advent of the digital computer has significantly increased our ability to carry out the numerical solution of such equations. In fact numerous books and papers have been published which deal, at least in part, with both the theory and practice of such solutions.

In the first chapter of this book we present a unified and direct development of many of the equations of interest in the solution of initial value ODEs. Later chapters will illustrate how these equations are used to obtain explicit numerical answers for a variety of engineering and scientific problems.

1.1. THE NUMERICAL PROBLEM AND NOMENCLATURE

We first consider the scalar ODE

$$dy/dx = y'(x) = f(x, y) \qquad (1.1\text{-}1)$$

with initial condition given by

$$y(x_0) = y_0 \qquad (1.1\text{-}2)$$

where $f(x, y)$ indicates any explicit functionality between the dependent variable y and the independent variable x. x_0 represents a specified value of x

1

and y_0 is then the initial value of $y(x)$ at $x = x_0$. By a solution to the system of (1.1-1) and (1.1-2) we mean a curve in the $y(x)$ versus x domain which passes through the point (y_0, x_0) and which satisfies (1.1-1). In a numerical sense we will be satisfied to obtain a discrete set of values of $y(x)$, called $\{y_n\}$, corresponding to a discrete set of x values, called $\{x_n\}$, which are an approximation, in some sense, to the equivalent continuous $y(x)$ versus x curve.

In addition to the first-order scalar equation it is possible to consider a set of simultaneous first-order equations or an equivalent high-order single equation. Thus we may write

$$dy_i/dx = y_i'(x) = f_i(x, y_1, y_2, \ldots, y_m) \qquad (1.1\text{-}3)$$

$$i = 1, 2, \ldots, m$$

$$y_i(x_0) = y_{0i} \qquad (1.1\text{-}4)$$

as representing a set of m simultaneous first-order ODEs. As long as the derivatives $y_1', y_2', \ldots, y_m'$ appear only on the left-hand side of the differential equations, then (1.1-3) is equivalent to one mth-order equation and vice versa. In general it is usually advantageous from a numerical point of view to consider the simultaneous equations rather than to deal with the single high-order equation; as a result we shall not specify the latter equation, although the case $m = 2$, the second-order equation, will be developed later since it represents a case with special properties.

Equations (1.1-3) and (1.1-4) can be written in an alternative form. If we define the vector $\mathbf{y} = [y_1, \ldots, y_m]$ we may write

$$dy/dx = \mathbf{y}' = \mathbf{f}(x, \mathbf{y}) \qquad (1.1\text{-}5)$$

$$\mathbf{y}(x_0) = \mathbf{y}_0 \qquad (1.1\text{-}6)$$

Further, we can remove the explicit dependence on x in the right-hand side by defining a new variable y_{m+1} in the form

$$y_{m+1}' = 1$$
$$y_{m+1}(x_0) = x_0 \qquad (1.1\text{-}7)$$

or by defining

$$\mathbf{z} = \begin{bmatrix} \mathbf{y} \\ y_{m+1} \end{bmatrix}, \qquad \mathbf{F} = \begin{bmatrix} \mathbf{f} \\ 1 \end{bmatrix}, \qquad \mathbf{z}_0 = \begin{bmatrix} \mathbf{y}_0 \\ x_0 \end{bmatrix}$$

there results

$$\mathbf{z}' = \mathbf{F}(\mathbf{z}) \qquad (1.1\text{-}8)$$

$$\mathbf{z}(x_0) = \mathbf{z}_0 \qquad (1.1\text{-}9)$$

Equation (1.1-8) is termed *autonomous* whereas (1.1-5) is called *nonautonomous*.

We shall assume that $f(x, y)$ is defined and continuous in the strip $a \leq x \leq b$, $-\infty < y < \infty$ where a and b are finite and that there exists a Lipschitz constant L such that for any $x \in [a, b]$ and any two numbers u and v,

$$|f(x, u) - f(x, v)| \leq L |u - v| \qquad (1.1\text{-}10)$$

If these two conditions are satisfied, then there exists exactly one function $y(x)$ which satisfies (1.1-1) and (1.1-2) on $[a, b]$, i.e., the initial value problem has a unique solution. In the nonscalar case vector norms replace absolute values in (1.1-10).

The above deals with the continuous system whose solution is $y(x)$ versus x. As we have pointed out, however, we shall develop numerical methods which yield, for a sequence $\{x_n\}$ of abscissas of $x_n > x_0$, a sequence of $\{y_n\}$ (or vectors) which approximate to the (exact) $y(x_n)$ (or vectors) of the desired solutions. To obtain these discrete sequences, we will consider the overall interval $(a = x_0, b)$ with a finite set of points $\{x_n\}$ which form a net, grid, or mesh along the x coordinate. Each point in the sequence will be related to a previous point by the relationship

$$x_{n+1} = x_n + h_n; \qquad n = 0, 1, 2, \ldots, N; \qquad x_0 = a; \quad x_N = b \qquad (1.1\text{-}11)$$

where h_n is the net, grid, or mesh spacing. In many cases $h_n = h = $ constant and the points $x_0, x_1, \ldots, x_N$ are equally spaced. Alternatively we see that

$$x_n = x_0 + nh; \qquad n = 0, 1, 2, \ldots, N; \qquad h_n = h \qquad (1.1\text{-}12)$$

relating x_n to the initial x_0. This may be further generalized to the form

$$x_\alpha = x_0 + \alpha h \qquad (1.1\text{-}13)$$

where α is a parameter; when α is a positive integer, (1.1-13) is equivalent to (1.1-12). However, when α is a fraction, x_α is a point within the discrete set $x_0, x_1, x_2, \ldots$. Equation (1.1-13) may be thought of as defining a new variable α which specifies a point along the x coordinate.

1.2. TAYLOR SERIES EXPANSION

Since we shall later be concerned with the use of the well-known Taylor series expansion of a function, a brief nonrigorous discussion is given at this point. If we assume that $f(x, y)$ is sufficiently differentiable or possesses continuous derivatives of all orders in both x and y, then a Taylor series for $y(x)$ about $x = x_0$ yields

$$y(x_0 + \alpha h) = y_0 + h\alpha y_0' + \frac{h^2 \alpha^2 y_0''}{2!} + \frac{h^3 \alpha^3 y_0'''}{3!} + \cdots \qquad (1.2\text{-}1)$$

for any α for which the series converges. For $\alpha = 1$, this becomes

$$y(x_1) = y_0 + h y_0' + \frac{h^2 y_0''}{2!} + \frac{h^3 y_0'''}{3!} + \cdots$$

or, equivalently,

$$y(x_{n+1}) = y(x_n) + h y'(x_n) + \frac{h^2}{2!} y''(x_n) + \frac{h^3}{3!} y'''(x_n) + \cdots \qquad (1.2\text{-}2)$$

To simplify the writing we adopt the convention

$$\begin{aligned}
y'(x) &= y^{[1]}(x) \\
y''(x) &= y^{[2]}(x) \\
y'''(x) &= y^{[3]}(x) \\
&\;\;\vdots
\end{aligned} \qquad (1.2\text{-}3)$$

Note that (1.2-2) states that given $y(x_n)$ and h we may calculate $y(x_{n+1})$ if we can evaluate $y'(x_n)$, $y''(x_n)$, ... However, since $f(x, y)$ is a function of two variables x and y the evaluation of the derivatives is complicated. In fact, from (1.1-1) using a simplified notation $y' = f$,

$$y'' = f' = f_x + f_y f$$

where f_x and f_y represent the derivatives of f with respect to x and y. In the same sense,

$$y''' = f'' = f_{xx} + 2f f_{xy} + f^2 f_{yy} + f_y[f_x + f_y f]$$

or

$$\begin{aligned}
f^{[0]} &= f \\
f^{[j+1]} &= f_x^{[j]} + f^{[j]} f, \qquad j = 0, 1, 2, \ldots
\end{aligned} \qquad (1.2\text{-}4)$$

Equation (1.2-4) is a recurrence relation for generating successive derivatives for $y(x)$ when $y(x)$ or $f(x, y)$ is a scalar. In the vector situation, equivalent results are obtained.

1.3. ASPECTS OF NUMERICAL INTERPOLATION

To lead into the development of certain equations of major importance in numerically solving ODEs, we shall now define a number of interpolation formulas. In general, these formulas start with a given sequence of the $y(x)$, $\{y_n\}$, at equally spaced $\{x_n\}$ ($h_n = h = $ constant). We specify the linear difference operators [6]:

1. Forward difference operator Δ such that

$$\Delta y_n = y_{n+1} - y_n$$
$$\Delta^2 y_n = \Delta y_{n+1} - \Delta y_n = y_{n+2} - 2y_{n+1} + y_n \qquad (1.3\text{-}1)$$
$$\Delta^3 y_n = \Delta^2 y_{n+1} - \Delta^2 y_n = y_{n+3} - 3y_{n+2} + 3y_{n+1} - y_n$$
$$\vdots$$
$$\Delta^q y_n = \Delta^{q-1} y_{n+1} - \Delta^{q-1} y_n$$

2. Backward difference operator ∇ such that

$$\nabla y_n = y_n - y_{n-1}$$
$$\nabla^2 y_n = \nabla y_n - \nabla y_{n-1} = y_n - 2y_{n-1} + y_{n-2}$$
$$\nabla^3 y_n = \nabla^2 y_n - \nabla^2 y_{n-1} = y_n - 3y_{n-1} + 3y_{n-2} - y_{n-3} \qquad (1.3\text{-}2)$$
$$\vdots$$
$$\nabla^q y_n = \nabla^{q-1} y_n - \nabla^{q-1} y_{n-1}$$

Using these operators it is possible to define the finite interpolation formulas of Newton

$$y_\alpha = y_0 + \alpha\,\Delta y_0 + \frac{\alpha(\alpha - 1)}{2!}\,\Delta^2 y_0 + \cdots$$

$$+ \frac{\alpha(\alpha - 1)\cdots(\alpha - n + 1)}{n!}\,\Delta^n y_0, \qquad \alpha = \frac{x_\alpha - x_0}{h} \qquad (1.3\text{-}3)$$

$$y_{n+\alpha} = y_n + \alpha\nabla y_n + \frac{\alpha(\alpha + 1)}{2!}\,\nabla^2 y_n + \cdots$$

$$+ \frac{\alpha(\alpha + 1)\cdots(\alpha + n - 1)}{n!}\,\nabla^n y_n, \qquad \alpha = \frac{x_\alpha - x_n}{h} \qquad (1.3\text{-}4)$$

These two formulas are termed Newton's forward formula (NFF) and Newton's backward formula (NBF) respectively. Each is obtained by fitting a polynomial to the sequence of points $\{y_n\}$ at equally spaced $\{x_n\}$. (Actually the formulas are polynomials in α because of the change of variable.) Fitting is taken to mean that the polynomial is made exact at the points themselves. Since there are $n + 1$ points $(y_0, y_1, \ldots, y_n)$ this means that if $y(x)$ is a polynomial of degree n or less, the formulas above will be exact and finite in length as shown. When $y(x)$ is not a polynomial or is a polynomial of degree greater than n, the formulas will not be exact (except at the points themselves). In other words an error will be made because a finite rather than an infinite series is used. In general form this error will appear as $T_\alpha = C_\alpha h^{n+1} y^{[n+1]}(\zeta)$ where C is a function of α and $y^{[n+1]}(\zeta)$ is evaluated at some $x_0 < \zeta < x_n$.

Equations (1.3-3) and (1.3-4) are in operator notation as stated. An alternative notation would be the Lagrangian form in which the operators are replaced by the points themselves using (1.3-1) and (1.3-2). Both forms of equations will be used in this book, although the Lagrangian form, because it is simpler, will be preferred.

1.3.1. Hermite Interpolation

Of further interest is the class of Hermite interpolation polynomials. Instead of fitting a polynomial to the $\{y_n\}$ at the discrete $\{x_n\}$, this approach fits a polynomial to the two sequences $\{y_n\}$ and $\{y_n'\}$ at the discrete $\{x_n\}$ [2]. This is equivalent to fitting the polynomial to $2n + 2$ points (rather than the $n + 1$ as above). In other words these polynomials, termed *osculating polynomials*, not only agree in value with a given function at specified arguments, but their derviatives up to some order also match the derivatives of the given function, usually at the same arguments. Higher-order osculation could also be used by requiring that the second derivatives match, etc.

An explicit form of the Hermite formula could be written as

$$y_\alpha = \sum_{i=0}^{n} a_i(\alpha)y_i + \sum_{i=0}^{n} b_i(\alpha)y_i' \tag{1.3-5}$$

where the $a_i(\alpha)$ and $b_i(\alpha)$ are polynomials of degree $2n + 1$ and are related to the Lagrange interpolation polynomial [2]. The error associated with a finite n in (1.3-5) is of the form

$$T_\alpha = C_\alpha h^{2n+2} y^{[2n+2]}(\zeta)$$

By actually evaluting the $a_i(\alpha)$ and $b_i(\alpha)$ (this may be done by the method of Section 1.7 or more specifically as given in Section 1.8) it is possible to obtain a considerable number of different formulas. Some of these are listed below along with the appropriate truncation errors. Note that because of the assumed structure, (1.3-5), the results are in Lagrangian form; both extrapolation and interpolation forms are given with α an integer; the lower limits in (1.3-5) have been set at $i = 1$ rather than $i = 0$ to conform to the literature results.

HERMITE EXTRAPOLATION

$$\begin{aligned} y_{n+1} &= 5y_{n-1} - 4y_n + 2h[y_{n-1}' + 2y_n'] \\ T_\alpha &= (h^4/6)y^{[4]}(\zeta) \end{aligned} \tag{1.3-6}$$

$$\begin{aligned} y_{n+1} &= 10y_{n-2} + 9y_{n-1} - 18y_n + 3h[y_{n-2}' + 6y_{n-1}' + 3y_n'] \\ T_\alpha &= (h^6/20)y^{[6]}(\zeta) \end{aligned} \tag{1.3-7}$$

$$\begin{aligned} y_{n+1} &= \tfrac{47}{3}y_{n-3} + 64y_{n-2} - 36y_{n-1} - \tfrac{128}{3}y_n \\ &\quad + 4h[y_{n-3}' + 12y_{n-2}' + 18y_{n-1}' + 4y_n'] \\ T_\alpha &= (h^8/70)y^{[8]}(\zeta) \end{aligned} \tag{1.3-8}$$

$$\begin{aligned} y_{n+1} &= \tfrac{131}{6}y_{n-4} + \tfrac{575}{3}y_{n-3} + 100y_{n-2} - \tfrac{700}{3}y_{n-1} - \tfrac{475}{6}y_n \\ &\quad + 5h[y_{n-4}' + 20y_{n-3}' + 60y_{n-2}' + 40y_{n-1}' + 5y_n'] \\ T_\alpha &= (h^{10}/252)y^{[10]}(\zeta) \end{aligned} \tag{1.3-9}$$

HERMITE INTERPOLATION

$$y_n = (1/2)[y_{n-1} + y_{n+1}] + (h/4)[y'_{n-1} - y'_{n+1}]$$
$$T_\alpha = (h^4/24)y^{[4]}(\zeta) \tag{1.3-10}$$

$$y_n = (1/27)[20y_{n-1} + 7y_{n+2}] + (2h/9)[2y'_{n-1} - y'_{n+2}]$$
$$T_\alpha = (h^4/6)y^{[4]}(\zeta) \tag{1.3-11}$$

$$y_n = (1/16)[9y_{n-1} + 9y_{n+1} - 2y_{n+3}]$$
$$+ (3h/32)[3y'_{n-1} + y'_{n+3}]$$
$$T_\alpha = -(3h^5/40)y^{[5]}(\zeta) \tag{1.3-12}$$

$$y_n = (1/81)[28y_{n-2} + 64y_{n+1} - 11y_{n+4}]$$
$$+ (4h/27)[2y'_{n-2} + y'_{n+4}]$$
$$T_\alpha = -(8h^5/15)y^{[5]}(\zeta) \tag{1.3-13}$$

$$y_n = (1/128)[45y_{n-1} + 72y_{n+1} + 11y_{n+3}]$$
$$+ (3h/64)[3y'_{n-1} - 12y'_{n+1} - y'_{n+3}]$$
$$T_\alpha = (h^6/80)y^{[6]}(\zeta) \tag{1.3-14}$$

Further and more extensive formulas can be given, but we shall not pursue this point.

1.4. DIFFERENTIATION FORMULAS

Once we have the various interpolation formulas we may differentiate to form $dy/dx = y'(x)$ (or y_n'), $y''(x)$, etc. Listed below are a few of these obtained from the linear operator equations. In all cases we treat the formulas as infinite series, with α taken as an integer.

$$hy_n' = (\Delta - \tfrac{1}{2}\Delta^2 + \tfrac{1}{3}\Delta^3 - \cdots)y_n \tag{1.4-1}$$
$$hy_n' = (\Delta + \tfrac{1}{2}\Delta^2 - \tfrac{1}{6}\Delta^3 + \cdots)y_{n-1} \tag{1.4-2}$$
$$h^2y_n'' = (\Delta^2 - \Delta^3 + \tfrac{11}{12}\Delta^4 - \tfrac{5}{6}\Delta^5 + \cdots)y_n \tag{1.4-3}$$
$$h^2y_n'' = (\Delta^2 - \tfrac{1}{12}\Delta^4 + \tfrac{1}{12}\Delta^5 - \cdots)y_{n-1} \tag{1.4-4}$$
$$hy_n' = (\nabla + \tfrac{1}{2}\nabla^2 + \tfrac{1}{3}\nabla^3 + \cdots)y_n \tag{1.4-5}$$
$$hy_n' = (\nabla - \tfrac{1}{2}\nabla^2 - \tfrac{1}{6}\nabla^3 + \cdots)y_{n+1} \tag{1.4-6}$$

It is possible to obtain many further formulas, but these are sufficient for our purposes.

1.5. SPECIFIC INTEGRATION FORMULAS

Just as the interpolation formulas were differentiated in Section 1.4 it is equally appropriate to write

$$\int_{x_n}^{x_n+\alpha h} y(x)\, dx \tag{1.5-1}$$

and by substituting any one of the polynomials for $y(x)$ obtain a set of integration formulas. We shall here list only a few of the possible integration formulas because in the next section these will be developed in further detail as applied to ODEs. In fact we shall there illustrate a method for developing almost tailor-made integration formulas, having a wide range of applicability.

Using NFF as an illustration in (1.5-1) with various limits on the integral it is easy to develop

$$\int_{x_n}^{x_{n+1}} y(x)\, dx = h[1 + \tfrac{1}{2}\Delta - \tfrac{1}{12}\Delta^2 + \tfrac{1}{24}\Delta^3 - \tfrac{19}{720}\Delta^4 + \cdots]y_n$$

$$(1.5\text{-}2)$$

$$\int_{x_n}^{x_{n+2}} y(x)\, dx = 2h[1 + \Delta + \tfrac{1}{6}\Delta^2 + 0\,\Delta^3 - \tfrac{1}{180}\Delta^4 + \cdots]y_n$$

$$(1.5\text{-}3)$$

$$\int_{x_{n-1}}^{x_n} y(x)\, dx = h[1 - \tfrac{1}{2}\Delta + \tfrac{5}{12}\Delta^2 - \tfrac{3}{8}\Delta^3 + \tfrac{251}{720}\Delta^4 - \cdots]y_n$$

$$(1.5\text{-}4)$$

Equivalent forms using NBF are

$$\int_{x_{n-1}}^{x_n} y(x)\, dx = h[1 - \tfrac{1}{2}\nabla - \tfrac{1}{12}\nabla^2 - \tfrac{1}{24}\nabla^3 - \tfrac{19}{720}\nabla^4 - \cdots]y_n \quad (1.5\text{-}5)$$

$$\int_{x_n}^{x_{n+1}} y(x)\, dx = h[1 + \tfrac{1}{2}\nabla + \tfrac{5}{12}\nabla^2 + \tfrac{3}{8}\nabla^3 + \tfrac{251}{720}\nabla^4 + \cdots]y_n \quad (1.5\text{-}6)$$

$$\int_{x_n}^{x_{n+2}} y(x)\, dx = h[2 + \tfrac{1}{3}\nabla^2 + \tfrac{1}{3}\nabla^3 + \tfrac{29}{90}\nabla^4 + \cdots]y_{n+1} \quad (1.5\text{-}7)$$

From (1.5-5) it also follows that

$$\int_{x_n}^{x_{n+1}} y(x)\, dx = h[1 - \tfrac{1}{2}\nabla - \tfrac{1}{12}\nabla^2 - \tfrac{1}{24}\nabla^3 - \tfrac{19}{720}\nabla^4 - \cdots]y_{n+1}$$

$$(1.5\text{-}8)$$

It is interesting to compare (1.5-6) and (1.5-8); both integrate over the same interval, but the coefficients in (1.5-8) decrease much faster than those in (1.5-6). This distinction will be mentioned again in Section 1.6.

As given above, the formulas are all infinite series. If the series are terminated after a finite number of terms, then an error is incurred. This error may be estimated by integrating the truncation error term of the original polynomial itself or, even more simply, by replacing the first neglected term, say $\Delta^n y_n$, by $h^{n+1} y^{[n]}(\zeta)$. Thus in (1.5-3) if we truncate the series *after* the Δ^2 term (note the Δ^3 term has a zero coefficient), we may write

$$\int_{x_n}^{x_{n+2}} y(x)\, dx = 2h[1 + \Delta + \tfrac{1}{6}\Delta^2]y_n - (h^5/90)y^{[4]}(\zeta) \qquad (1.5\text{-}9)$$

where $x_n < \zeta < x_{n+2}$.

We will finish this section by specifying the most celebrated group of integration formulas, namely the Newton–Cotes closed and open formulas. These formulas are obtained by integrating NFF over an increasing set of points, i.e., (x_n, x_{n+1}), (x_n, x_{n+2}), (x_n, x_{n+3}), ... The closed formulas are obtained by retaining the same number of differences as h intervals in the integration. The open formulas do not include the end point values of the integration intervals.

NEWTON–COTES CLOSED FORMULAS

$$\int_{x_n}^{x_{n+1}} y(x)\,dx = (h/2)[y_n + y_{n+1}] - (h^3/12)y^{[2]}(\zeta) \tag{1.5-10}$$

$$\int_{x_n}^{x_{n+2}} y(x)\,dx = (h/3)[y_n + 4y_{n+1} + y_{n+2}] - (h^5/90)y^{[4]}(\zeta) \tag{1.5-11}$$

$$\int_{x_n}^{x_{n+3}} y(x)\,dx = (3h/8)[y_n + 3y_{n+1} + 3y_{n+2} + y_{n+3}] - (3h^5/80)y^{[4]}(\zeta) \tag{1.5-12}$$

$$\int_{x_n}^{x_{n+4}} y(x)\,dx = (2h/45)[7y_n + 32y_{n+1} + 12y_{n+2} + 32y_{n+3} + 7y_{n+4}]$$
$$- (8h^7/945)y^{[6]}(\zeta) \tag{1.5-13}$$

$$\int_{x_n}^{x_{n+5}} y(x)\,dx = (5h/288)[19y_n + 75y_{n+1} + 50y_{n+2} + 50y_{n+3}$$
$$+ 75y_{n+4} + 19y_{n+5}] - (275h^7/12096)y^{[6]}(\zeta) \tag{1.5-14}$$

Note that (1.5-11) is merely (1.5-9) with the linear operators in (1.5-9) expanded. Equation (1.5-10) is the *trapezoidal rule* and (1.5-11) is *Simpson's rule*.

NEWTON–COTES OPEN FORMULAS

$$\int_{x_n}^{x_{n+2}} y(x)\,dx = 2h[y_{n+1}] + (h^3/3)y^{[2]}(\zeta) \tag{1.5-15}$$

$$\int_{x_n}^{x_{n+3}} y(x)\,dx = (3h/2)[y_{n+1} + y_{n+2}] + (3h^3/4)y^{[2]}(\zeta) \tag{1.5-16}$$

$$\int_{x_n}^{x_{n+4}} y(x)\,dx = (4h/3)[2y_{n+1} - y_{n+2} + 2y_{n+3}] + (14h^5/45)y^{[4]}(\zeta) \tag{1.5-17}$$

$$\int_{x_n}^{x_{n+5}} y(x)\,dx = (5h/24)[11y_{n+1} + y_{n+2} + y_{n+3} + 11y_{n+4}]$$
$$+ (95h^5/144)y^{[4]}(\zeta) \tag{1.5-18}$$

$$\int_{x_n}^{x_{n+6}} y(x)\,dx = (3h/10)[11y_{n+1} - 14y_{n+2} + 26y_{n+3} - 14y_{n+4} + 11y_{n+5}]$$
$$+ (41h^7/140)y^{[6]}(\zeta) \tag{1.5-19}$$

1.6. INTEGRATION FORMULAS FOR ODE

If we integrate both sides of (1.1-1) between (x_n, y_n) and (x_{n+1}, y_{n+1}), there results

$$
\begin{aligned}
y_{n+1} &= y_n + \int_{x_n}^{x_{n+1}} f(x, y)\, dx \\
&= y_n + \int_{x_n}^{x_{n+1}} y'(x)\, dx \\
&= y_n + h \int_0^1 y'(\alpha)\, d\alpha
\end{aligned}
\tag{1.6-1}
$$

Attaching the index sequence of $n = 0, 1, 2, \ldots$ to this equation and noting that y_0 is known at x_0, (1.1-2), it becomes apparent that (1.6-1) can be used recursively to generate y_1, y_2, y_3, $\ldots$, as long as we have some way to evalute the integral. In other words the solution of the ODE is converted to evaluating an integral. But this is exactly the problem attacked in Section 1.5 with the one change that now we have $y'(x)$ as the integrand whereas before we had $y(x)$. Thus all of the formulas in Section 1.5 can be used directly in (1.6-1), but with the replacement of $y(x)$ with $y'(x)$ or with $f(x, y)$ and of y_n with y_n' or $f(x_n, y_n) = f_n$.

Let us now proceed to derive many of the equations used to represent (1.6-1). These derivations follow from the use of NBF in two forms, the first an extrapolation form and the second an interpolation form. Further, we generalize by retaining q differences in NBF. Thus from (1.3-4) we write [in terms of $y'(x)$ and not $y(x)$] the finite series

$$
y'_{n+\alpha} = y_n' + \alpha \nabla y_n' + \frac{\alpha(\alpha + 1)}{2!} \nabla^2 y_n' + \cdots
$$

$$
+ \frac{\alpha(\alpha + 1) \cdots (\alpha + q - 1)}{q!} \nabla^q y_n'
\tag{1.6-2}
$$

$$
y'_{n+\alpha} = y'_{n+1} + (\alpha - 1)\nabla y'_{n+1} + \frac{(\alpha - 1)(\alpha)}{2!} \nabla^2 y'_{n+1} + \cdots
$$

$$
+ \frac{(\alpha - 1)(\alpha)(\alpha + 1) \cdots (\alpha + q - 2)}{q!} \nabla^q y'_{n+1}
\tag{1.6-3}
$$

If we substitute (1.6-2) into (1.6-1) there results [replace $y'(x)$ with $y'_{n+\alpha}$]

$$
y_{n+1} = y_n + h \sum_{i=0}^{q} a_i \nabla^i y_n', \qquad a_0 = 1
\tag{1.6-4}
$$

where

$$a_i = \int_0^1 \frac{\alpha(\alpha + 1) \cdots (\alpha + i - 1)}{i!} \, d\alpha, \qquad i > 0$$

The error term associated with truncating after the qth ∇ is

$$T_\alpha = h^{q+2} \int_0^1 \frac{\alpha(\alpha + 1) \cdots (\alpha + q)}{(q + 1)!} \, y^{[q+2]}(\zeta) \, d\zeta$$

But since the coefficient of $y^{[q+2]}$ does not change sign in $(0, 1)$, it is possible to write

$$T_\alpha = a_{q+1} h^{q+2} y^{[q+2]}(\zeta) \tag{1.6-5}$$

By actually calculating the a_i in (1.6-4) one arrives at

$$y_{n+1} = y_n + h[1 + \tfrac{1}{2}\nabla + \tfrac{5}{12}\nabla^2 + \tfrac{3}{8}\nabla^3 + \tfrac{251}{720}\nabla^4 + \cdots]y_n' \tag{1.6-6}$$

which is merely (1.5-6). Note that if we truncate this series after the ∇^2 term, then $T_\alpha = \tfrac{3}{8}h^4 y^{[4]}(\zeta)$; if we truncate before the first ∇ then (for $q = 0$)

$$y_{n+1} = y_n + h y_n' \qquad \text{with} \qquad T_\alpha = \tfrac{1}{2}h^2 y^{[2]}(\zeta) \tag{1.6-7}$$

Equation (1.6-7) is the celebrated *Euler's formula*. Let us also adopt the convention that for $T_\alpha = \tfrac{1}{2}h^2 y^{[2]}(\zeta)$, we write

$$T_\alpha = O(h^2) \tag{1.6-8}$$

indicating that the truncation error is of *order* h^2.

Actually we can generalize further by rewriting (1.6-1) in the form

$$y_{n+1} = y_{n-r} + h \int_{-r}^1 y'_{n+\alpha} \, d\alpha \tag{1.6-9}$$

where r is any positive integer. The case $r = 0$ is, of course, the one already discussed. Substituting (1.6-2) in (1.6-9) the result is for $r = 1, 3$, and 5,

$$y_{n+1} = y_{n-1} + h[2 + 0\nabla + \tfrac{1}{3}\nabla^2 + \tfrac{1}{3}\nabla^3 + \tfrac{29}{90}\nabla^4 + \cdots]y_n', \qquad r = 1 \tag{1.6-10}$$

$$y_{n+1} = y_{n-3} + h[4 - 4\nabla + \tfrac{8}{3}\nabla^2 + 0\nabla^3 + \tfrac{14}{45}\nabla^4 + \cdots]y_n', \qquad r = 3 \tag{1.6-11}$$

$$y_{n+1} = y_{n-5} + h[6 - 12\nabla + 15\nabla^2 - 9\nabla^3 + \tfrac{33}{10}\nabla^4 + 0\nabla^5 + \cdots]y_n' \qquad r = 5 \tag{1.6-12}$$

The error term associated with truncation of these formulas is more complicated to calculate because the integrands in $(-r, 1)$ change sign. We shall not list these here except for certain special cases, which arise by noting that the coefficient of the rth difference is always zero. If we retain the rth differences only (the accuracy of the result is identical for retaining the rth or the $r-1$ difference):

$$y_{n+1} = y_{n-1} + 2h[y_n']$$
$$T_\alpha = (h^3/3)y^{[3]}(\zeta)$$

(1.6-13)

$$y_{n+1} = y_{n-3} + 4h[y_n' - \nabla y_n' + \tfrac{2}{3}\nabla^2 y_n']$$
$$T_\alpha = (14h^5/45)y^{[5]}(\zeta)$$

(1.6-14)

$$y_{n+1} = y_{n-5} + 6h[y_n' - 2\nabla y_n' + \tfrac{5}{2}\nabla^2 y_n' - \tfrac{3}{2}\nabla^3 y_n' + \tfrac{11}{20}\nabla^4 y_n']$$
$$T_\alpha = (41h^7/140)y^{[7]}(\zeta)$$

(1.6-15)

Actually these last three equations, with the differences expanded, are nothing more than a version of three of the Newton–Cotes open formulas, namely (1.5-15), (1.5-17), and (1.5-19) respectively. Since $y'(x)$ is involved rather than $y(x)$, the error terms are one derivative higher in the present section.

Having used the exrapolation form of NBF (1.6-2) we can repeat the procedure with the interpolation form (1.6-3). The analogous results for $r = 0, 1, 3,$ and 5 are

$$y_{n+1} = y_n + h[1 - \tfrac{1}{2}\nabla - \tfrac{1}{12}\nabla^2 - \tfrac{1}{24}\nabla^3 - \tfrac{19}{720}\nabla^4 - \cdots]y_{n+1}', \quad r = 0$$

(1.6-16)

$$y_{n+1} = y_{n-1} + h[2 - 2\nabla + \tfrac{1}{3}\nabla^2 + 0\nabla^3 - \tfrac{1}{90}\nabla^4 - \cdots]y_{n+1}', \quad r = 1$$

(1.6-17)

$$y_{n+1} = y_{n-3} + h[4 - 8\nabla + \tfrac{20}{3}\nabla^2 - \tfrac{8}{3}\nabla^3 + \tfrac{14}{45}\nabla^4 - 0\nabla^5 - \cdots]y_{n+1}'$$

$$r = 3 \quad (1.6\text{-}18)$$

$$y_{n+1} = y_{n-5} + h[6 - 18\nabla + 27\nabla^2 - 24\nabla^3 + \tfrac{123}{10}\nabla^4$$
$$- \tfrac{33}{10}\nabla^5 + \cdots]y_{n+1}', \qquad r = 5 \quad (1.6\text{-}19)$$

Some interesting and useful results may be obtained by truncating after a certain number of differences. In the $r = 0$ case truncation after the first difference yields

$$y_{n+1} = y_n + (h/2)[y_{n+1}' + y_n']$$
$$T_\alpha = -(h^3/12)y^{[3]}(\zeta)$$

(1.6-20)

which is called the *modified Euler formula*, the *trapezoidal rule*, and the *Crank–Nicholson* method. For $r = 1$ and 3, truncating after the second and the fourth differences (note that the third and fifth differences are zero) respectively yields

$$y_{n+1} = y_{n-1} + (h/3)[y_{n+1}' + 4y_n' + y_{n-1}']$$
$$T_\alpha = -(h^5/90)y^{[5]}(\zeta)$$

(1.6-21)

and

$$y_{n+1} = y_{n-3} + (2h/45)[7y'_{n+1} + 32y_n' + 12y'_{n-1} + 32y'_{n-2} + 7y'_{n-3}]$$
$$T_\alpha = -(8h^7/945)y^{[7]}(\zeta)$$

$$(1.6\text{-}22)$$

The three formulas (1.6-20)–(1.6-22) are forms of the Newton–Cotes closed formulas and equivalent to (1.5-10), (1.5-11), and (1.5-13).

1.6.1 Adams Forms

For convenience in later use we shall now group the $r = 0$ forms, then the $r = 1$ forms, etc. The $r = 0$ equations, (1.6-6) and (1.6-16), are called the *Adams–Bashforth* (A–B) and *Adams–Moulton* (A–M) formulas respectively. These are shown in operator notation through ∇^6 in (1.6-23) and (1.6-24).

Adams–Bashforth, $r = 0$

$$y_{n+1} = y_n + h[y_n' + \tfrac{1}{2}\nabla y_n' + \tfrac{5}{12}\nabla^2 y_n' + \tfrac{3}{8}\nabla^3 y_n' + \tfrac{251}{720}\nabla^4 y_n'$$
$$+ \tfrac{475}{1440}\nabla^5 y_n' + \tfrac{19087}{60480}\nabla^6 y_n' + \cdots]$$

$$(1.6\text{-}23)$$

Adams–Moulton, $r = 0$

$$y_{n+1} = y_n + h[y'_{n+1} - \tfrac{1}{2}\nabla y'_{n+1} - \tfrac{1}{12}\nabla^2 y'_{n+1} - \tfrac{1}{24}\nabla^3 y'_{n+1} - \tfrac{19}{720}\nabla^4 y'_{n+1}$$
$$- \tfrac{3}{160}\nabla^5 y'_{n+1} - \tfrac{863}{60480}\nabla^6 y'_{n+1} + \cdots]$$

$$(1.6\text{-}24)$$

Note that the coefficients of the A–M equation decrease faster than those in the A–B equation. These are again (1.5-6) and (1.5-8) on which we previously commented.

By suitable truncation of these formulas we may get a whole series of new equations. We have already mentioned that for (1.6-23) the truncation before any differences ($q = 0$) yields Euler's method (1.6-7); if we truncate after the first ($q = 1$), then the second ($q = 2$), etc. difference, we get the equations below and Table 1.1. The Lagrangian forms are shown here and

TABLE 1.1

ADAMS–BASHFORTH FORMS

q	Coefficient of h	y_n'	y'_{n-1}	y'_{n-2}	y'_{n-3}	y'_{n-4}	y'_{n-5}
0	1	1					
1	1/2	3	−1				
2	1/12	23	−16	5			
3	1/24	55	−59	37	−9		
4	1/720	1901	−2774	2616	−1274	251	
5	1/1440	4277	−7923	9982	−7298	2877	−475

in Chapter 4 we present an extended form of Table 1.1, including all the truncation errors.

$$y_{n+1} = y_n + hy_n' \qquad (q = 0) \tag{1.6-25}$$

$$y_{n+1} = y_n + (h/2)[3y_n' - y_{n-1}'] \qquad (q = 1) \tag{1.6-26}$$

$$y_{n+1} = y_n + (h/12)[23y_n' - 16y_{n-1}' + 5y_{n-2}'] \qquad (q = 2) \tag{1.6-27}$$

$$y_{n+1} = y_n + (h/24)[55y_n' - 59y_{n-1}' + 37y_{n-2}' - 9y_{n-3}'] \qquad (q = 3) \tag{1.6-28}$$

In a completely equivalent manner we could list the first few equations resulting from truncating the Adams–Moulton equation. However, here we merely list the values in Table 1.2, with an extended version given in Chapter

TABLE 1.2

ADAMS–MOULTON FORMS

q	Coefficient of h	y_{n+1}'	y_n'	y_{n-1}'	y_{n-2}'	y_{n-3}'	y_{n-4}'
0	1	1					
1	1/2	1	1				
2	1/12	5	8	−1			
3	1/24	9	19	−5	1		
4	1/720	251	646	−264	106	−19	
5	1/1440	475	1427	−798	482	−173	27

4. Note that the Adams formulas involve y_{n+1}, y_n and weighted derivative values.

1.6.2. Nystrom Forms

By contrast or as a direct extension we may consider the set of two $r = 1$ equations (1.6-10) and (1.6-17). These are called the *Nystrom* formulas and use values of y_{n+1} and y_{n-1} and weighted derivative values. Thus

$$y_{n+1} = y_{n-1} + h[2y_n' + 0\nabla y_n' + \tfrac{1}{3}\nabla^2 y_n' + \tfrac{1}{3}\nabla^3 y_n' \\ + \tfrac{29}{90}\nabla^4 y_n' + \tfrac{14}{45}\nabla^5 y_n' + \tfrac{1139}{3780}\nabla^6 y_n' + \cdots] \tag{1.6-29}$$

$$y_{n+1} = y_{n-1} + h[2y_{n+1}' - 2\nabla y_{n+1}' + \tfrac{1}{3}\nabla^2 y_{n+1}' + 0\nabla^3 y_{n+1}' \\ - \tfrac{1}{90}\nabla^4 y_{n+1}' - \tfrac{1}{90}\nabla^5 y_{n+1}' - \tfrac{37}{3780}\nabla^6 y_{n+1}' + \cdots] \tag{1.6-30}$$

As in the Adams forms, the coefficients in the second equation (the implicit equation based originally on an interpolation polynomial) decrease faster than those in the first equation (the explicit equation based originally on an extrapolation polynomial). We could tabulate all the different cases, but we merely present in Table 1.3 the coefficients for the explicit method. Note that the case $q = 0$ in (1.6-29) yields

$$y_{n+1} = y_{n-1} + 2hy_n' \tag{1.6-31}$$

which is (1.6-13) and termed the *midpoint rule*. Because the first term discarded from (1.6-29) has a zero coefficient, this formula is more accurate than would appear, the error being $O(h^3)$.

TABLE 1.3

NYSTROM EXPLICIT FORM

q	Coefficient of h	y_n'	y_{n-1}'	y_{n-2}'	y_{n-3}'	y_{n-4}'
0	1	2				
1	1	2	0			
2	1/3	7	-2	1		
3	1/3	8	-5	4	-1	
4	1/90	269	-266	294	-146	29

The same idea of coupling the $r = 0$ and $r = 1$ equations can be extended to the cases $r = 3$ and $r = 5$. The Hermite formulas (1.3-6)–(1.3-14) are already in the form of the formulas of this section, i.e., they involve values of $y_{n+1}, y_n, \ldots$ and weighted first derivatives, and can be considered as equivalent to the formulas developed in this section.

1.7. GENERALIZED INTEGRATION FORMULAS FOR ODE

In the previous section we derived a number of equations relating y_{n+1} to $y_n, y_{n-1}, \ldots$ and to weighted derivative values at these points. In each case the equations were obtained directly from manipulating interpolation or extrapolation polynomials. In the present section we proceed via an entirely different route which yields all of the previous formulas and at the same time can be used to obtain many other important formulas.

To start this discussion we first define the generalized linear, k-step differential–difference equation with constant coefficients

$$y_{n+1} = \alpha_1 y_n + \alpha_2 y_{n-1} + \cdots + \alpha_k y_{n+1-k}$$
$$+ h[\beta_0 y_{n+1}' + \beta_1 y_n' + \cdots + \beta_k y_{n+1-k}'] \tag{1.7-1}$$

or

$$y_{n+1} = \sum_{i=1}^{k} \alpha_i y_{n+1-i} + h \sum_{i=0}^{k} \beta_i y_{n+1-i}' \tag{1.7-2}$$

Alternative forms of (1.7-1) and (1.7-2) may be given by revising the datum point on the differences. In (1.7-1) and (1.7-2), y_{n+1} is used on the left side and only equal or lower values of y_i or y_i' are used on the right side. In other words the basis is y_{n+1} and all lower values. The analogous equations which use y_n and all higher values, i.e., $y_{n+1}, y_{n+2}, \ldots$ are given by

$$y_{n+k} = \alpha_1 y_{n+k-1} + \alpha_2 y_{n+k-2} + \cdots + \alpha_k y_n$$
$$+ h[\beta_0 y'_{n+k} + \cdots + \beta_k y_n'] \qquad (1.7\text{-}1')$$

or

$$y_{n+k} = \sum_{i=1}^{k} \alpha_i y_{n+k-i} + h \sum_{i=0}^{k} \beta_i y'_{n+k-i} \qquad (1.7\text{-}2')$$

While both forms are basically equivalent we prefer to generally use (1.7-1) and (1.7-2).

It is apparent that all of the previous formulas can be encompassed within (1.7-1) by a suitable choice of the $2k + 1$ parameters $\alpha_1, \ldots, \alpha_k, \beta_0, \beta_1, \ldots, \beta_k$. While we show the upper limit in each summation of (1.7-2) to be the integer k, this is only for convenience; the upper limits can be different if desired and all that we subsequently say will follow with obvious modifications.

Of importance is the fact that a constant spacing h is assumed, that the α_i and β_i are constants independent of h (see Section 1.7.5), that only y_i and y_i' values are allowed (no y_i'', y_i''', $\ldots$) and that only function values or derivatives at the equally spaced x_i are used. In later sections we will remove many of these assumptions on the specific form of (1.7-1).

While we are discussing (1.7-1) a number of definitions can be stated. First, k can be related directly to the integer r used in Section 1.6. We make the change to k [see discussion for (1.7-3)] to conform to the literature usage. If $k = 1$, y_{n+1} relates only to y_n and y'_{n+1} and y_n'; we refer to all such equations as forming a *single-step* method [a step from (x_n, y_n) to (x_{n+1}, y_{n+1})]. If $k > 1$, y_{n+1} relates to at least y_n and y_{n-1}. We refer to all such equations as forming a *multiple-step* method.

Further terminology relates to the explicit value for β_0. Thus, if $\beta_0 = 0$, the resulting equation is referred to as an *open, explicit,* or *predictor* equation because y_{n+1} only occurs on the left-hand side of the equation. In other words, y_{n+1} can be calculated directly from the right-hand side values. If $\beta_0 \neq 0$, the resulting equation is referred to as a *closed, implicit,* or *corrector* equation since y_{n+1} occurs on both sides of the equation. In other words the unknown y_{n+1} cannot be calculated directly since it is contained within y'_{n+1}.

Having specified the form of (1.7-1), we now seek the explicit numerical values of the coefficients $\alpha_1, \alpha_2, \ldots, \alpha_k, \beta_0, \beta_1, \beta_2, \ldots, \beta_k$. To determine these coefficients we will embed a finite polynomial in x in the equation to yield an exact solution to the differential–difference equation. If (1.7-1) is made exact for a polynomial of degree p, we say the resulting formula is of order p. Now we proceed to illustrate how this exact solution (and the determination of the coefficients) can be developed by the method of undetermined coefficients.

1.7.1. The Method of Undetermined Coefficients

To illustrate the procedure called the method of undetermined coefficients, we pick a simple case, namely $k = 1$, in (1.7-1). This yields the single-step formula

$$y_{n+1} = \alpha_1 y_n + h[\beta_0 y'_{n+1} + \beta_1 y'_n] \qquad (1.7\text{-}3)$$

which has the form of (1.6-20). There are three unknown parameters α_1, β_0, and β_1 and we can make (1.7-3) exact if $y(x)$ is a polynomial of degree $2(p = 2)$. Thus we substitute $y(x) = 1$, x, and x^2 in (1.7-3) yielding three equations in the three unknown coefficients. Explicitly, we use

$$y_i = 1, \qquad y_i' = 0$$
$$y_i = x, \qquad y_i' = 1 \qquad i = n, \quad n+1$$
$$y_i = x_i^2, \qquad y_i' = 2x_i$$

Using $y_i = 1$, $y_i' = 0$ in (1.7-3) yields immediately

$$y_{n+1} = x_{n+1}^0 = 1 = \alpha_1 x_n^0 + h[\beta_0 0 + \beta_1 0]$$

or $\alpha_1 = 1$. Substituting $y_i = x_i$, $y_i' = 1$ in (1.7-3) yields $1 = \beta_0 + \beta_1$ where we have also used $(x_{n+1} - x_n)/h = 1$. Finally, substituting $y_i = x_i^2$, $y_i' = 2x_i$ yields (using $1 = \beta_0 + \beta_1$) $\beta_0 = \frac{1}{2}$ and $\beta_1 = \frac{1}{2}$. Thus we have derived the second-order $(p = 2)$ equation

$$y_{n+1} = y_n + (h/2)[y'_{n+1} + y'_n] \qquad (p = 2) \qquad (1.7\text{-}4)$$

which is exactly (1.6-20).

At this point we may ask why we use this method to derive an equation which we already have. The reason lies in the flexibility of the approach. Here we satisfied the condition that the number of coefficients to be determined was $p + 1 = 3$. Suppose instead we use a polynomial of degree less than 2; obviously, if we use $p = 1$, we can at most determine two of the coefficients with a third left free or unspecified. In fact, we can write immediately that

$$y_{n+1} = y_n + h[(1 - \beta_1)y'_{n+1} + \beta_1 y'_n] \qquad (p = 1) \qquad (1.7\text{-}5)$$

as a first-order $(p = 1)$ formula. We may now specify β_1 in any manner desired; thus we may choose a specific value for β_1 because of truncation error considerations or to satisfy alternative requirements such as maintaining stability (Chapter 3). The important point to realize is that we may leave certain of the parameters "free" or undetermined to be specified later to satisfy conditions other than the order of the equation.

To show how β_1 in (1.7-5) affects the truncation error and how to estimate the truncation error via the undetermined coefficient method, let us proceed by the following crude analysis. We have made (1.7-3) exact for $p = 2$; if $y(x)$ is not a polynomial of degree two, then the main error in (1.7-4) is related to the first neglected polynomial x^3. On this basis we write (1.7-4) as

$$y_{n+1} = y_n + (h/2)[y_{n+1}' + y_n'] + Cy^{[3]} \qquad (1.7\text{-}6)$$

indicating that the truncation error is a constant times $y^{[3]}$. But if the main error is x^3, then $y^{[3]} = 6$ and substituting $y_i = x_i^3$, $y_i' = 3x_i^2$ in (1.7-6) yields, after some manipulation, that

$$C = -h^3/12 \qquad (1.7\text{-}7)$$

Thus the truncation error is $T_\alpha = -(h^3/12)y^{[3]}(\zeta)$ exactly as before. If we repeat this procedure in (1.7-5) but now use $y_i = x_i^2$, $y_i' = 2x_i$, and $T_\alpha = Cy^{[2]}$ ($y^{[2]} = 2$) there results

$$C = -h^2[\tfrac{1}{2} - \beta_1] \qquad (1.7\text{-}8)$$

or

$$y_{n+1} = y_n + h[(1 - \beta_1)y_{n+1}' + \beta_1 y_n'] - h^2[\tfrac{1}{2} - \beta_1]y^{[2]}(\zeta) \qquad (1.7\text{-}9)$$

Equation (1.7-9) is of order 1 ($p = 1$), but we still can choose β_1 in any fashion desired. If we choose $\beta_1 = \tfrac{1}{2}$, then the result is to make a second-order equation; if we choose $\beta_1 = 1$, then we get Euler's method or (1.6-7) which is first order.

Carrying out the substitution of the pth degree polynomial in the difference equation and the subsequent manipulations is really not as complicated as might be thought. The same results are obtained if one uses $h = 1$ and $x_n = 0$ in the manipulations. Further, for the $p + 1$ equations which result there are summation relations which may be written for each substitution.

We also wish to point out that the results of the method of undetermined coefficients can be accomplished in a seemingly different manner. If each y_i and y_i' is expanded in a Taylor series about x_n, the coefficients of the powers of h in (1.7-1) will vanish up through the integer p. This number p is obviously the order of the polynomial approximating the function and the product of h^{p+1} and its coefficient is the first term of the truncation error. For example, we may write

$$y_{n+1} = y_n + hy_n' + \tfrac{1}{2}h^2 y_n'' + \tfrac{1}{6}h^3 y_n''' + \cdots$$
$$hy_{n+1}' = hy_n' + h^2 y_n'' + \tfrac{1}{2}h^3 y_n''' + \cdots$$

and for (1.7-4), the following may be constructed:

		y_n	$hy_n{}'$	h^2y_n''	h^3y_n'''
From	y_{n+1}	1	1	1/2	1/6
	$-y_n$	-1	0	0	0
	$-\frac{1}{2}hy_{n+1}'$	0	$-1/2$	$-1/2$	$-1/4$
	$-\frac{1}{2}hy_n{}'$	0	$-1/2$	0	0
Sum of columns		0	0	0	1/12

Thus the coefficients of the powers of h up through the integer 2 will vanish and the truncation error will be $-\frac{1}{12}h^3y^{[3]}$. This is, of course, identical to the result obtained previously.

1.7.2. Adams Forms

As a further illustration of the method let us select the $k = 4$ step equation

$$y_{n+1} = \alpha_1 y_n + \alpha_2 y_{n-1} + \alpha_3 y_{n-2} + \alpha_4 y_{n-3}$$
$$+ h[\beta_0 y_{n+1}' + \beta_1 y_n{}' + \beta_2 y_{n-1}' + \beta_3 y_{n-2}' + \beta_4 y_{n-3}']$$
$$(1.7\text{-}10)$$

We have nine coefficients and we could evaluate them so as to form an eighth-order process. Instead, we arbitrarily select $\alpha_2 = \alpha_3 = \alpha_4 = \beta_0 = 0$ leaving five coefficients, and the best possible order is $p = 4$. If we make (1.7-10) exact for $y(x) = 1$, x, x^2, x^3, and x^4, we obtain

$$\alpha_1 = 1, \quad \beta_1 = \tfrac{55}{24}, \quad \beta_2 = -\tfrac{59}{24}, \quad \beta_3 = \tfrac{37}{24}, \quad \beta_4 = -\tfrac{9}{24}$$

or

$$y_{n+1} = y_n + (h/24)[55y_n{}' - 59y_{n-1}' + 37y_{n-2}' - 9y_{n-3}'] \quad (1.7\text{-}11)$$

This is one of the Adams–Bashforth equations, (1.6-28). By selecting $\alpha_2 = \alpha_3 = \alpha_4 = 0$, we forced the result to become one of the Adams equations; the choice $\beta_0 = 0$ then picked the Bashforth form.

1.7.3. Higher-Order Derivative Forms

Next we consider two cases which involve higher-order derivatives. First, we select the equation

$$y_{n+1} = \alpha_1 y_n + \alpha_2 y_{n-1} + \alpha_3 y_{n-2}$$
$$+ h^2[\gamma_0 y_{n+1}'' + \gamma_1 y_n'' + \gamma_2 y_{n-1}'' + \gamma_3 y_{n-2}''] \quad (1.7\text{-}12)$$

Note this is not of the form (1.7-1) previously considered, since the second derivative terms are included and the first derivative terms are excluded. Nevertheless, the same approach can be used profitably. Since we have seven coefficients, we could make this exact for an order 6. Instead we will only make the formula of order 5 leaving one free parameter which we will take as α_3. The truncation error will be of the form $T_\alpha = Cy^{[6]}$. Substituting $y(x) = 1, x, \ldots, x^5$ yields the following coefficient values:

$$\alpha_1 = 2 + \alpha_3, \qquad \alpha_2 = -(1 + 2\alpha_3), \qquad \alpha_3 = \alpha_3$$
$$\gamma_0 = 1/12, \qquad\qquad \gamma_1 = (10 - \alpha_3)/12$$
$$\gamma_2 = (1 - 10\alpha_3)/12, \qquad \gamma_3 = -\alpha_3/12$$

Selecting the case $\alpha_3 = 0$ as an example there results

$$y_{n+1} = 2y_n - y_{n-1} + (h^2/12)[y''_{n+1} + 10y''_n + y''_{n-1}] \qquad (1.7\text{-}13)$$

with $T_\alpha = -(h^6/240)y^{[6]}(\zeta)$. This equation is termed *Numerov's* equation [see (1.9-11)].

Next we select the equation

$$y_{n+1} = \alpha_1 y_n + h[\beta_0 y'_{n+1} + \beta_1 y_n'] + h^2[\gamma_0 y''_{n+1} + \gamma_1 y''_n] \qquad (1.7\text{-}14)$$

with five parameters. We can make (1.7-14) exact for $p = 4$ or for $y(x) = 1, x, \ldots, x^4$ to yield

$$\alpha_1 = 1, \qquad \beta_0 = \beta_1 = \tfrac{1}{2}$$
$$\gamma_0 = -\tfrac{1}{12}, \qquad \gamma_1 = \tfrac{1}{12}$$

Thus we obtain

$$y_{n+1} = y_n + (h/2)[y'_{n+1} + y_n'] + (h^2/12)[-y''_{n+1} + y''_n] \qquad (1.7\text{-}15)$$

with $T_\alpha = (h^5/720)y^{[5]}(\zeta)$. This result will be renumbered as (1.10-4) later.

1.7.4. Gaussian Forms

Next we choose the special equation

$$y_{n+1} = \alpha_1 y_n + h\beta_1 f(x_{n+\beta_1}, y_{n+\beta_1}) + h\beta_2 f(x_{n+\beta_2}, y_{n+\beta_2}) \qquad (1.7\text{-}16)$$

Here we have the usual parameters α_1, β_1, and β_2, but in addition the values $x_{n+\beta_1}$ and $x_{n+\beta_2}$ within x_n and x_{n+1} at which to evaluate f must be determined. In other words we do not have an equal-step type formula, but rather a variable-step formula with the derivatives to be determined from the undetermined coefficient procedure.

Since we have five parameters we can make (1.7-16) exact for $p = 4$ or for $y(x) = 1, \ldots, x^4$. Substituting consecutively, there results

$$\alpha_1 = 1, \qquad \beta_1 + \beta_2 = 1$$
$$\beta_1 x_{n+\beta_1} + \beta_2 x_{n+\beta_2} = \tfrac{1}{2}$$
$$\beta_1 x_{n+\beta_1}^2 + \beta_2 x_{n+\beta_2}^2 = \tfrac{1}{3}$$
$$\beta_1 x_{n+\beta_1}^3 + \beta_2 x_{n+\beta_2}^3 = \tfrac{1}{4}$$

Regarding the last four of these equations as containing the 4 unknowns β_1, β_2, $x_{n+\beta_1}$, and $x_{n+\beta_2}$, these can be solved to yield

$$\alpha_1 = 1, \qquad \beta_1 = \beta_2 = \tfrac{1}{2}$$

$$x_{n+\beta_1} = \frac{3 - \sqrt{3}}{6}, \qquad x_{n+\beta_2} = \frac{3 + \sqrt{3}}{6} \qquad (1.7\text{-}17)$$

Equation (1.7-16) with the parameters of (1.7-17) will be referred to in Section 1.12 and in subsequent chapters of this book as the 2-point Gauss formula. As can be seen the formula is fourth order ($p = 4$).

1.7.5. Variable Step Forms

Finally in this section we consider the coefficients of (1.7-1) as functions of h (nonconstant coefficients). Using (1.7-10) as a starting formula we allow for $h = h_n$ and define

$$\tau_1 = (x_n - x_{n-1})/h, \qquad \tau_2 = (x_n - x_{n-2})/h$$

$$\tau_3 = (x_n - x_{n-3})/h \qquad (1.7\text{-}18)$$

Further, we select the explicit form of (1.7-10), $\beta_0 = 0$ and allow $\alpha_2 = \alpha_3 = \alpha_4 = 0$. Obviously, if $\tau_1 = 1$, $\tau_2 = 2$, and $\tau_3 = 3$, we would then merely obtain the constant h Adams formula of (1.7-11). Instead, we do not fix τ_1, τ_2, and τ_3 at these integer values, but allow them to be free in the undetermined coefficient procedure.

Proceeding as before (the algebra is now more complicated) we arrive at the recursive relations

$$\beta_4 = \frac{2(2 + 3\tau_1)(\tau_2 + \tau_1) + 3(1 - 2\tau_1{}^2)}{12\tau_3(\tau_3 - \tau_1)(\tau_2 - \tau_3)}$$

$$\beta_3 = \frac{2 + 3\tau_1 - 6\tau_3(\tau_3 - \tau_1)\beta_4}{6\tau_2(\tau_2 - \tau_1)} \qquad (1.7\text{-}19)$$

$$\beta_2 = -\frac{1}{2\tau_1}(1 + 2\tau_3\beta_4 + 2\tau_2\beta_3)$$

$$\beta_1 = 1 - \beta_2 - \beta_3 - \beta_4$$

and a truncation error of

$$T_\alpha = \{1 - \tfrac{5}{12}[3(\tau_1 + \tau_2 + \tau_3) + 4(\tau_1\tau_2 + \tau_1\tau_3 + \tau_2\tau_3)] + 6\tau_1\tau_2\tau_3\}$$
$$\times h^5 y^{[5]}(\zeta)/5! \qquad (1.7\text{-}20)$$

The result is obviously of order 4, and allows for a variable step calculation since τ_1, τ_2, and τ_3 can now be chosen in any manner desired. The special case $\tau_1 = 1$, $\tau_2 = 2$, and $\tau_3 = 3$ yields (1.7-11). Further details on this variable step approach are given by Van Wyk [9].

1.8. COMPILATION OF VARIOUS MULTIPLE-STEP INTEGRATION FORMULAS INCLUDING y_i AND y_i'

In this section we shall use the ideas presented above to develop and compile a large number of different possible integration formulas involving y_n, y_{n-1}, ... and y_{n+1}', y_n', y_{n-1}', ... at the various points in the discrete set $\{x_n\}$. None of the derivations will be given. The formulas will be divided into the two areas of $\beta_0 = 0$ (the explicit type equation) and $\beta_0 \neq 0$ (the implicit type equation). Within each area, the number of back points k and the order of the result p will be important. Many of these results were first given by Hamming [2].

1.8.1. Explicit Forms

Consider the explicit $k = 4$ case

$$y_{n+1} = \alpha_1 y_n + \alpha_2 y_{n-1} + \alpha_3 y_{n-2} + \alpha_4 y_{n-3}$$
$$+ h[\beta_1 y_n' + \beta_2 y_{n-1}' + \beta_3 y_{n-2}' + \beta_4 y_{n-3}'] \qquad (1.8\text{-}1)$$

but with the proviso that three of the four α_i are zero. Thus we have five coefficients and we can make the formula exact for $p = 4$. By the method of undetermined coefficients, there results Table 1.4. Note that the equation

TABLE 1.4

EXPLICIT EQUATION (1.8-1)

			$k=4$,	$p=4$			
α_1	α_2	α_3	α_4	β_1	β_2	β_3	β_4
1	0	0	0	55/24	$-59/24$	37/24	$-9/24$
0	1	0	0	8/3	$-5/3$	4/3	$-1/3$
0	0	1	0	21/8	$-9/8$	15/8	$-3/8$
0	0	0	1	9/24	$-4/3$	8/3	0

with $\alpha_1 = 1$ is an Adams–Bashforth formula (1.6-28), that the $\alpha_2 = 1$ case is a Nystrom formula (Table 1.3), etc.

Next we consider the case $k = 3$

$$y_{n+1} = \alpha_1 y_n + \alpha_2 y_{n-1} + \alpha_3 y_{n-2} + h[\beta_1 y_n' + \beta_2 y_{n-1}' + \beta_3 y_{n-2}']$$
$$(1.8\text{-}2)$$

Since there are six coefficients, (1.8-2) can be made of order 5. Instead, only order 4, $p = 4$, will be specified by allowing α_3 as a free parameter. The result is shown in Table 1.5 for a few selected values of α_3. Note that the case $\alpha_3 = 10$ yields the Hermite formula (1.3-7).

TABLE 1.5

EXPLICIT EQUATION (1.8-2)[a]

$k = 3, \quad p = 4$				
$\alpha_1 = -8 - \alpha_3$	-18	-9	-8	-7
$\alpha_2 = 9$	9	9	9	9
$\alpha_3 = \alpha_3$	10	1	0	-1
$\beta_1 = (17 + \alpha_3)/3$	9	6	17/3	16/3
$\beta_2 = (14 + 4\alpha_3)/3$	18	6	14/3	10/3
$\beta_3 = (-1 + \alpha_3)/3$	3	0	$-1/3$	$-2/3$
[b] $T_\alpha = (40 - 4\alpha_3)/3$	0	12	40/3	44/3

[a] From "Numerical Methods for Scientists and Engineers," R. W. Hamming. Copyright © 1962 by McGraw-Hill, Inc. Used by permission of McGraw-Hill Book Company.
[b] Multiply by $h^5 y^{[5]}(\zeta)/5!$.

Using (1.8-2), but allowing an order of only 3 there exist two free parameters. Calling these α_2 and α_3 we list the coefficients in Table 1.6.

TABLE 1.6

EXPLICIT EQUATION (1.8-2)

$k = 3, \quad p = 3$	
$\alpha_1 = 1 - \alpha_2 + \alpha_3$	$\beta_1 = (23 + 5\alpha_2 + 4\alpha_3)/12$
$\alpha_2 = \alpha_2$	$\beta_2 = (-16 + 8\alpha_2 + 16\alpha_3)/12$
$\alpha_3 = \alpha_3$	$\beta_3 = (5 - \alpha_2 + 4\alpha_3)/12$
$T_\alpha = (9 - \alpha_2) h^4 y^{[4]}(\zeta)/4!$	

Another case of interest is the $k = 3$ equation

$$y_{n+1} = \alpha_1 y_n + \alpha_2 y_{n-1} + \alpha_3 y_{n-2} + h[\beta_1 y_n' + \beta_2 y_{n-1}'] \qquad (1.8\text{-}3)$$

With α_3 a free parameter, we can make (1.8-3) of order $p = 3$ and in Table 1.7 are the results for selected α_3. Of interest is that the case $\alpha_3 = 0$ yields the Hermite formula (1.3-6) and that $\alpha_3 = 1$ yields another Hermite formula we had not previously listed, namely

$$y_{n+1} = -9y_n + 9y_{n-1} + y_{n-2} + 6h[y_n' + y_{n-1}'] \qquad (1.8\text{-}4)$$

with $T_\alpha = (h^5/10)y^{[5]}(\zeta)$.

TABLE 1.7

EXPLICIT EQUATION (1.8-3)[a]

$k = 3, \quad p = 3$			
$\alpha_1 = -4 - 5\alpha_3$	-54	-9	-4
$\alpha_2 = 5 + 4\alpha_3$	45	9	5
$\alpha_3 = \alpha_3$	10	1	0
$\beta_1 = 4 + 2\alpha_3$	24	6	4
$\beta_2 = 2 + 4\alpha_3$	42	6	2
[b] $T_\alpha = 4 - 4\alpha_3$	-36	0^c	4

[a] From "Numerical Methods for Scientists and Engineers," R. W. Hamming. Copyright © 1962 by McGraw-Hill, Inc. Used by permission of McGraw-Hill Book Company.
[b] Multiply by $h^4 y^{[4]}(\zeta)/4!$.
[c] $T_\alpha = 12h^5 y^{[5]}(\zeta)/5!$.

Finally, we select the $k = 4$ equation

$$y_{n+1} = \alpha_1 y_n + \cdots + \alpha_4 y_{n-3} + h[\beta_1 y_n' + \cdots + \beta_4 y_{n-3}'] \quad (1.8\text{-}5)$$

Table 1.8 shows the result for $\beta_4 = 0$, α_3 and α_4 free, and $p = 4$. Note that

TABLE 1.8

EXPLICIT EQUATION (1.8-5)

$\beta_4 = 0, \quad k = 4, \quad p = 4$				
$\alpha_1 = -8 - \alpha_3 + 8\alpha_4$	-8	0	-9	-1
$\alpha_2 = 9 - 9\alpha_4$	9	0	9	0
$\alpha_3 = \alpha_3$	0	0	1	1
$\alpha_4 = \alpha_4$	0	1	0	1
$\beta_1 = (17 + \alpha_3 - 9\alpha_4)/3$	$17/3$	$8/3$	6	3
$\beta_2 = (14 + 4\alpha_3 - 18\alpha_4)/3$	$14/3$	$-4/3$	6	0
$\beta_3 = (-1 + \alpha_3 + 9\alpha_4)/3$	$-1/3$	$8/3$	0	3
[a] $T_\alpha = (40 - 4\alpha_3 + 72\alpha_4)/3$	$40/3$	$122/3$	9	36

[a] Multiply by $h^5 y^{[5]}(\zeta)/5!$.

the case $\alpha_3 = 0$, $\alpha_4 = 1$ in Table 1.8 yields an adaptation of the Newton–Cotes open end formula (1.5-17),

$$y_{n+1} = y_{n-3} + (4h/3)(2y_n' - y_{n-1}' + 2y_{n-2}')$$

Equation (1.8-4) is also in this set.

1.8.2. Implicit Forms

First we start with the $k = 2$ form

$$y_{n+1} = \alpha_1 y_n + \alpha_2 y_{n-1} + h[\beta_0 y'_{n+1} + \beta_1 y'_n + \beta_2 y'_{n-1}] \qquad (1.8\text{-}6)$$

and for α_2 as the free parameter make (1.8-6) of order $p = 3$. Table 1.9

TABLE 1.9

IMPLICIT EQUATION $(1.8\text{-}6)^a$

$k = 2,\quad p = 3$			
$\alpha_1 = 1 - \alpha_2$	1	0	4/5
$\alpha_2 = \alpha_2$	0	1	1/5
$\beta_0 = (5 - \alpha_2)/12$	5/12	1/3	2/5
$\beta_1 = (8 + 8\alpha_2)/12$	8/12	4/3	4/5
$\beta_2 = (-1 + 5\alpha_2)/12$	$-1/12$	1/3	0
$^b T_\alpha = -1 + \alpha_2$	-1	0^c	$-4/5$

^{a} From "Numerical Methods for Scientists and Engineers," R. W. Hamming. Copyright © 1962 by McGraw-Hill, Inc. Used by permission of McGraw-Hill Book Company.

^{b} Multiply by $h^4 y^{[4]}(\zeta)/4!$.

$^c T_\alpha = -h^5 y^{[5]}(\zeta)/90$.

illustrates some results. Clearly the case $\alpha_2 = 1$ yields an adaptation of Simpson's rule (1.5-11) which is frequently termed *Milne's equation*

$$y_{n+1} = y_{n-1} + (h/3)[y'_{n+1} + 4y'_n + y'_{n-1}] \qquad (1.8\text{-}7)$$

$T_\alpha = -(h^5/90)y^{[5]}(\zeta)$ and (1.8-7) is of order 4.

Next we use the $k = 3$ form

$$y_{n+1} = \alpha_1 y_n + \cdots + \alpha_3 y_{n-2} + h[\beta_0 y'_{n+1} + \cdots + \beta_3 y'_{n-2}] \qquad (1.8\text{-}8)$$

and select α_2 and α_3 as free. By specifying that $p = 4$ we then obtain the results of Table 1.10. There are a number of interesting results in this table, namely, $\alpha_2 = \alpha_3 = 0$ [an Adams–Moulton formula]; $\alpha_2 = 1$, $\alpha_3 = 0$ [Equation (1.8-7) of Milne]; $\alpha_2 = 0$, $\alpha_3 = 1$ [called the $\frac{3}{8}$ rule and is an adaptation of (1.5-12)]; and $\alpha_2 = 0$, $\alpha_3 = -\frac{1}{8}$ [called *Hamming's* formula and to be used in Chapter 4].

Using (1.8-8) but with $\beta_3 = 0$ and α_2 as the free parameter there results Table 1.11 with $p = 4$. Note that we generate Milne's equation again, Hamming's again, and a number of other possible formulas.

In summary, we are able to generate an enormous number of explicit and implicit formulas by the method of undetermined coefficients.

TABLE 1.10

IMPLICIT EQUATION (1.8-8)[a]

$k = 3, \quad p = 4$

$\alpha_1 = 1 - \alpha_2 - \alpha_3$	1	0	0	1/3	1/2	0	9/8
$\alpha_2 = \alpha_2$	0	1	0	1/3	1/2	2/3	0
$\alpha_3 = \alpha_3$	0	0	1	1/3	0	1/3	-1/8
$\beta_0 = (1/24)(9 - \alpha_2)$	9/24	1/3	3/8	13/36	17/48	25/72	3/8
$\beta_1 = (1/24)(19 + 13\alpha_2 + 8\alpha_3)$	19/24	4/3	9/8	39/36	51/48	91/72	6/8
$\beta_2 = (1/24)(-5 + 13\alpha_2 + 32\alpha_3)$	-5/24	1/3	9/8	15/36	3/48	43/72	3/8
$\beta_3 = (1/24)(1 - \alpha_2 + 8\alpha_3)$	1/24	0	3/8	5/36	1/48	9/72	0
$^b\,T_x = (1/6)(-19 + 11\alpha_2 - 8\alpha_3)$	-19/6	-4/3	-9/2	-3	-9/4	-43/18	-3

[a] From "Numerical Methods for Scientists and Engineers," R. W. Hamming. Copyright © 1962 by McGraw-Hill, Inc. Used by permission of McGraw-Hill Book Company.

[b] Multiply by $h^5 y^{[5]}(\zeta)/5!$

TABLE 1.11

IMPLICIT EQUATION (1.8-8)

$\beta_3 = 0, \quad k = 3, \quad p = 4$

$\alpha_1 = (1/8)(9 - 9\alpha_2)$	0	9/17	1	9/8	9/5
$\alpha_2 = \alpha_2$	1	9/17	1/9	0	-3/5
$\alpha_3 = -(1/8)(1 - \alpha_2)$	0	-1/17	-1/9	-1/8	-1/5
$\beta_0 = (1/24)(9 - \alpha_2)$	1/3	6/17	10/27	3/8	2/5
$\beta_1 = (1/12)(9 + 7\alpha_2)$	4/3	18/17	22/27	3/4	2/5
$\beta_2 = (1/24)(-9 + 17\alpha_2)$	1/3	0	-8/27	-3/8	-4/5
$T_x \times h^5 y^{[5]}(\zeta)$	-1/90	-3/170	-19/810	-1/40	-1/30

1.9. COMPILATION OF VARIOUS MULTIPLE-STEP FORMULAS INCLUDING y_i AND y_i''

Because the special second-order equation $y'' = f(x, y)$ occurs frequently in physical representations, there is sufficient motivation to investigate various k-step formulas which use only y_n, y_{n-1}, ... and y_{n+1}'', y_n'', y_{n-1}'', ... In fact, in Section 1.7.3 we already derived one case, (1.7-13), which fits into this category. Here we wish to develop this approach further.

1.9.1. Explicit Forms

Consider the explicit form

$$y_{n+1} = \alpha_1 y_n + \alpha_2 y_{n-1} + h^2[\gamma_1 y_n'' + \gamma_2 y_{n-1}''] \tag{1.9-1}$$

This is a 2-step process ($k = 2$) with four parameters and it can be made exact for $p = 3$. If we do so we find

$$y_{n+1} = 2y_n - y_{n-1} + h^2 y_n''$$
$$T_\alpha = (h^4/12)y^{[4]}(\zeta) \tag{1.9-2}$$

An alternative $k = 4$, $p = 5$ formula which finds application is

$$y_{n+1} = 2y_{n-1} - y_{n-3} + (4h^2/3)(y_n'' + y_{n-1}'' + y_{n-2}'')$$
$$T_\alpha = -(16h^6/240)y^{[6]}(\zeta) \tag{1.9-3}$$

while others are given by

$$y_{n+1} = 2y_n - y_{n-1} + (h^2/12)[13y_n'' - 2y_{n-1}'' + y_{n-2}''] \tag{1.9-4}$$

$$y_{n+1} = 2y_n - y_{n-1} + (h^2/12)[14y_n'' - 5y_{n-1}'' + 4y_{n-2}'' - 14y_{n-3}''] \tag{1.9-5}$$

$$y_{n+1} = 2y_n - y_{n-1} + (h/240)$$
$$\times [29y_n'' - 176y_{n-1}'' + 194y_{n-2}'' - 96y_{n-3}'' + 19y_{n-4}''] \tag{1.9-6}$$

where (1.9-4) is called *Stormer's equation*.

1.9.2. Implicit Forms

An equivalent approach can be made in the implicit form. If we select the $k = 2$ form

$$y_{n+1} = \alpha_1 y_n + \alpha_2 y_{n-1} + h^2[\gamma_0 y_{n+1}'' + \gamma_2 y_{n-1}''] \tag{1.9-7}$$

and make it exact for $p = 3$, there results

$$y_{n+1} = 2y_n - y_{n-1} + (h^2/2)[y_{n+1}'' + y_{n-1}'']$$
$$T_\alpha = -(h^4/24)y^{[4]}(\zeta) \tag{1.9-8}$$

Alternatively we might select $k = 3$ in the form

$$y_{n+1} = \alpha_1 y_n + \alpha_2 y_{n-1} + \alpha_3 y_{n-2}$$
$$+ h^2[\gamma_0 y''_{n+1} + \gamma_1 y''_n + \gamma_2 y''_{n-1} + \gamma_3 y''_{n-2}] \tag{1.9-9}$$

and make exact for $p = 5$ with α_3 as a free parameter. The results are shown in Table 1.12.

TABLE 1.12

IMPLICIT EQUATION $(1.9-9)^a$

	$k = 3, \quad p = 5$		
$\alpha_1 = 2 + \alpha_3$	3	2	1
$\alpha_2 = -1 - 2\alpha_3$	-3	-1	1
$\alpha_3 = \alpha_3$	1	0	-1
$\gamma_0 = 1/12$	1/12	1/12	1/12
$\gamma_1 = (10 - \alpha_3)/12$	9/12	10/12	11/12
$\gamma_2 = (1 - 10\alpha_3)/12$	$-9/12$	1/12	11/12
$\gamma_3 = -\alpha_3/12$	$-1/12$	0	1/12
$^b T_\alpha = -3(1 - \alpha_3)$	0	-3	-6

^{a} From "Numerical Methods for Scientists and Engineers," R. W. Hamming. Copyright © 1962 by McGraw-Hill, Inc. Used by permission of McGraw-Hill Book Company.

^{b} Multiply by $h^6 y^{[6]}/6!$.

The case $\alpha_3 = 0$ yields

$$y_{n+1} = 2y_n - y_{n-1} + (h^2/12)[y''_{n+1} + 10y''_n + y''_{n-1}]$$
$$T_\alpha = -(h^6/240)y^{[6]}(\zeta) \tag{1.9-10}$$

which is known as *Numerov's* or the *royal road* formula. This is (1.7-13) as derived previously. An additional implicit form is

$$y_{n+1} = 2y_n - y_{n-1} + (h^2/240)[19y''_{n+1} + 204y''_n + 14y''_{n-1} + 4y''_{n-2} - y''_{n-3}] \tag{1.9-11}$$

1.10. MULTIPLE-STEP INTERGRATION FORMULAS: HIGHER DERIVATIVE TERMS

As a final set of formulas of the specific type under consideration, we now consider first those formulas which involve y_i, y_i' and y_i'' and then y_i, y_i', y_i'' and y_i'''. The first set are like (1.7-14) previously derived. Further higher-order derivatives can be used, but the details should be evident from the present development.

1.10.1. Explicit Forms

We start with the explicit form

$$y_{n+1} = \alpha_1 y_n + \alpha_2 y_{n-1} + h[\beta_1 y_n' + \beta_2 y_{n-1}'] + h^2[\gamma_1 y_n'' + \gamma_2 y_{n-1}'']$$

$$(1.10\text{-}1)$$

with six parameters. This is a 2-step $(k = 2)$ formula and with $p = 4$ one free parameter can be used; we select α_2 as the free parameter. The result of the undetermined coefficient approach is shown in Table 1.13.

TABLE 1.13

EXPLICIT EQUATION $(1.10\text{-}1)^a$

	$k = 2$,	$p = 4$	
$\alpha_1 = 1 - \alpha_2$	1	1/2	0
$\alpha_2 = \alpha_2$	0	1/2	1
$\beta_1 = (-1 + \alpha_2)/2$	$-1/2$	$-1/4$	0
$\beta_2 = (3 + \alpha_2/12$	3/2	7/4	2
$\gamma_1 = (17 - \alpha_2)/12$	17/12	33/24	4/3
$\gamma_2 = (7 + \alpha_2)/12$	7/12	15/24	2/3
$^b T_\alpha = (31 + \alpha_2)/6$	31/6	21/4	16/3

^{a} From "Numerical Methods for Scientists and Engineers," R. W. Hamming. Copyright © 1962 by McGraw-Hill, Inc. Used by permission of McGraw-Hill Book Company.
^{b} Multiply by $h^5 y^{[5]}/5!$.

The cases $\alpha_2 = 0$ and 1 yield the formulas

$$y_{n+1} = y_n + (h/2)[-y_n' + 3y_{n-1}'] + (h^2/12)[17y_n'' + 7y_{n-1}'']$$
$$T_\alpha = (31h^5/720)y^{[5]}(\zeta)$$

$$(1.10\text{-}2)$$

and

$$y_{n+1} = y_{n-1} + 2hy_{n-1}' + (2h^2/3)[2y_n'' + y_{n-1}'']$$
$$T_\alpha = (16h^5/360)y^{[5]}(\zeta)$$

$$(1.10\text{-}3)$$

1.10.2. Implicit Forms

We can do the same for the implicit forms but we merely write here two equations which find utility:

$$y_{n+1} = y_n + (h/2)[y_{n+1}' + y_n'] + (h^2/12)[-y_{n+1}'' + y_n'']$$
$$T_\alpha = (h^5/720)y^{[5]}(\zeta)$$

$$(1.10\text{-}4)$$

which is (1.7-15) as previously derived and

$$y_{n+1} = 2y_n - y_{n-1} + (3h/8)[y'_{n+1} - y'_{n-1}] + (h^2/24)[y''_{n+1} - 8y''_n + y''_{n-1}]$$
$$\text{(1.10-5)}$$

with $T_\alpha = (h^8/60480)y^{[8]}(\zeta)$.

1.10.3. Higher-Derivative Forms

It would seem apparent that we might continue the procedure of including higher-order derivatives in different k-step equations. In particular, Lambert and Mitchell [5] have developed the necessary equations and presented a table for all the forms. Typical explicit formulas which result are

$$y_{n+1} = y_n + hy_n' + (h^2/2)y''_n + (h^3/6)y'''_n$$
$$T_\alpha = (h^4/24)y^{[4]}(\zeta)$$
$$\text{(1.10-6)}$$

$$y_{n+1} = y_n + (h/2)[15y_n' - 13y'_{n-1}] - (h^2/10)[31y''_n + 29y''_{n-1}]$$
$$+ (h^3/120)[111y'''_n - 49y'''_{n-1}]$$
$$T_\alpha = (209h^7/100,800)y^{[7]}(\zeta)$$
$$\text{(1.10-7)}$$

and a typical implicit formula is

$$y_{n+1} = y_n + (h/2)[y'_{n+1} + y_n'] - (h^2/10)[y''_{n+1} - y''_n]$$
$$+ (h^3/120)[y'''_{n+1} + y'''_n]$$
$$T_\alpha = (h^7/100,800)y^{[7]}(\zeta)$$
$$\text{(1.10-8)}$$

Many other formulas such as the Adams–Bashforth, Adams–Moulton, etc., can also be ascertained in the Lambert–Mitchell table.

The question arises immediately as to whether a higher accuracy (a lower truncation error) can more easily be obtained by increasing k or by increasing the order of the derivatives. Since either procedure involves the determination of further coefficients or parameters, both paths will increase the accuracy. As Lambert and Mitchell show, the truncation error is more strongly influenced by the order of the derivatives than the number of steps k. Thus to achieve a minimum truncation error, the highest-order derivative should be used with an associated small k. The use of a small k also tends to remove stability problems (see Chapter 3); however, it may frequently be difficult to calculate the higher derivatives if $f(x, y)$ is a complicated function.

1.11. FURTHER DEFINITIONS

1.11.1. The Local Truncation Error

In polynomial approximation at $n + 1$ data points the difference or error between the true value of $y(x)$ and the calculated approximation is given by the truncation error. If $y(x)$ is known analytically, then this error or at least its maximum value or bound can be calculated. For numerical differentiation or integration, one can differentiate or integrate the error function to estimate the error. As already indicated, the first neglected term in the operational form can be used to specify the truncation error.

The same concept can be carried over directly to the solution of ODEs. By analyzing the first neglected term in a finite series or, alternatively, in the method of undetermined coefficients, by substituting a $p + 1$ degree polynomial into a formula of order p we can develop an error term of the form

$$T(x, h) = Ch^{p+1} y^{[p+1]}(\zeta) \tag{1.11-1}$$

Now we term $T(x, h)$ (rather than T_α) the *local truncation error*, since the algorithm will be applied over and over as the calculation proceeds from x_0 to $x_1, \ldots, x_n, x_{n+1}, \ldots$. In a broad sense we can define a general single-step method by

$$y_{n+1} = y_n + h\Phi(x_n, y_n; h), \qquad n = 0, 1, 2, \ldots \tag{1.11-2}$$

where Φ is the *increment function* and depends on x_n, y_n, and the h used. All the quantities in (1.11-2) (except h) can be scalars or vectors. As before, we define $y(x)$ as the exact solution for the ODE and then for any x we call

$$T(x, h) = y(x) + h\Phi(x, y(x); h) - y(x + h) \tag{1.11-3}$$

as the local truncation (or discretization) error. If p is the largest integer p' with the property that

$$T(x, h) = O(h^{p'+1}) \qquad \text{as} \quad h \to 0 \tag{1.11-4}$$

then p is called the order or order of accuracy of the single-step method. This conforms to the definition of p already used. The terminology here is that for any two scalar quantities, say $u(h)$ and $g(h)$, depending on h, the notation $u(h) = O(g(h))$ means that there exists some positive constant C_1, independent of h, such that

$$\lim_{h \to 0} |u(h)/g(h)| \le C_1$$

The order of a method is regarded as a measure of the accuracy of the method, since the local truncation error is comparable to h^{p+1} but not to any higher power of h.

As an illustration of these concepts, let us pick the simplest case in which $\Phi = f(x_n, y_n)$. This means that (1.11-2) becomes

$$y_{n+1} = y_n + hf(x_n, y_n)$$

which is merely Euler's method. From our previous discussion (see (1.6-7)), we know that $T(x, h) = \frac{1}{2}h^2 y^{[2]}(\zeta)$ and thus $p = 1$, the algorithm is first order, and the local truncation error is $O(h^2)$. Using (1.11-3) we can write

$$\begin{aligned} T(x, h) &= y(x) + hf(x, y(x)) - y(x + h) \\ &= y(x) + hy'(x) - y(x + h) \end{aligned} \tag{1.11-5}$$

as long as one continuous derivative exists. In the same sense it is not difficult to show that a Taylor series expansion through p terms is of order p as long as p continuous derivatives exist.

Of further interest is the question of bounds on the errors. There are basically two types; the first is a strict bound, related to (1.11-1), such as

$$|\text{error}| < h^{p+1}C \tag{1.11-6}$$

and the second is an asymptotic bound such as

$$\text{error} \sim h^{p+1}C$$

or

$$h^{-(p+1)}[\text{error}] \to C \quad \text{as} \quad h \to 0 \tag{1.11-7}$$

As a typical illustration, for Euler's method and (1.11-5) if

$$|y^{[2]}(x)| \le M$$

then by using a Taylor series it is possible to show directly that

$$|T(x, h)| = |y(x) + hy'(x) - y(x + h)| \le (h^2/2)M$$

This is a strict bound on the local truncation error of Euler's method.

While the form of (1.11-1) is usually correct, there are cases when erroneous results can be obtained. We wish to develop here the conditions under which the form is correct.

Let us start the discussion by considering the linear operator $L[y(x)]$, where the operator is of order p if

$$\begin{aligned} L[x^j] &= 0, \quad j = 0, 1, \dots, p \\ L[x^j] &\ne 0, \quad j = p + 1 \end{aligned} \tag{1.11-8}$$

Using Taylor's theorem and assuming the derivatives of y exist, we may write for $y(x)$

$$y(x) = \sum_{j=0}^{p} \frac{y^{[j]}(0)}{j!} x^j + \frac{1}{p!} \int_0^k (x - v)_+^p y^{[p+1]}(v) \, dv \tag{1.11-9}$$

where

$$(x - v)_+ = \max(0, x - v) \tag{1.11-10}$$

Combining terms after using L, there results

$$L[y(x)] = \frac{1}{p!} \int_0^k G_p(v) y^{[p+1]}(v) \, dv \tag{1.11-11}$$

where $G_p(v)$, called the *influence function*, is given by

$$G_p(v) = L[(x - v)_+^p] \tag{1.11-12}$$

If $f(x)$ is continuous in $a \le x \le b$ and $g(x)$ does not change sign in that interval, then there exists at least one number $x = \zeta$ between a and b such that

$$\int_a^b f(x)g(x)\,dx = f(\zeta)\int_a^b g(x)\,dx \tag{1.11-13}$$

provided $g(x)$ is integrable on $[a, b]$ [2]. But this means that if $G_p(v)$ is of constant sign (and only if) we may write (1.11-11) as

$$L[y(x)] = \frac{y^{[p+1]}(\zeta)}{p!}\int_0^k G_p(v)\,dv \tag{1.11-14}$$

or, from (1.11-8), as

$$L[y(x)] = \frac{L[x^{p+1}]}{(p+1)!}\,y^{[p+1]}(\zeta) \tag{1.11-15}$$

It then follows that if we define $T(x, h)$ to be the difference $y_{n+1} - y(x_{n+1})$ that

$$T(x, h) = \frac{L[x^{p+1}]}{(p+1)!}\,h^{p+1}y^{[p+1]}(\zeta) \tag{1.11-16}$$

The key point is that we get the form of (1.11-1) if and only if $G_p(v)$ remains of constant sign over the interval of integration. Hamming has illustrated the case of a general k-step formula with variable parameters. By leaving some of the parameters free (note that this is exactly what has been done to derive many of the formulas) it is possible to examine plots of $G_p(v)$ versus the free parameters. On this basis, one finds certain regions or sets of explicit values of the free parameters for which $G_p(v)$ remain of constant sign. Only within these regions (which may, of course, be infinite) will the truncation expressions of the form of (1.11-1) hold correctly. Fortunately this holds for the forms we shall analyze.

1.11.2 The Accumulated Truncation Error

While the local truncation error specifies the error at one step in a calculation, the *accumulated truncation error* indicates how the local errors propagate over the entire calculation. Since this error is extremely important, we here illustrate how one may estimate it using Euler's method as an example. We specify that the exact value of $y(x)$ is given by

$$y(x_{n+1}) = y(x_n) + h\Phi(x_n, y(x_n); h) - T(x_n, h) \tag{1.11-17}$$

This equation determines the local truncation error $T(x_n, h)$ with the negative sign used for convenience in later manipulation. Note that as $h \to 0$, $T(x_n, h) \to 0$ as normally desired. Defining the error ε_n

$$\varepsilon_n = y_n - y(x_n) \tag{1.11-18}$$

subtracting (1.11-2) and (1.11-17) yields

$$\varepsilon_{n+1} = \varepsilon_n + h[\Phi(x_n, y_n; h) - \Phi(x_n, y(x_n); h)] + T(x_n, h) \tag{1.11-19}$$

For Euler's method, $\Phi(x_n, y_n; h) = f(x_n, y_n)$ and (1.11-19) becomes

$$\varepsilon_{n+1} = \varepsilon_n + h[f(x_n, y_n) - f(x_n, y(x_n))] + T(x_n, h) \tag{1.11-20}$$

Assuming $f(x, y)$ satisfies a Lipschitz condition such that

$$|f(x_n, y_n) - f(x_n, y(x_n))| \le L|y_n - y(x_n)| \le L|\varepsilon_n| \tag{1.11-21}$$

it follows that

$$|\varepsilon_{n+1}| \le (1 + hL)|\varepsilon_n| + T(x, h) \tag{1.11-22}$$

where $T(x, h)$ is the maximum local truncation error,

$$T(x, h) = \max_n |T(x_n, h)| \tag{1.11-23}$$

But we may get a further and more explicit form of this equation by using $n = 0, 1, 2, \ldots$ and letting $T(x, h) = T$. Thus

$$|\varepsilon_1| \le (1 + hL)|\varepsilon_0| + T$$
$$|\varepsilon_2| \le (1 + hL)|\varepsilon_1| + T$$
$$\le (1 + hL)^2|\varepsilon_0| + [1 + (1 + hL)]T$$
$$|\varepsilon_3| \le (1 + hL)|\varepsilon_2| + T$$
$$\le (1 + hL)^3|\varepsilon_0| + [1 + (1 + hL) + (1 + hL)^2]T$$
$$\vdots$$

By noting that the coefficient on T is forming a geometric series, we now write that, by induction,

$$|\varepsilon_{n+1}| \le (1 + hL)^{n+1}|\varepsilon_0| + \left[\frac{(1 + hL)^{n+1} - 1}{hL}\right]T \tag{1.11-24}$$

This equation gives us a bound on the accumulated truncation error at the $(n + 1)$st step in the Euler method calculations.

If the initial condition for the differential equation has zero error, then $\varepsilon_0 = 0$ and

$$|\varepsilon_{n+1}| \le \left[\frac{(1 + hL)^{n+1} - 1}{hL}\right]T \tag{1.11-25}$$

Alternatively we could have approached this same problem using the mean value theorem. (This requires certain smoothness conditions for $f(x, y)$, but these are usually fulfilled.) In this case we would have replaced (1.11-20) with

$$
\begin{aligned}
\varepsilon_{n+1} &= \varepsilon_n + hf_{\bar{y}}\varepsilon_n + T(x_n, h) \\
&= (1 + hf_{\bar{y}})\varepsilon_n + T(x_n, h)
\end{aligned}
\tag{1.11-26}
$$

with $f_{\bar{y}} = [\partial f/\partial y]_{\bar{y}}$, $y_n \le \bar{y} \le y(x_n)$. This is a linear, first-order difference equation whose solution we can find easily. In fact if we assume $|f_{\bar{y}}| \le L$ and $|T(x_n, h)| \le T$, then we can obtain (1.11-24) in a direct manner.

We see that (1.11-22) allows a calculation of the bound on the accumulation of the local truncation error assuming ε_n is known. Equations (1.11-24) or (1.11-25), by contrast, give an independent error bound without any knowledge of the previous errors. By being more explicit about T it is possible to gain further information. For example, for Euler's method we know that

$$
T(x_n, h) = (h^2/2)y''(\zeta)
$$

with $y'' = f' = f_x + f_y f$. If we then bound $y''(\zeta)$ by

$$
M = \max_{a \le \zeta \le b} y''(\zeta)
\tag{1.11-27}
$$

assuming f_x and f_y are continuous,

$$
T = (h^2/2)M
\tag{1.11-28}
$$

But noting that

$$
1 + u \le e^u, \qquad u \ge 0
$$

we see that

$$
(1 + hL)^{n+1} \le \exp[(n + 1)hL] = \exp[L(x_{n+1} - x_0)]
\tag{1.11-29}
$$

When (1.11-28) and (1.11-29) are used in (1.11-24), there results

$$
|\varepsilon_{n+1}| \le \exp[L(x_{n+1} - x_0)]|\varepsilon_0| + (hM/2L)\,[\exp[L(x_{n+1} - x_0)] - 1]
\tag{1.11-30}
$$

or, for $\varepsilon_0 = 0$,

$$
|\varepsilon_{n+1}| \le (hM/2L)\,[\exp[L(x_{n+1} - x_0)] - 1]
\tag{1.11-31}
$$

Obviously as $h \to 0$, the truncation error at x_{n+1} goes to zero.

1.11.3. Other Items

All of the above is based on the proposition of exact arithmetic, i.e., all numbers are evaluated with infinite precision. Obviously this is almost never the case when fixed word length computers are used and we must mention the *local rounding error*. Thus we really do not have (1.11-2), but instead have

$$
\tilde{y}_{n+1} = \tilde{y}_n + h\Phi(x_n, \tilde{y}_n; h) + e_n
\tag{1.11-32}
$$

where e_n is the local rounding error. The tilde over the terms in (1.11-32) indicate that these are rounded and not exact numbers. It would seem fairly obvious that this rounding error decreases as the computer word length is increased (e.g., double precision) but increases as h decreases, since more arithmetic calculations are required to get from $x = x_0$ to $x = x_n$.

If the local round-off error is included in the derivation of the truncation error in Euler's method, then instead of (1.11-31) there results

$$E_{n+1} \leq \left[\frac{hM}{2} + \frac{e}{h} \right] \left[\frac{\exp[L(x_{n+1} - x_0)] - 1}{L} \right] \qquad (1.11\text{-}33)$$

where $e = \max_n e_n$, $E_0 = 0$, and $E_n = \tilde{y}_n - y(x_n)$. This equation tends to confirm our intuitive idea that as $h \to 0$, the truncation error decreases toward zero whereas the round-off error increases toward infinity. Obviously there is an optimum value of h, h_{opt}, for which the total error is the smallest. Below this value the main contribution to the overall error is round-off, while above this value it is truncation. By minimizing the sum of $(hM/2) + (e/h)$, it is possible to determine that optimum h for Euler's method as

$$h_{opt} = (2Me)^{1/2} \qquad (1.11\text{-}34)$$

For this value of h_{opt}, $(hM/2) = (e/h)$ and the errors are equal. Even when other numerical integration methods are used, it is still a good rule to select an h for which the truncation error and the round-off error have equal orders of magnitude. At this point maximum accuracy will be achieved.

To briefly test out these ideas, we have selected the system $y' = -y$, $y_0 = 1.0$, whose solution is $y(x) = e^{-x}$. Using a series of h values from $h = 10^{-6}$ to $h = 1.0$, we computed the numerical total error at $x = 1.0$ using Euler's method and a fourth-order Runge–Kutta method to be derived in Chapter 2. Both single precision (SP) and double precision (DP) results were obtained on an IBM 360 computer for which e is approximately $1_{10^{-7}}$ and $1_{10^{-14}}$, respectively. Since $M = 1$ in this case, (1.11-34) would predict h_{opt} of approximately h_{opt} (SP) $= 5_{10^{-4}}$ and h_{opt} (DP) $= 1_{10^{-7}}$. Table 1.14 shows the actual numerical data obtained.

These data show that the minimum error in Euler (SP) is approximately where predicted while the (DP) minimum is beyond the data obtained (as it roughly should be). We can see further that at large h the SP and DP results are essentially identical, but that as h decreases the round-off becomes important in the SP results but not in the DP. The fourth-order integration results are roughly the same in a qualitative fashion, but since the truncation error is much smaller than for Euler's method, the SP minimum occurs at a larger h and the minimum even occurs in the DP results. In general we can see the actual advantage to decreasing values of h in terms of accuracy; but also we can see that beyond a certain value of h this feature actually turns around.

TABLE 1.14

INFLUENCE OF TRUNCATION AND ROUND-OFF ERRORS
FOR $y' = -y$ AT $x = 1.0$

h	Euler		Fourth-order Runge–Kutta	
	SP	DP	SP	DP
$1_{10^{-6}}$	—	$1.8_{10^{-7}}$	—	—
$1_{10^{-5}}$	$1.6_{10^{-3}}$	$1.8_{10^{-6}}$	$1.6_{10^{-3}}$	$5.2_{10^{-13}}$
$5_{10^{-5}}$	$3.3_{10^{-4}}$	$9.2_{10^{-6}}$	$2.3_{10^{-4}}$	$1.1_{10^{-13}}$
$1_{10^{-4}}$	$1.8_{10^{-4}}$	$1.8_{10^{-5}}$	$1.6_{10^{-4}}$	$7.6_{10^{-14}}$
$2.5_{10^{-4}}$	$1.4_{10^{-4}}$	$4.6_{10^{-5}}$	$9.2_{10^{-4}}$	$3.4_{10^{-14}}$
$5_{10^{-4}}$	$1.5_{10^{-4}}$	$9.2_{10^{-5}}$	$5.8_{10^{-5}}$	$1.7_{10^{-14}}$
$1_{10^{-3}}$	$2.0_{10^{-4}}$	$1.8_{10^{-4}}$	$2.1_{10^{-5}}$	$5.1_{10^{-15}}$
$2.5_{10^{-3}}$	$4.7_{10^{-4}}$	$4.6_{10^{-4}}$	$6.9_{10^{-6}}$	$1.2_{10^{-13}}$
$5_{10^{-3}}$	$9.2_{10^{-4}}$	$9.2_{10^{-4}}$	$3.3_{10^{-6}}$	$1.9_{10^{-12}}$
$1_{10^{-2}}$	$1.8_{10^{-3}}$	$1.8_{10^{-3}}$	$1.7_{10^{-6}}$	$3.0_{10^{-11}}$
$5_{10^{-2}}$	$9.4_{10^{-3}}$	$9.4_{10^{-3}}$	$3.6_{10^{-7}}$	$2.0_{10^{-8}}$
$1_{10^{-1}}$	$1.9_{10^{-2}}$	$1.9_{10^{-2}}$	$5.3_{10^{-7}}$	$3.3_{10^{-7}}$
$5_{10^{-1}}$	$1.2_{10^{-1}}$	$1.2_{10^{-1}}$	$2.9_{10^{-4}}$	$2.9_{10^{-4}}$
1.0	—	—	$7.1_{10^{-3}}$	$7.1_{10^{-3}}$

Other terms of interest for later sections of this book are the *stability* and the *principal error function*. The stability, as with the accumulated error, is qualitatively related to how errors (local truncation or round-off) propagate through n steps of the algorithm, i.e., they relate to the sensitivity of the algorithm to error propagation. The principal error function is merely ϕ in the local truncation error written as

$$T(x, h) = h^{p+1}\phi + O(h^{p+2}) \qquad (1.11\text{-}35)$$

For Euler's method, $\phi = \frac{1}{2}y^{[2]}(\zeta)$. ϕ may be thought of as all the terms in the local truncation error except for the h^{p+1} term.

REFERENCES

1. Davis, P. J., and Rabinowitz, P., "Numerical Integration." Blaisdell, New York, 1967.
2. Hamming, R. W., "Numerical Methods for Scientists and Engineers." McGraw-Hill, New York, 1962.
3. Henrici, P., "Discrete Variable Methods in Ordinary Differential Equations." Wiley, New York, 1962.
4. Kopal, Z., "Numerical Analysis." Wiley, New York, 1955.
5. Lambert, J. D., and Mitchell, A. R., On the solution of $y' = f(x, y)$ by a class of high accuracy difference formulae of low order, *Z. Angew Math. Phys.* **13**, 223 (1962).

6. Lapidus, L. "Digital Computation for Chemical Engineers." McGraw-Hill, New York, 1962.
7. Salzer, H. E., Osculatory extrapolation and a new method for the numerical integration of differential equations, *J. Franklin Inst.* **262**, 111 (1956).
8. Salzer, H. E., Numerical integration of $y'' = \phi(x, y, y')$ using osculatory interpolation, *J. Franklin Inst.* **263**, 401 (1957).
9. Van Wyk, R., Variable mesh methods for differential equations, *NASA Report*, CR-1247 (Nov., 1968).

2

Runge–Kutta and Allied Single-Step Methods

In this chapter the principles of Runge–Kutta type single-step methods will be developed and a large number of specific integration formulas will be presented. However, aspects of stability for the wide variety of explicit and implicit formulas will be postponed until Chapter 3. Further, in Chapter 4 certain formulas will be presented combining the Runge–Kutta principles with those of the multiple-step equations.

2.1. DEVELOPMENT OF SINGLE-STEP RUNGE–KUTTA FORMULAS

We have already mentioned one single-step formula, Euler's formula as given by (1.6-7), in which the values (x_{n+1}, y_{n+1}) are obtained from a knowledge of only (x_n, y_n). In this section we want to develop the basis for an entire class of single-step methods within the broad classification of Runge–Kutta formulas.

Runge was the first to point out that it was possible to avoid the successive differentiation in the Taylor series while preserving the accuracy. The new feature is to set up a problem with undetermined parameters and make the result as high order as possible by using evaluations of $f(x, y)$ *within* the interval (x_n, y_n) and (x_{n+1}, y_{n+1}). In other words the derivatives in the Taylor series are bypassed by requiring $f(x, y)$ to be evaluated a number of additional times within the interval mentioned.

Thus we set up the general single-step equations

$$y_{n+1} = y_n + \sum_{i=1}^{v} w_i k_i \qquad (2.1\text{-}1)$$

with the w_i as weighting coefficients to be determined, v as the number of $f(x, y)$ substitutions, and the k_i satisfying the explicit sequence

$$k_i = hf\left(x_n + c_i h, \; y_n + \sum_{j=1}^{i-1} a_{ij} k_j\right), \qquad c_1 = 0, \qquad i = 1, 2, \ldots, v \qquad (2.1\text{-}2)$$

or

$$\begin{aligned}
k_1 &= hf(x_n, y_n) \\
k_2 &= hf(x_n + c_2 h, \; y_n + a_{21} k_1) \\
k_3 &= hf(x_n + c_3 h, \; y_n + a_{31} k_1 + a_{32} k_2) \\
&\vdots
\end{aligned} \qquad (2.1\text{-}2')$$

Note immediately that if $v = 1$, then only k_1 is needed and (2.1-1) yields Euler's formula. It can be seen that there are parameters $w_1, \ldots, w_v$ in (2.1-1) and $c_2, \ldots, c_v, a_{2j}, \ldots$ in (2.1-2) that must be determined. Each set of parameters (when determined) will specify the points (x, y) at which $f(x, y)$ is to be evaluated. Thus while the overall calculation yields y_{n+1} from y_n, it is necessary to evaluate $f(x, y)$ at points between x_{n+1} and x_n.

To obtain specific values for the unknown parameters, we expand y_{n+1} in powers of h such that it agrees with the solution of the differential equations to a specified number of terms in a Taylor series. For illustration purposes, let us choose $v = 3$ and note that we then have eight parameters w_1, w_2, w_3, c_2, c_3, a_{21}, a_{31}, and a_{32} to evaluate.

Since $f(x, y)$ is a function of both x and y, we see that [see (1.2-3) and (1.2-4)]

$$\begin{aligned}
y' &= f(x, y) = f \\
y'' &= f_x + f_y f \\
y''' &= f_{xx} + 2ff_{xy} + f^2 f_{yy} + f_y(f_x + ff_y)
\end{aligned} \qquad (2.1\text{-}3)$$

Using $f(x_n, y_n) = f_n$ a Taylor series through third-order ($v = 3$) terms yields

$$\begin{aligned}
y_{n+1} = y_n &+ hf_n + \tfrac{1}{2}h^2(f_x + ff_y)_n \\
&+ \tfrac{1}{6}h^3(f_{xx} + 2ff_{xy} + f^2 f_{yy} + f_y f_x + f_y ff_y)_n
\end{aligned} \qquad (2.1\text{-}4)$$

Expanding the terms in (2.1-2) in the same manner yields

$$\begin{aligned}
k_1 =\; & hf_n \\
k_2 =\; & hf_n + h^2(c_2 f_x + a_{21} ff_y)_n \\
& + h^3((c_2{}^2/2)f_{xx} + c_2 a_{21} ff_{xy} + \tfrac{1}{2}a_{21} f^2 f_{yy})_n \\
k_3 =\; & hf_n + h^2(c_3 f_x + a_{31} ff_y + a_{32} ff_y)_n \\
& + h^3((c_3{}^2/2)f_{xx} + c_3(a_{31} + a_{32}) ff_{xy} + \tfrac{1}{2}(a_{31} + a_{32})^2 f^2 f_{yy} \\
& + a_{32}(c_3 f_x + a_{21} ff_y) f_y)_n
\end{aligned} \qquad (2.1\text{-}5)$$

Substituting (2.1-5) into (2.1-1) and comparing term by term with (2.1-4) we obtain the eight identities

$$
\begin{aligned}
w_1 + w_2 + w_3 &= 1 & c_2 w_2 + c_3 w_3 &= \tfrac{1}{2} \\
a_{21} w_2 + (a_{31} + a_{32}) w_3 &= \tfrac{1}{2} & \tfrac{1}{2} c_2{}^2 w_2 + \tfrac{1}{2} c_3{}^2 w_3 &= \tfrac{1}{6} \\
c_2 a_{21} w_2 + c_3 (a_{31} + a_{32}) w_3 &= \tfrac{1}{3} & \tfrac{1}{2} a_{21}^2 w_2 + \tfrac{1}{2} (a_{31} + a_{32})^2 w_3 &= \tfrac{1}{6} \\
c_2 a_{31} w_3 &= \tfrac{1}{6} & a_{21} a_{31} w_3 &= \tfrac{1}{6}
\end{aligned}
\tag{2.1-6}
$$

However, an examination of these equations reveals that

$$
c_2 = a_{21} \quad \text{and} \quad c_3 = a_{31} + a_{32} \tag{2.1-7}
$$

and that there are only four independent relations left. This is connected with six unknown parameters since (2.1-7) removes two unknowns from the original eight mentioned above. Thus if we leave two free parameters, say c_2 and c_3, we may determine all the others uniquely. This is equivalent to the case $p = 3$ in our previous terminology and we thus call the result a third-order Runge–Kutta system ($p = v = 3$).

2.1.1. Specific Runge–Kutta Forms

Let us now examine some possible results for different order Runge–Kutta systems. Of specific interest is that any analysis as above *always* leads to the following identities

$$
\sum_{i=1}^{v} w_i = 1
$$
$$
c_i = \sum_{j=1}^{i-1} a_{ij}, \quad i \neq 1, \quad i = 2, \ldots, v
\tag{2.1-8}
$$

If we had carried the analysis through to only h^2 terms, there would result:

$$
w_1 + w_2 = 1, \quad c_2 w_2 = \tfrac{1}{2}, \quad c_2 = a_{21} \tag{2.1-9}
$$

If we leave c_2 as a free parameter, we determine w_1 and w_2. Thus for $c_2 = \tfrac{1}{2}, \tfrac{1}{3}$, and 1 there results (w_1, w_2) of $(0, 1)$, $(\tfrac{1}{4}, \tfrac{3}{4})$ and $(\tfrac{1}{2}, \tfrac{1}{2})$ respectively. These in turn yield the final three equations

$$
y_{n+1} = y_n + h f(x_n + \tfrac{1}{2} h, y_n + \tfrac{1}{2} h f_n) \tag{2.1-10}
$$

$$
y_{n+1} = y_n + (h/4)[f(x_n, y_n) + 3 f(x_n + \tfrac{2}{3} h, y_n + \tfrac{2}{3} h f_n)] \tag{2.1-11}
$$

$$
y_{n+1} = y_n + (h/2)[f(x_n, y_n) + f(x_n + h, y_n + h f_n)] \tag{2.1-12}
$$

as typical $p = v = 2$ Runge–Kutta formulas.

This same approach can be carried further for $v = 3$ and $v = 4$, but we shall merely present a few of the more well-known forms of each here. Thus three typical third-order formulas which come from (2.1-6) are

$$
\begin{aligned}
y_{n+1} &= y_n + \tfrac{1}{6}[k_1 + 4k_2 + k_3] \\
k_1 &= hf(x_n, y_n) \\
k_2 &= hf(x_n + \tfrac{1}{2}h, y_n + \tfrac{1}{2}k_1) \\
k_3 &= hf(x_n + h, y_n - k_1 + 2k_2)
\end{aligned}
\tag{2.1-13}
$$

and

$$
\begin{aligned}
y_{n+1} &= y_n + \tfrac{1}{4}[k_1 + 3k_3] \\
k_1 &= hf(x_n, y_n) \\
k_2 &= hf(x_n + \tfrac{1}{3}h, y_n + \tfrac{1}{3}k_1) \\
k_3 &= hf(x_n + \tfrac{2}{3}h, y_n + \tfrac{2}{3}k_2)
\end{aligned}
\tag{2.1-14}
$$

and

$$
\begin{aligned}
y_{n+1} &= y_n + \tfrac{1}{9}[2k_1 + 3k_2 + 4k_3] \\
k_1 &= hf(x_n, y_n) \\
k_2 &= hf(x_n + \tfrac{1}{2}h, y_n + \tfrac{1}{2}k_1) \\
k_3 &= hf(x_n + \tfrac{3}{4}h, y_n + \tfrac{3}{4}k_2)
\end{aligned}
\tag{2.1-15}
$$

Typical fourth-order formulas are given by

$$
\begin{aligned}
y_{n+1} &= y_n + \tfrac{1}{6}[k_1 + 2k_2 + 2k_3 + k_4] \\
k_1 &= hf(x_n, y_n) \\
k_2 &= hf(x_n + \tfrac{1}{2}h, y_n + \tfrac{1}{2}k_1) \\
k_3 &= hf(x_n + \tfrac{1}{2}h, y_n + \tfrac{1}{2}k_2) \\
k_4 &= hf(x_n + h, y_n + k_3)
\end{aligned}
\tag{2.1-16}
$$

and

$$
\begin{aligned}
y_{n+1} &= y_n + \tfrac{1}{8}[k_1 + 3k_2 + 3k_3 + k_4] \\
k_1 &= hf(x_n, y_n) \\
k_2 &= hf(x_n + \tfrac{1}{3}h, y_n + \tfrac{1}{3}k_1) \\
k_3 &= hf(x_n + \tfrac{2}{3}h, y_n + \tfrac{1}{3}k_1 + k_2) \\
k_4 &= hf(x_n + h, y_n + k_1 - k_2 + k_3)
\end{aligned}
\tag{2.1-17}
$$

By a suitable choice of the free parameters, it is possible to obtain a large number of second-, third-, fourth- and higher-order formulas.

Finally we point out that the fourth-order equation can be directly adapted to the situation of a second-order differential equation not containing a first derivative explicitly. Thus we have

$$
\begin{aligned}
y_{n+1} &= y_n + hy_n' + \tfrac{1}{6}[k_1 + 2k_2] \\
hy_{n+1}' &= hy_n' + \tfrac{1}{6}[k_1 + 4k_2 + k_3]
\end{aligned}
\tag{2.1-18}
$$

where

$$k_1 = h^2 f(x_n, y_n)$$
$$k_2 = h^2 f[x_n + \tfrac{1}{2}h, y_n + \tfrac{1}{2}h y_n' + \tfrac{1}{8}k_1]$$
$$k_3 = h^2 f[x_n + h, y_n + h y_n' + \tfrac{1}{2}k_2]$$

2.1.2. Alternative Approaches to Generation of Formulas

In the previous sections we have developed both single- and multiple-step formulas using the method of undetermined parameters. We wish to present here an alternative approach to this technique based on Gaussian quadrature.

To illustrate the aspects of the repeated Gaussian quadrature approach, we recall that the single-step Runge–Kutta formulas advanced from the point x_n to the point x_{n+1} by evaluating $f(x, y)$ at a number of ordinates within (x_n, x_{n+1}). Thus if we consider (2.1-14)

$$y_{n+1} = y_n + \tfrac{1}{4}[k_1 + 3k_3] \tag{2.1-19}$$

where

$$k_1 = h f(x_n, y_n)$$
$$k_2 = h f(x_n + \tfrac{1}{3}h, y_n + \tfrac{1}{3}k_1)$$
$$k_3 = h f(x_n + \tfrac{2}{3}h, y_n + \tfrac{2}{3}k_2)$$

we may visualize the sequence of calculation steps to yield y_{n+1} as being those of Table 2.1. In Step 1, $y_{n+1/3}$ is evaluated from y_n using

$$y_{n+1/3} = y_n + (h/3)y_n' \tag{2.1-20}$$

which is an Euler formula, $p = 1$, (1.6-7); in Step 2, $y_{n+2/3}$ is evaluated from y_n and $y_{n+1/3}$ using

$$y_{n+2/3} = y_n + (2h/3)y_{n+1/3}' \tag{2.1-21}$$

which is a midpoint formula, $p = 2$, (1.6-31); in Step 3, y_{n+1} is evaluated from y_n and $y_{n+2/3}$ using

$$y_{n+1} = y_n + \tfrac{1}{4}y_n' + \tfrac{3}{4}y_{n+2/3}' \tag{2.1-22}$$

TABLE 2.1

SERIES OF STEPS FOR THIRD-ORDER RUNGE–KUTTA

Step	Evaluate	Weights used	Order	Name
1	$y_{n+1/3}$	1	1	Euler
2	$y_{n+2/3}$	1	2	Midpoint
3	y_{n+1}	1/4, 3/4	3	2-point Radau

which is a Radau 2-point formula, $p = 3$, to be developed shortly. The h actually used in (2.1-20) and (2.1-21) is one-third the normal sized h because of the step in x. This series of steps which uses successively more accurate formulas generates the exact results as the third-order Runge–Kutta formula.

The important feature in the above analysis is the location or specification of the intermediate points. Actually the Runge–Kutta forms themselves allowed this specification for the above example, but in the more general sense to be discussed shortly we will not have any such guides. Instead we shall turn to the use of Gauss, Radau, and Lobatto quadratures for evaluating integrals.

As distinct from the polynomial or difference methods of Section 1.5, which used equally spaced abscissas, these quadrature methods use unequally spaced points, but the points are spaced symmetrically around the midpoint $x_{n+1/2}$. A high order of accuracy is achieved by allowing the location of the points themselves to be coefficients in the undetermined coefficient process. An example of the procedure used and the results obtained have already been given in Section 1.7.4 and (1.7-16) and (1.7-17). Table 2.2 represents a summary of the information we desire for present purposes. This can be

TABLE 2.2

GAUSS, RADAU, AND LOBATTO QUADRATURE

Abscissa	Weights	Order	Common name
Gauss quadrature			
$1/2$	1	2	midpoint
$(3 \pm \sqrt{3})/6$	$1/2$	4	2-point Gauss
Radau quadrature			
0	1	1	Euler
0	$1/4$	3	2-point Radau
$2/3$	$3/4$		
0	$1/9$	5	3-point Radau
$(6 - \sqrt{6})/10$	$(16 + \sqrt{6})/36$		
$(6 + \sqrt{6})/10$	$(16 - \sqrt{6})/36$		
Lobatto quadrature			
0, 1	$1/2$	2	trapezoidal rule
0, 1	$1/6$	4	Milne
$1/2$	$2/3$		
0, 1	$1/12$	6	4-point Lobatto
$(5 \pm \sqrt{5})/10$	$5/12$		

found in a number of references, for example [33], but with the change that the integration interval is now $(0, 1)$ (or x_n to x_{n+1}) and not $(-1, +1)$. Note that in Table 2.2 the fourth-order, 2-point Gauss corresponds to the points of $(1.7\text{-}17)$.

It is important to see from Table 2.2 that in the Gauss forms no end points (abscissa of 0 or 1) are involved, in the Radau forms one end point (abscissa of 0) is involved, and in the Lobatto forms both end points (abscissa of 0 and 1) are involved.

Gates [27] (see also Stoller and Morrison [70] and Day [21]) used this information to develop high-order single-step formulas using increasingly higher-order intermediate formulas. We do not wish to repeat this material here although two points are of interest. First, the calculations are explicit proceeding from x_n directly to x_{n+1} through the intermediate points. Later we shall look at implicit methods using the Gauss, Radau, and Lobatto forms which fall directly into the Runge–Kutta form. Second, as distinct from the Runge–Kutta forms in which all the weights of the different equations are interconnected, the current method determines the weights essentially independently of one equation to the next. But this freedom requires a penalty and it is in the number of evaluations of $f(x, y)$ that the current method requires. As we shall emphasize, the Runge–Kutta formulas of fourth-, fifth-, and sixth-order require four, six, and eight evaluations, i.e., for $(r, v) = \{(4, 4), (5, 6), \text{ and } (6, 8)\}$. By contrast, Gates' method requires $\{(4, 5), (5, 7), \text{ and } (6,11)\}$ respectively. Since the number of times that $f(x, y)$ is calculated may, in a large calculation, determine the overall computation time, this increase may be significant for $r > 4$.

2.2. CONDENSED NOMENCLATURE FOR RUNGE–KUTTA METHODS

We will now introduce a condensed representation of the Runge–Kutta methods developed by Butcher [8]. To illustrate this representation, consider the main Runge–Kutta equations in the form of $(2.1\text{-}2)$ and $(2.1\text{-}2')$, and arrange the coefficients in the array form

$$
\begin{array}{c|cccc}
0 & & & & \\
c_2 & a_{21} & & & \\
c_3 & a_{31} & a_{32} & & \\
c_4 & a_{41} & a_{42} & a_{43} & \\
\hline
 & w_1 & w_2 & w_3 & w_4
\end{array}
$$

or

$$
\begin{array}{c|c}
\mathbf{c} & \mathbf{A_L} \\
\hline
 & \mathbf{w}^T
\end{array}
$$

In other words, we form a lower triangular matrix $\mathbf{A}_L$ containing the multiplying factors a_{ij} of the k_i in the increments of y_n, a vector $\mathbf{c}$ containing the increments on x_n and a vector transposed $\mathbf{w}^T$ containing the v weighting factors $w_1, \ldots, w_v$. If we select the third-order case of (2.1-13) this may be symbolically represented as

$$
\begin{array}{c|ccc}
0 & & & \\
\tfrac{1}{2} & \tfrac{1}{2} & & \\
1 & -1 & 2 & \\
\hline
& \tfrac{1}{6} & \tfrac{4}{6} & \tfrac{1}{6}
\end{array}
$$

From the identities of (2.1-8) we have a convenient means for quickly checking the validity of these symbolic diagrams.

Finally, we note that because $\mathbf{A}_L$ is a lower triangular matrix the calculation for k_{i+1} requires only $k_i, k_{i-1}, \ldots, k_1$. As such, these formulas will be called *explicit* or *open-ended Runge–Kutta*. If we modify the equations so that elements appear on the main diagonal or in the upper right triangle, then the k_{i+1} will be functions of $k_{i+1}, k_i, \ldots, k_1$ or $k_v, k_{v-1}, \ldots, k_{i+1}, k_i, \ldots, k_1$. Such forms are then *implicit* or *closed-end Runge–Kutta* formulas.

2.3. EXPLICIT RUNGE–KUTTA EQUATIONS OF DIFFERENT ORDER

In this section we shall tabulate a variety of different explicit Runge–Kutta formulas which hold for single scalar ODE. The results can be extended directly to sets of first-order equations.

Let us start the discussion by quoting a number of interesting and important results. With p the order of the Runge–Kutta process and v the number of *substitutions, derivative evaluations,* or *stages,* it is possible to suggest that for a given integer v the order p be as large as possible. In fact, for each v there is a largest value of p for which the nonlinear algebraic equations associated with the Runge–Kutta formulas have a solution. If $N(v)$ is the largest value of p for a v stage process, then

$$N(v) = v, \qquad 1 \le v \le 4 \tag{2.3-1}$$

and

$$
\begin{aligned}
N(5) &= 4, \qquad N(6) = 5, \qquad N(7) = 6 \\
N(8) &= 6, \qquad N(9) = 7
\end{aligned}
\tag{2.3-2}
$$

In other words, an order 1, $\ldots$, 4 formula can be obtained with an equal number of stages or function evaluations $p = v \le 4$; however, as the order is increased a larger number of substitutions is required. There is no fifth-order, five stage process, but instead a fifth-order process requires six substitutions; the use of nine stages will generate at most a seventh-order process.

In order to show this relationship in subsequent discussion, we shall adopt the symbolism of specifying the (p, v) pair for the different formulas.

Butcher [5] has tabulated all the Taylor series coefficients and error terms for the Runge–Kutta processes of order $p \leq 8$. Sarafyan and Brown [66] have also detailed a computer algorithm for deriving the various formulas.

2.3.1. First-Order Formulas (1, 1)

The only $(1, 1)$ formula is Euler's

$$y_{n+1} = y_n + hf(x_n, y_n) \qquad (2.3\text{-}3)$$

which we have mentioned previously.

2.3.2. Second-Order Formulas (2, 2)

Here we have the Runge–Kutta formulas of (2.1-9)–(2.1-12) with c_2 as the free parameter. The symbolic diagram for the $p = 2$ case as a function of c_2 can be specified as

$$
\begin{array}{c|cc}
0 & & \\
c_2 & c_2 & \\
\hline
& 1 - \tfrac{1}{2}c_2 & \tfrac{1}{2}c_2
\end{array}
$$

with the local truncation error given by

$$T(x, h) = h^3[\tfrac{1}{6} - (c_2/4)](f_{xx} + 2f_{xy}f + f_{yy}f^2)$$
$$+ (h^3/6)(f_x f_y + f_y^2 f)$$

or, using (1.2-4) and $Df = f' = f_x + ff_y$, as

$$
\begin{aligned}
T(x, h) &= h^3[\tfrac{1}{6} - (c_2/4)]D^2f + (h^3/6)f_y Df \\
&= h^3[\tfrac{1}{6} - (c_2/4)]f'' + (h^3/4)c_2 f_y f'
\end{aligned}
\qquad (2.3\text{-}4)
$$

For $c_2 = \tfrac{1}{2}, \tfrac{2}{3}$, and 1, we obtain the following forms

$$
\begin{array}{c|cc}
0 & & \\
\tfrac{1}{2} & \tfrac{1}{2} & \\
\hline
& 0 & 1
\end{array}
\qquad c_2 = \tfrac{1}{2} \qquad (2.3\text{-}5)
$$

Heun form

$$
\begin{array}{c|cc}
0 & & \\
\tfrac{2}{3} & \tfrac{2}{3} & \\
\hline
& \tfrac{1}{4} & \tfrac{3}{4}
\end{array}
\qquad c_2 = \tfrac{2}{3} \qquad (2.3\text{-}6)
$$

Improved Euler

$$
\begin{array}{c|cc}
0 & & \\
1 & 1 & \\
\hline
& \frac{1}{2} & \frac{1}{2}
\end{array}
\qquad c_2 = 1 \qquad\qquad (2.3\text{-}7)
$$

We note the following about these three second-order formulas:

1. In (2.3-5), one weight is zero, $w_1 = 0$, and the formula is a midpoint formula.
2. In (2.3-6), the use of $c_2 = \frac{2}{3}$ has the feature of canceling the first term in $T(x, h)$ of (2.3-4).
3. Equation (2.3-7) is termed the improved Euler method. It is not the trapezoidal rule because

$$
\begin{aligned}
k_2 &= hf(x_n + h, y_n + k_1) \\
&= hf(x_n + h, y_n + hf_n)
\end{aligned}
$$

When $f(x, y)$ simplifies to $f(x)$, however, we do get the trapezoidal rule.

2.3.3. Third-Order Formulas (3, 3)

Here we have (2.1-6) and (2.1-13)–(2.1-15). There are two free parameters c_2 and c_3 resulting from eight unknowns with six equations and the local truncation error is given by

$$
\begin{aligned}
T(x, h) = (h^4/4!)\{ &[1 - 4(c_2{}^3 w_2 + c_3{}^3 w_3)]D^3 f \\
&+ (1 - 12c_2{}^2 a_{32} w_3)f_y D^2 f + (3 - 24c_2 c_3 w_3)Df\,Df_y + f_y{}^2 Df \}
\end{aligned}
$$

$$(2.3\text{-}8)$$

The following formulas are frequently used
   Classic form

$$
\begin{array}{c|ccc}
0 & & & \\
\frac{1}{2} & \frac{1}{2} & & \\
1 & -1 & 2 & \\
\hline
& \frac{3}{8} & \frac{2}{3} & \frac{1}{6}
\end{array}
\qquad c_2 = \tfrac{1}{2}, \quad c_3 = 1 \qquad (2.3\text{-}9)
$$

Nystrom form

$$
\begin{array}{c|ccc}
0 & & & \\
\frac{2}{3} & \frac{2}{3} & & \\
\frac{2}{3} & 0 & \frac{2}{3} & \\
\hline
& \frac{1}{4} & \frac{3}{8} & \frac{3}{8}
\end{array}
\qquad c_2 = c_3 = \tfrac{2}{3} \qquad (2.3\text{-}10)
$$

Heun form

$$
\begin{array}{c|ccc}
0 \\
\frac{1}{3} & \frac{1}{3} \\
\frac{2}{3} & 0 & \frac{2}{3} \\
\hline
& \frac{1}{4} & 0 & \frac{3}{4}
\end{array}
\qquad c_2 = \tfrac{1}{3}, \quad c_3 = \tfrac{2}{3}
\qquad (2.3\text{-}11)
$$

Equation (2.3-9) is termed the classic third-order Runge–Kutta form and becomes Simpson's rule when $f(x, y) = f(x)$.

2.3.4. Fourth-Order Formulas (4, 4)

Here we again have two free parameters resulting from thirteen unknowns with eleven equations, those parameters being c_2 and c_3. The local truncation error is given by

$$
\begin{aligned}
T(x, h) = h^5 \Bigg\{ & \left[\frac{1}{120} - \frac{w_2 c_2{}^4 + w_3 c_3{}^4 + w_4 c_4{}^4}{24} \right] D^4 f \\
&+ \left[\frac{1}{20} - \frac{w_3 c_2 c_3{}^2 a_{32} + w_4 c_4{}^2 (c_2 a_{42} + c_3 a_{43})}{2} \right] D^2 f_y \, Df \\
&+ \left[\frac{1}{30} - \frac{w_3 a_{32} c_2{}^2 c_3 + w_4 c_4 (a_{42} c_2{}^2 + a_{43} c_3{}^2)}{2} \right] Df_y D^2 f \\
&+ \left[\frac{1}{120} - \frac{w_4 a_{43} a_{32} c_2{}^2}{2} \right] f_y{}^2 D^2 f \\
&+ \left[\frac{1}{40} - \frac{w_3 a_{32}^2 c_2{}^2 + w_4 (a_{43} c_3 + a_{42} c_2)^2}{2} \right] f_{yy} D^2 f \\
&+ \left[\frac{1}{120} - \frac{w_3 a_{32} c_2{}^3 + w_4 (a_{43} c_3{}^3 + a_{42} c_2{}^3)}{6} \right] f_y D^3 f \\
&+ \left[\frac{7}{120} - w_4 a_{43} a_{32} c_2 (c_3 + c_4) \right] f_y Df_y \, Df \\
&+ \frac{1}{120} f_y{}^3 Df \Bigg\}
\end{aligned}
\qquad (2.3\text{-}12)
$$

It is apparent that $T(x, h)$ is already so complicated that we shall not attempt to write it down for higher-order cases. Specific formulas of interest are:

Classic form

$$
\begin{array}{c|cccc}
0 \\
\frac{1}{2} & \frac{1}{2} \\
\frac{1}{2} & 0 & \frac{1}{2} \\
1 & 0 & 0 & 1 \\
\hline
& \frac{1}{6} & \frac{1}{3} & \frac{1}{3} & \frac{1}{6}
\end{array}
\qquad c_2 = c_3 = \tfrac{1}{2}
\qquad (2.3\text{-}13)
$$

Kutta form

$$
\begin{array}{c|cccc}
0 \\
\frac{1}{3} & \frac{1}{3} \\
\frac{2}{3} & -\frac{1}{3} & 1 \\
1 & 1 & -1 & 1 \\
\hline
& \frac{1}{8} & \frac{3}{8} & \frac{3}{8} & \frac{1}{8}
\end{array}
\qquad c_2 = \tfrac{1}{3}, \quad c_3 = \tfrac{2}{3} \qquad (2.3\text{-}14)
$$

Gill form

$$
\begin{array}{c|cccc}
0 \\
\frac{1}{2} & \frac{1}{2} \\
\frac{1}{2} & (\sqrt{2}-1)/2 & (2-\sqrt{2})/2 \\
1 & 0 & -\sqrt{2}/2 & 1+\sqrt{2}/2 \\
\hline
& \frac{1}{6} & (2-\sqrt{2})/6 & (2+\sqrt{2})/6 & \frac{1}{6}
\end{array}
\qquad c_2 = c_3 = \tfrac{1}{2} \qquad (2.3\text{-}15)
$$

A variety of other formulas are possible with some of the $w_i = 0$, but we shall not present them here. The interested reader should consult Kopal [40]. Equation (2.3-13) is termed the classic fourth-order Runge–Kutta form while (2.3-15), which is due to Gill, has the feature of minimizing round-off error. Fyfe [26] has extended Gill's procedure to other fourth-order cases.

2.3.5. Fifth-Order Formulas (5, 6)

As soon as we consider $p \geq 5$ the complexity of the algebraic equations and the freedom of parameter choice leads to a wide set of possible Runge–Kutta formulas. For $p = 5$, there are sixteen equations and twenty-one unknowns and thus five free parameters. Here we will quote some of the resulting formulas and indicate which seem to have special features of interest. The most well known is

Nystrom form

$$
\begin{array}{c|cccccc}
0 \\
\frac{1}{3} & \frac{1}{3} \\
\frac{2}{5} & \frac{4}{25} & \frac{6}{25} \\
1 & \frac{1}{4} & -\frac{12}{4} & \frac{15}{4} \\
\frac{2}{3} & \frac{6}{81} & \frac{90}{81} & -\frac{50}{81} & \frac{8}{81} \\
\frac{4}{5} & \frac{6}{75} & \frac{36}{75} & \frac{10}{75} & \frac{8}{75} & 0 \\
\hline
& \frac{23}{192} & 0 & \frac{125}{192} & 0 & -\frac{81}{192} & \frac{125}{192}
\end{array}
\qquad (2.3\text{-}16)
$$

Luther [46], in particular, has made the point that many fifth-order formulas may be considered as falling within certain families. Thus, from (1.5-13) we see that a Newton–Cotes closed end formula has weights $\frac{7}{90}$, $\frac{32}{90}$, $\frac{12}{90}$, $\frac{32}{90}$, and $\frac{7}{90}$, and is of order six ($p = 6$). Luther thus suggests that all

Runge–Kutta formulas for which $f(x, y) = f(x)$ yields these weights or others like them be classified within the Newton–Cotes family of Runge–Kutta formulas (the order is now one less, namely $p = 5$).

Typical examples are

Luther, Newton–Cotes family

$$
\begin{array}{c|cccccc}
0 & & & & & & \\
1 & 1 & & & & & \\
1 & \frac{1}{2} & \frac{1}{2} & & & & \\
\frac{1}{4} & \frac{14}{64} & \frac{5}{64} & -\frac{3}{64} & & & \\
\frac{1}{2} & -\frac{12}{96} & -\frac{12}{96} & \frac{8}{96} & \frac{64}{96} & & \\
\frac{3}{4} & 0 & -\frac{9}{64} & \frac{5}{64} & \frac{16}{64} & \frac{36}{64} & \\
\hline
 & \frac{7}{90} & 0 & \frac{7}{90} & \frac{32}{90} & \frac{12}{90} & \frac{32}{90}
\end{array}
\tag{2.3-17}
$$

Luther, Newton–Cotes family

$$
\begin{array}{c|cccccc}
0 & & & & & & \\
1 & 1 & & & & & \\
\frac{1}{2} & \frac{3}{8} & \frac{1}{8} & & & & \\
1 & -\frac{1}{2} & -\frac{1}{2} & 2 & & & \\
\frac{1}{4} & \frac{4}{64} & -\frac{5}{64} & \frac{20}{64} & -\frac{3}{64} & & \\
\frac{3}{4} & \frac{12}{64} & \frac{9}{64} & -\frac{12}{64} & \frac{7}{64} & \frac{32}{64} & \\
\hline
 & \frac{7}{90} & 0 & \frac{12}{90} & \frac{7}{90} & \frac{32}{90} & \frac{32}{90}
\end{array}
\tag{2.3-18}
$$

Butcher [7] has also presented a series of fifth-order formulas, the first three of which are in the Newton–Cotes family. These are

Butcher forms

$$
\begin{array}{c|cccccc}
0 & & & & & & \\
\frac{1}{8} & \frac{1}{8} & & & & & \\
\frac{1}{4} & 0 & \frac{1}{4} & & & & \\
\frac{1}{2} & \frac{1}{2} & -1 & 1 & & & \\
\frac{3}{4} & \frac{3}{16} & 0 & 0 & \frac{9}{16} & & \\
1 & -\frac{5}{7} & \frac{4}{7} & \frac{12}{7} & -\frac{12}{7} & \frac{8}{7} & \\
\hline
 & \frac{7}{90} & 0 & \frac{32}{90} & \frac{12}{90} & \frac{32}{90} & \frac{7}{90}
\end{array}
\tag{2.3-19}
$$

$$
\begin{array}{c|cccccc}
0 & & & & & & \\
\frac{1}{4} & \frac{1}{4} & & & & & \\
\frac{1}{4} & \frac{1}{8} & \frac{1}{8} & & & & \\
\frac{1}{2} & 0 & -\frac{1}{2} & 1 & & & \\
\frac{3}{4} & \frac{3}{16} & 0 & 0 & \frac{9}{16} & & \\
1 & -\frac{3}{7} & \frac{2}{7} & \frac{12}{7} & -\frac{12}{7} & \frac{8}{7} & \\
\hline
 & \frac{7}{90} & 0 & \frac{32}{90} & \frac{12}{90} & \frac{32}{90} & \frac{7}{90}
\end{array}
\tag{2.3-20}
$$

$$
\begin{array}{c|cccccc}
0 \\
-\frac{1}{2} & -\frac{1}{2} \\
\frac{1}{4} & \frac{5}{16} & -\frac{1}{16} \\
\frac{1}{2} & -\frac{3}{4} & \frac{1}{4} & 1 \\
\frac{3}{4} & \frac{3}{16} & 0 & 0 & \frac{9}{16} \\
1 & 0 & -\frac{1}{7} & \frac{12}{7} & -\frac{12}{7} & \frac{8}{7} \\
\hline
 & \frac{7}{90} & 0 & \frac{32}{90} & \frac{12}{90} & \frac{32}{90} & \frac{7}{90}
\end{array}
\qquad (2.3\text{-}21)
$$

$$
\begin{array}{c|cccccc}
0 \\
\frac{1}{5} & \frac{1}{5} \\
\frac{2}{5} & 0 & \frac{2}{5} \\
\frac{1}{3} & \frac{7}{36} & 0 & \frac{5}{36} \\
\frac{4}{5} & 0 & 0 & \frac{4}{5} & 0 \\
1 & \frac{1}{4} & 0 & -\frac{35}{4} & \frac{54}{7} & \frac{25}{14} \\
\hline
 & \frac{5}{48} & 0 & 0 & \frac{27}{56} & \frac{125}{336} & \frac{1}{24}
\end{array}
\qquad (2.3\text{-}22)
$$

$$
\begin{array}{c|cccccc}
0 \\
-\frac{1}{5} & -\frac{1}{5} \\
\frac{2}{5} & \frac{4}{5} & -\frac{2}{5} \\
\frac{1}{3} & \frac{7}{36} & 0 & \frac{5}{36} \\
\frac{4}{5} & 0 & 0 & \frac{4}{5} & 0 \\
1 & \frac{1}{4} & 0 & -\frac{35}{4} & \frac{54}{7} & \frac{25}{14} \\
\hline
 & \frac{5}{48} & 0 & 0 & \frac{27}{56} & \frac{125}{336} & \frac{1}{24}
\end{array}
\qquad (2.3\text{-}23)
$$

There are some interesting features associated with (2.3-19)–(2.3-23). Form (2.3-20) will find application later (Section 2.8) as a convenient means of estimating the truncation error. Further, in (2.3-21) in all steps after the first, k_2 may be replaced by k_4 from the preceding step (as we go from x_{n-1} to x_n to x_{n+1}) with no loss in order. Similarly, (2.3-22) may be used as a first step and (2.3-23) for all following steps, with the replacement of k_2 by k_5 from the preceding step after starting. In this manner, two of the five Butcher forms require, effectively, only five stages, i.e., they behave as (5, 5) processes.

Just as there are fifth-order Runge–Kutta formulas which are in the Newton–Cotes family, Luther has also presented formulas which are in the Gauss, Radau, and Lobatto families (see Table 2.2 for weights, etc.). Some of these are presented on page 53. Note the large number of zeros in these formulas indicating they are computationally effective.

Sarafyan [62] has also developed a number of fifth-order Runge–Kutta formulas. We present one of these here for use later in an interesting truncation error analysis.

Luther, Gauss family

$$
\begin{array}{c|ccccc}
0 & & & & & \\
1 & 1 & & & & \\
1/2 & 3/8 & 1/8 & & & \\
1 & -1/2 & -1/2 & 2 & & \\
(5-\sqrt{15})/10 & -\sqrt{15}/100 & -10/100 & (60-8\sqrt{15})/100 & -\sqrt{15}/100 & \\
(5+\sqrt{15})/10 & (-6-\sqrt{15})/20 & -2/20 & 12/20 & (6-\sqrt{15})/20 & 4\sqrt{15}/20 \\
\hline
& 0 & 0 & 8/18 & 5/18 & 5/18
\end{array}
\tag{2.3-24}
$$

Luther, Radau family

$$
\begin{array}{c|ccccc}
0 & & & & & \\
4/11 & 4/11 & & & & \\
2/5 & 9/50 & 11/50 & & & \\
1 & 0 & -11/4 & 15/4 & & \\
(6-\sqrt{6})/10 & (81+9\sqrt{6})/600 & 0 & (255-55\sqrt{6})/600 & (24-14\sqrt{6})/600 & \\
(6+\sqrt{6})/10 & (81-9\sqrt{6})/600 & 0 & (255+55\sqrt{6})/600 & (24+14\sqrt{6})/600 & 0 \\
\hline
& 4/36 & 0 & 0 & (16+\sqrt{6})/36 & (16-\sqrt{6})/36
\end{array}
\tag{2.3-25}
$$

Luther, Lobatto family

$$
\begin{array}{c|ccccc}
0 & & & & & \\
1 & 1 & & & & \\
1/2 & 3/8 & 1/8 & & & \\
1 & -1/2 & -1/2 & 2 & & \\
(5-\sqrt{5})/10 & (25-7\sqrt{5})/100 & (5-5\sqrt{5})/100 & (20+4\sqrt{5})/100 & -2\sqrt{5}/100 & \\
(5+\sqrt{5})/10 & (3+\sqrt{5})/20 & (1+\sqrt{5})/20 & (4-4\sqrt{5})/20 & 2/20 & 4\sqrt{5}/20 \\
\hline
& 1/2 & 0 & 0 & 1/12 & 5/12 & 5/12
\end{array}
\tag{2.3-26}
$$

Sarafyan form

$$
\begin{array}{c|cccccc}
0 \\
\frac{1}{2} & \frac{1}{2} \\
\frac{1}{2} & \frac{1}{4} & \frac{1}{4} \\
1 & 0 & -1 & 2 \\
\frac{2}{3} & \frac{7}{27} & \frac{10}{27} & 0 & \frac{1}{27} \\
\frac{2}{10} & \frac{28}{625} & -\frac{125}{625} & \frac{546}{625} & \frac{54}{625} & -\frac{378}{625} \\
\hline
 & \frac{14}{336} & 0 & 0 & \frac{35}{336} & \frac{162}{336} & \frac{125}{336}
\end{array}
\qquad (2.3\text{-}27)
$$

Finally, we mention and exhibit the fifth-order Runge–Kutta forms developed by Fehlberg [24], Shanks [69], and Lawson [43], each of which has some special feature associated with it. The Fehlberg form, just as (2.3-27), can be used for truncation error analysis, the Shanks form is (5, 5) because the algebraic equations are not solved exactly and the Lawson form has an extended region of stability (see Chapter 3 for further details).

Fehlberg form

$$
\begin{array}{c|ccccccc}
0 \\
\frac{1}{6} & \frac{1}{6} \\
\frac{4}{15} & \frac{4}{75} & \frac{16}{75} \\
\frac{2}{3} & \frac{5}{6} & -\frac{8}{3} & \frac{5}{2} \\
\frac{4}{5} & -\frac{8}{5} & \frac{144}{25} & -4 & \frac{16}{25} \\
1 & \frac{361}{320} & -\frac{18}{5} & \frac{407}{128} & -\frac{11}{80} & \frac{55}{128} \\
\hline
 & \frac{31}{384} & 0 & \frac{1125}{2816} & \frac{9}{32} & \frac{125}{768} & \frac{5}{66}
\end{array}
\qquad (2.3\text{-}28)
$$

Shanks form (5, 5)

$$
\begin{array}{c|cccccc}
0 \\
\frac{1}{9000} & \frac{1}{9000} \\
\frac{3}{10} & -\frac{4047}{10} & \frac{4050}{10} \\
\frac{3}{4} & \frac{20241}{8} & -\frac{20250}{8} & \frac{15}{8} \\
1 & -\frac{931041}{81} & \frac{931500}{81} & -\frac{490}{81} & \frac{112}{81} \\
\hline
 & \frac{105}{1134} & 0 & \frac{500}{1134} & \frac{448}{1134} & \frac{81}{1134}
\end{array}
\qquad (2.3\text{-}29)
$$

Lawson form, Newton–Cotes family

$$
\begin{array}{c|cccccc}
0 \\
\frac{1}{2} & \frac{1}{2} \\
\frac{1}{4} & \frac{3}{16} & \frac{1}{16} \\
\frac{1}{2} & 0 & 0 & \frac{1}{2} \\
\frac{3}{4} & 0 & -\frac{3}{16} & \frac{6}{16} & \frac{9}{16} \\
1 & \frac{1}{7} & \frac{4}{7} & \frac{6}{7} & -\frac{12}{7} & \frac{8}{7} \\
\hline
 & \frac{7}{90} & 0 & \frac{32}{90} & \frac{12}{90} & \frac{32}{90} & \frac{7}{90}
\end{array}
\qquad (2.3\text{-}30)
$$

2.3.6. Sixth-Order Formulas (6, 7) and (6, 8)

From (2.3-2) we can see that sixth-order Runge–Kutta formulas can be obtained from seven stages (6, 7) and from eight stages (6, 8). The former is generally to be preferred.

The most famous sixth-order formula is due to Huta and was analyzed by Butcher [6]. The formula as shown below is (6, 8).

Butcher form (6, 8)

$$
\begin{array}{c|cccccccc}
0 \\
\frac{1}{9} & \frac{1}{9} \\
\frac{1}{6} & \frac{1}{24} & \frac{3}{24} \\
\frac{1}{3} & \frac{1}{6} & -\frac{3}{6} & \frac{4}{6} \\
\frac{1}{2} & -\frac{5}{8} & \frac{27}{8} & -\frac{24}{8} & \frac{48}{8} \\
\frac{2}{3} & \frac{221}{9} & -\frac{981}{9} & \frac{867}{9} & -\frac{102}{9} & \frac{1}{9} \\
\frac{5}{6} & -\frac{783}{48} & \frac{678}{48} & -\frac{472}{48} & -\frac{66}{48} & \frac{80}{48} & \frac{3}{48} \\
1 & \frac{761}{82} & -\frac{2079}{82} & \frac{1002}{82} & \frac{834}{82} & -\frac{454}{82} & -\frac{9}{82} & \frac{72}{82} \\
\hline
& \frac{41}{840} & \frac{216}{840} & 0 & \frac{27}{840} & \frac{272}{840} & \frac{27}{840} & \frac{216}{840} & \frac{41}{840}
\end{array}
\tag{2.3-31}
$$

More useful formulas which are (6, 7) are due to Butcher [7], to Luther [44], and to Fehlberg [24]; Shanks has also presented an "almost" (6, 6) formula. We list certain of these below.

Butcher form (6, 7)

$$
\begin{array}{c|ccccccc}
0 \\
\frac{1}{3} & \frac{1}{3} \\
\frac{2}{3} & 0 & \frac{2}{3} \\
\frac{1}{3} & \frac{1}{12} & \frac{1}{3} & -\frac{1}{12} \\
\frac{1}{2} & -\frac{1}{16} & \frac{9}{8} & -\frac{3}{16} & -\frac{3}{8} \\
\frac{1}{2} & 0 & \frac{9}{8} & -\frac{3}{8} & -\frac{3}{4} & \frac{1}{2} \\
1 & \frac{9}{44} & -\frac{9}{11} & \frac{63}{44} & \frac{18}{11} & 0 & -\frac{16}{11} \\
\hline
& \frac{11}{120} & 0 & \frac{27}{40} & \frac{27}{40} & -\frac{4}{15} & -\frac{4}{15} & \frac{11}{120}
\end{array}
\tag{2.3-32}
$$

Shanks form (6, 6)

$$
\begin{array}{c|cccccc}
0 \\
\frac{1}{300} & \frac{1}{300} \\
\frac{1}{5} & -\frac{29}{5} & \frac{30}{5} \\
\frac{3}{5} & \frac{323}{5} & -\frac{330}{5} & \frac{10}{5} \\
\frac{14}{15} & -\frac{510104}{810} & \frac{521640}{810} & -\frac{12705}{810} & \frac{1925}{810} \\
1 & -\frac{417923}{77} & \frac{427350}{77} & -\frac{10605}{77} & \frac{1309}{77} & -\frac{54}{77} \\
\hline
& \frac{198}{3698} & 0 & \frac{1225}{3698} & \frac{1540}{3698} & \frac{810}{3698} & -\frac{77}{3698}
\end{array}
\tag{2.3-33}
$$

2.3.7. Higher-Order Formulas

Higher-order Runge–Kutta forms have been developed, but we shall not quote any of these here. Shanks has specified $(7, 7)$, $(7, 9)$, $(8, 10)$, and $(8, 12)$ formulas and Fehlberg a $(7, 8)$ and a $(8, 9)$ form.

2.4. RUNGE–KUTTA FORMULAS DERIVED FROM TRUNCATION ERROR ANALYSIS

We now extend the previous analysis to find some Runge–Kutta forms for $p \leq 4$ which minimize the truncation error. As will be seen the analysis seems exceedingly difficult to develop for $p > 4$. We shall largely follow Ralston's work [56] although we need to mention that King [38] has extended some of these results.

In much of the analysis to follow we assume that the local truncation error has the form,

$$T(x, h) = h^{p+1}\phi \qquad (2.4\text{-}1)$$

where ϕ is the principal error function, and that certain bounds exist, namely,

$$|\phi| < CMK^p, \qquad C = \text{constant} \qquad (2.4\text{-}2)$$

$$|f(x, y)| < M, \qquad M, K = \text{constants}$$

$$\frac{\partial^{i+j}f}{\partial x^i \, \partial y^j} < \frac{K^{i+j}}{M^{j-1}}, \qquad i + j \leq p \qquad (2.4\text{-}3)$$

There does not seem to be any physical reason for defining these bounds, other than the simplicity of the results.

2.4.1 Second-Order Formulas

For the second-order case, we note from the equation before (2.3-4) that the local truncation error is given by

$$T(x, h) = h^3[\tfrac{1}{6} - (c_2/4)](f_{xx} + 2f_{xy}f + f_{yy}f^2) + (h^3/6)(f_x f_y + f_y^2 f) \qquad (2.4\text{-}4)$$

But from (2.4-3)

$$f < M, \qquad f_{xy} < K^2$$
$$f_x < KM, \qquad f_{xx} < K^2M$$
$$f_y < K, \qquad f_{yy} < K^2/M$$

and thus (2.4-4) becomes

$$|T(x, h)/h^3| = |\phi| < [4|\tfrac{1}{6} - (c_2/4)| + \tfrac{1}{3}]MK^2 \qquad (2.4\text{-}5)$$

Clearly the smallest bound on ϕ occurs when $c_2 = \frac{2}{3}$ in which case $|\phi| < \frac{1}{3} MK^2$. This is, of course, the case previously called the Heun form, (2.3-6). In fact, we may now tabulate as Ralston [56] has done (no new formulas) the following conditions for the second-order formulas of Section 2.3

$$c_2 = \tfrac{1}{2}, \qquad |\phi| < \tfrac{1}{2}MK^2 \tag{2.3-5}$$

$$c_2 = \tfrac{2}{3}, \qquad |\phi| < \tfrac{1}{3}MK^2 \qquad \text{[Optimum]} \tag{2.3-6}$$

$$c_2 = 1, \qquad |\phi| < \tfrac{2}{3}MK^2 \tag{2.3-7}$$

2.4.2. Third-Order Formulas

In the third-order case we have two free parameters c_2 and c_3 and a local truncation error given by (2.3-8). By going through the same procedure as above, Ralston was able to show that the case $c_2 = \frac{1}{2}$, $c_3 = \frac{3}{4}$ leads to the optimum $|\phi|$. On this basis we have the optimum third-order formula:

Optimum third-order

$$
\begin{array}{c|ccc}
0 & & & \\
\frac{1}{2} & \frac{1}{2} & & \\
\frac{3}{4} & 0 & \frac{3}{4} & \\
\hline
 & \frac{2}{9} & \frac{1}{3} & \frac{4}{9}
\end{array}
\tag{2.4-6}
$$

and we may summarize the third-order cases as

$$c_2 = \tfrac{1}{2}, \quad c_3 = 1, \qquad |\phi| < \tfrac{13}{26}MK^3 \tag{2.3-9}$$

$$c_2 = c_3 = \tfrac{1}{4}, \qquad\qquad |\phi| < \tfrac{1}{4}MK^3 \tag{2.3-10}$$

$$c_2 = \tfrac{1}{2}, \quad c_3 = \tfrac{3}{4}, \qquad |\phi| < \tfrac{1}{9}MK^3 \qquad \text{[Optimum]} \tag{2.4-6}$$

$$c_2 = \tfrac{1}{3}, \quad c_3 = \tfrac{2}{3}, \qquad |\phi| < \tfrac{2}{3}MK^3 \tag{2.3-11}$$

Kuntzmann [41] has also derived an optimum third-order formula by minimizing the sum of the absolute values of the local truncation error. His formula uses the parameters $c_2 = (10 - 2\sqrt{13})/6 = 0.4648162$ and $c_3 = (1 + \sqrt{13})/6 = 0.7675919$. Note how close these are to the optimum of (2.4-6); the actual form is

Kuntzmann optimum

$$
\begin{array}{c|ccc}
0 & & & \\
0.4648162 & 0.4648162 & & \\
0.7675919 & -0.0581020 & 0.8256939 & \\
\hline
 & 0.2071768 & 0.3585646 & 0.4342585
\end{array}
\qquad
\begin{array}{l}
c_2 \simeq 0.46 \\
c_3 \simeq 0.77
\end{array}
\tag{2.4-7}
$$

2.4.3. Fourth-Order Formulas

Once again we have a two free parameter case with the local truncation error given by (2.3-12). After some difficult algebraic manipulations, Ralston was able to show the following optimum formula and results for bounds on the other previously shown formulas.

$$c_2 = 0.4, \quad c_3 = \tfrac{7}{8} - (3\sqrt{5})/16, \qquad |\phi| < 5.46_{10^{-2}} M K^4 \qquad \text{[Optimum]} \tag{2.4-8}$$

0			
0.4	0.4		
0.45573725	0.29697761	0.15875964	
1	0.21810040	−3.05096516	3.83286476
	0.17476028	−0.55148066	1.20553560	0.17118478

$$c_2 = c_3 = \tfrac{1}{2}, \qquad |\phi| < \tfrac{73}{720} M K^4 \tag{2.3-13}$$

$$c_2 = \tfrac{1}{3}, \quad c_3 = \tfrac{2}{3}, \qquad |\phi| < \tfrac{107}{1080} M K^4 \tag{2.3-14}$$

$$\text{Gill,} \qquad |\phi| < \left[\frac{53}{360} - \frac{1}{12\sqrt{2}} \right] M K^4 \tag{2.3-15}$$

Kuntzmann has suggested an optimal formula in which $c_2 = \tfrac{2}{5}$, $c_3 = \tfrac{3}{5}$ which is very close to (2.4-8). This formula is explicitly given by

Kuntzmann optimum

0				
$\tfrac{2}{5}$	$\tfrac{2}{5}$			
$\tfrac{3}{5}$	$-\tfrac{3}{20}$	$\tfrac{3}{4}$		
1	$\tfrac{19}{44}$	$-\tfrac{15}{44}$	$\tfrac{40}{44}$	
	$\tfrac{55}{360}$	$\tfrac{125}{360}$	$\tfrac{125}{360}$	$\tfrac{55}{360}$

$$c_2 = \tfrac{2}{5}, \quad c_3 = \tfrac{3}{5} \tag{2.4-9}$$

Finally we mention the work of Hull and Johnson [35] who set up three measures of the local truncation error including the absolute value of the error, the absolute square of the error, and the procedure used by Ralston. For the fourth-order case, these measures all lead to optimum values of $c_2 \simeq 0.35$ and $c_3 \simeq 0.45$ which correspond closely to those of Ralston. In other words none of the criteria are particularly sensitive to the choice of parameter values, as long as these values are near to the optimum ones. However, by applying those small variations in the parameters to some simple appearing differential equations the truncation error itself varied widely (a factor of 25) with the parameters. This points out that the strict bounds on the local truncation error are a function of the particular differential equation studied and that the bounds are probably very conservative. As an indication of this feature note that the bound on the truncation error for the classical fourth-order Runge–Kutta is given by

$$|T(x, h)| < 10.14_{10^{-2}} M K^4 h^5 \tag{2.4-10}$$

In the case $f(x, y) = f(x)$ this Runge–Kutta formula becomes Simpson's rule and the error term is given by

$$|T(x, h)| \le (h^5/2880)y^{[5]}(\zeta) \qquad (2.4\text{-}11)$$

This comes directly from (1.6-21) where the error is given as $-(h^5/90)y^{[5]}(\zeta)$. However, since we are using an h which is one-half that for (1.6-21) we get $(1/90)(1/2)^5 = (1/2880)$. Comparing (2.4-10) and (2.4-11), we see that the bound used in this section is about 300 times too large.

2.5. QUADRATURE AND IMPLICIT FORMULAS

In the previous sections of this chapter we have presented a large number of explicit Runge–Kutta formulas. These formulas have the feature, as discussed in detail in Chapter 3, of always being stable for a small enough h. Stable is taken to mean that errors committed in the calculation do not propagate and increase in amplitude, but rather decrease as n increases. As a result, valid computational solutions are obtained. However, there are two defects in the formulas even assuming stability for small h. The first is that the h may need to be so small for stability as to require an excessive amount of computation time while the second is that the numerical solution may not approximate the true solution of the differential equation even for this small finite h.

It is primarily to remove these two defects (as will be shown in Chapter 3) that we now consider another class of Runge–Kutta formulas, the semi-implicit and the implicit forms. For these cases we shall find that any h can be used and still have a stable solution and that the numerical solution approximates the true solution of the differential equation.

The basic idea is to fill in the main diagonal or the upper triangular array of $\mathbf{A}_L$ with nonzero a_{ij}. The result is then a k_i which is a function at any line of $k_1, k_2, \ldots, k_{i-1}, k_i, k_{i+1}, \ldots, k_v$; an iterative solution is then required because all the k_i appear in all the equations.

As will be seen, many of the resulting implicit Runge–Kutta formulas fall within the Gauss, Radau, or Lobatto families as already outlined in Section 2.1.

2.5.1. Basic Forms

Let us start the discussion by recalling the basic structure of the Runge–Kutta formulas, namely,

$$y_{n+1} = y_n + \sum_{i=1}^{v} w_i k_i \qquad (2.5\text{-}1)$$

where

$$k_i = hf\left(x_n + c_i h, y_n + \sum_{j=1}^{i-1} a_{ij} k_j\right), \qquad c_1 = 0, \quad i = 1, 2, \ldots, v \qquad (2.5\text{-}2)$$

v is the number of substitutions and $c_i = \sum_{j=1}^{v} a_{ij}$. There are two points of interest in (2.5-1). First we note that by imposing $c_1 = 0$, we automatically specify that the first k_i, k_1, is evaluated at $x = x_n$. If we had not imposed this constraint k_1 might be evaluated at another value of x. Second, we have written the upper limit in the summation for k_i as $i - 1$. This automatically means that the lower triangular matrix $\mathbf{A}_L$ results for the a_{ij}. If we raised this limit to $j = v$ then we would fill in the elements of the matrix and at the same time increase the number of parameters. In this way we obtain the implicit Runge–Kutta formulas we are going to present shortly. In fact, following Butcher [8, 9] we may define the three arrangements:

Explicit	Total of $v(v + 1)/2$ parameters to choose	In any k_i, only use $k_1, \ldots, k_{i-1}$. Upper limit on summation is $i - 1$. Only $a_{i,i-1} \neq 0$.
Semi-Implicit	Total of $v(v + 3)/2$ parameters to to choose	In any k_i, only use $k_1, \ldots, k_i$. Upper limit on summation is i
Implicit	Total of $v(v + 1)$ parameters to choose	In any k_i, use $k_1, \ldots, k_v$. No restriction placed on the a_{ij}. Upper limit on summation is v.

We can proceed to develop the formulas in either of two ways: (1) by augmenting the system we may write the system differential equation as $\mathbf{y}' = \mathbf{f}(\mathbf{y})$ with x implicit in $\mathbf{y}$ (while we shall show most of our equations in scalar form, this formulation contains the equivalent vector or multi-dimensional case). This equation can then be solved by a quadrature formula in the form

$$I = \int_{x_0}^{x_1} f(x)\, dx$$

or

$$I = h \sum_{i=1}^{v} w_i f(x_0 + c_i h).$$

Butcher shows that for each Runge–Kutta process, there corresponds such a quadrature form characterized by the values $w_1, \ldots, w_v$ and $c_1, \ldots, c_v$, i.e., a Gauss–Legendre (or Gauss) quadrature. (2) Alternatively we can expand the Runge–Kutta formulation of (2.5-2) directly to include more of the a_{ij} and then follow the normal Taylor series approach.

2.5.2. Gauss Forms

In the case of a Gauss quadrature Butcher shows that $c_1, c_2, \ldots, c_v$ are roots of $P_v(2c - 1) = 0$ where $P_v(x)$ is a Legendre polynomial of degree v. Note that this merely specifies the x_i with the particular polynomials (shifted polynomials) transformed to $(0, 1)$ instead of the usual $(-1, 1)$. An algorithm for v substitutions and order $2v$, $p = 2v$, is given by

  a.   Find roots of $P_v(2c - 1) = 0$. This yields $c_1, \ldots, c_v$.
  b.   For each i, $i = 1, 2, \ldots, v$, find a_{ij} $(j = 1, 2, \ldots, v)$ as solutions to the linear system

$$\sum_{j=1}^{v} a_{ij} c_j^{k-1} = (1/k)c_i^k, \qquad k = 1, 2, \ldots, v$$

  c.   Find w_j $(j = 1, 2, \ldots, v)$ as solutions to the linear equations

$$\sum_{j=1}^{v} w_j c_j^{k-1} = 1/k, \qquad k = 1, 2, \ldots, v$$

$c_1 \neq 0$ in these equations. On this basis Butcher presents a series of Runge–Kutta implicit Gauss formulas with $v = 1, 2, \ldots, 5$ and $p = 2v$, We show below the cases $v = 1, 2, 3$ equivalent also to those given by Ceshchino and Kuntzmann [15]

Gauss implicit. (Equivalent to trapezoidal rule.)

$$\frac{\frac{1}{2} \ \Big| \ \frac{1}{2}}{1} \qquad v = 1, \quad p = 2 \qquad (2.5\text{-}3)$$

Gauss implicit. (See Table 2.2 for 2-point Guass.)

$$
\begin{array}{c|cc}
(3 - \sqrt{3})/6 & 1/4 & (3 - 2\sqrt{3})/12 \\
(3 + \sqrt{3})/6 & (3 + 2\sqrt{3})/12 & 1/4 \\
\hline
& 1/2 & 1/2
\end{array}
\qquad v = 2, \quad p = 4 \qquad (2.5\text{-}4)
$$

Gauss implicit

$$
\begin{array}{c|ccc}
(5 - \sqrt{15})/10 & 5/36 & (10 - 3\sqrt{15})/45 & (25 - 6\sqrt{15})/180 \\
1/2 & (10 + 3\sqrt{15})/72 & 2/9 & (10 - 3\sqrt{15})/72 \\
(5 + \sqrt{15})/10 & (25 + 6\sqrt{15})/180 & (10 + 3\sqrt{15})/45 & 5/36 \\
\hline
& \frac{5}{18} & \frac{4}{9} & \frac{5}{18}
\end{array}
$$

$$v = 3, \quad p = 6 \qquad (2.5\text{-}5)$$

Note the high order achieved by these implicit formulas for the number of stages or substitutions.

2.5.3. Radau and Lobatto Forms

Next we turn to use of the Radau and Lobatto forms and specify the three cases:

Case 1. $c_1 = 0; c_2, \ldots, c_v$ free parameters
Case 2. $c_v = 1; c_1, \ldots, c_{v-1}$ free parameters
Case 3. $c_1 = 0, c_v = 1; c_2, \ldots, c_{v-1}$ free parameters

In Case 1, $x = 0$ (or x_n) is specified as one of the points to evaluate $f(x, y)$; in Case 2, $x = 1$ (or x_{n+1}) is specified as one of the points to evaluate $f(x, y)$. These are then Radau quadratures. In Case 3, $x = 0$ (or x_n) and $x = 1$ (or x_{n+1}) are both specified as points to evaluate $f(x, y)$. This is then a Lobatto quadrature. The order of the processes is, because of the number of specified points, $p = 2v - 1$ for Cases 1 and 2 and $p = 2v - 2$ for Case 3.

Because of the relationship that $c_i = \sum_{j=1}^{v} a_{ij}$ it follows that

1 and 3: Since $c_1 = 0$, $a_{11} = a_{12} = \cdots = a_{1v} = 0$
2 and 3: Since $c_v = 1$, $a_{1v} = a_{2v} = \cdots = a_{vv} = 0$

and for 1 and 3 the first row of the a_{ij} matrix will be zero while for 2 and 3 the last column of the a_{ij} matrix will be zero.

On this basis we may now present Runge–Kutta implicit forms for Cases 1, 2, and 3.

CASE 1 FORMULAS

Radau implicit. (Euler formula.)

$$\begin{array}{c|c} 0 & 0 \\ \hline & 1 \end{array} \qquad v = 1, \quad p = 1 \qquad (2.5\text{-}6)$$

Radau implicit. (See Table 2.2 for 2-point Radau.)

$$\begin{array}{c|cc} 0 & 0 & 0 \\ \frac{2}{3} & \frac{1}{3} & \frac{1}{3} \\ \hline & \frac{1}{4} & \frac{3}{4} \end{array} \qquad v = 2, \quad p = 3 \qquad (2.5\text{-}7)$$

Radau implicit. (See Table 2.2 for 3-point Radau.)

$$\begin{array}{c|ccc} 0 & 0 & 0 & 0 \\ (6 - \sqrt{6})/10 & (9 + \sqrt{6})/75 & (24 + \sqrt{6})/120 & (168 - 73\sqrt{6})/600 \\ (6 + \sqrt{6})/10 & (9 - \sqrt{6})/75 & (168 + 73\sqrt{6})/600 & (24 - \sqrt{6})/120 \\ \hline & 1/9 & (16 + \sqrt{6})/36 & (16 - \sqrt{6})/36 \end{array}$$
$$v = 3, \quad p = 5 \qquad (2.5\text{-}8)$$

CASE 2 FORMULAS

Radau implicit. (See Table 2.2 for 2-point Radau.)

$$\begin{array}{c|cc} \frac{1}{3} & \frac{1}{3} & 0 \\ 1 & 1 & 0 \\ \hline & \frac{3}{4} & \frac{1}{4} \end{array} \qquad v = 2, \quad p = 3 \qquad (2.5\text{-}9)$$

Radau implicit. (See Table 2.2 for 3-point Radau.)

$$
\begin{array}{c|ccc}
(4 - \sqrt{6})/10 & (24 - \sqrt{6})/120 & (24 - 11\sqrt{6})/120 & 0 \\
(4 + \sqrt{6})/10 & (24 + 11\sqrt{6})/120 & (24 + \sqrt{6})/120 & 0 \\
1 & (6 - \sqrt{6})/12 & (6 + \sqrt{6})/12 & 0 \\
\hline
 & (16 - \sqrt{6})/36 & (16 + \sqrt{6})/36 & 1/9
\end{array}
$$

$$v = 3, \quad p = 5 \qquad (2.5\text{-}10)$$

<small>CASE 3 FORMULAS</small>
Lobatto implicit (trapezoidal rule)

$$
\begin{array}{c|cc}
0 & 0 & 0 \\
1 & 1 & 0 \\
\hline
 & \frac{1}{2} & \frac{1}{2}
\end{array}
\qquad v = 2, \quad p = 2 \qquad (2.5\text{-}11)
$$

Lobatto implicit (Milne form)

$$
\begin{array}{c|ccc}
0 & 0 & 0 & 0 \\
\frac{1}{2} & \frac{1}{4} & \frac{1}{4} & 0 \\
1 & 0 & 1 & 0 \\
\hline
 & \frac{1}{6} & \frac{2}{3} & \frac{1}{6}
\end{array}
\qquad v = 3, \quad p = 4 \qquad (2.5\text{-}12)
$$

Lobatto implicit (See Table 2.2 for 4-point Lobatto.)

$$
\begin{array}{c|cccc}
0 & 0 & 0 & 0 & 0 \\
(5 - \sqrt{5})/10 & (5 + \sqrt{5})/60 & 1/6 & (15 - 7\sqrt{5})/60 & 0 \\
(5 + \sqrt{5})/10 & (5 - \sqrt{5})/60 & (15 + 7\sqrt{5})/60 & 1/6 & 0 \\
1 & 1/6 & (5 - \sqrt{5})/12 & (5 + \sqrt{5})/12 & 0 \\
\hline
 & 1/12 & 5/12 & 5/12 & 1/12
\end{array}
$$

$$v = 4, \quad p = 6 \qquad (2.5\text{-}13)$$

Lobatto implicit

$$
\begin{array}{c|ccccc}
0 & 0 & 0 & 0 & 0 & 0 \\
(7 - \sqrt{21})/14 & 1/14 & 1/9 & (13 - 3\sqrt{21})/63 & (14 - 3\sqrt{21})/126 & 0 \\
1/2 & 1/32 & (91 + 21\sqrt{21})/576 & 11/72 & (97 - 21\sqrt{21})/576 & 0 \\
(7 + \sqrt{21})/14 & 1/14 & (14 + 3\sqrt{21})/126 & (13 + 3\sqrt{21})/63 & 1/9 & 0 \\
7 & 0 & 7/18 & 2/9 & 7/18 & 0 \\
\hline
 & 1/20 & 49/180 & 16/45 & 49/180 & 1/20
\end{array}
$$

$$v = 5, \quad p = 8 \qquad (2.5\text{-}14)$$

We thus have available a whole collection of new Runge–Kutta formulas which, while requiring iteration, yield a high order for the number of stages.

2.5.4. Semi-Implicit Forms

Intermediate between the explicit and the implicit formulas is the formulation due to Rosenbrock [59] and refined further by Haines [30]. The motivation behind this approach is based upon obtaining stable Runge–Kutta processes as well as maintaining computational efficiency, i.e., avoiding the necessity to iterate. The basic equations are given by

$$y_{n+1} = y_n + w_1 k_1 + w_2 k_2 + w_3 k_3 \qquad (2.5\text{-}15)$$

with

$$
\begin{aligned}
k_1 &= h[f(y_n) + a_1 A(y_n)k_1] \\
k_2 &= h[f(y_n + b_1 k_1) + a_2 A(y_n + c_1 k_1)k_2] \\
k_3 &= h[f(y_n + b_2 k_1 + d_1 k_2) + a_3 A(y_n + c_2 k_1 + e_1 k_2)k_3]
\end{aligned}
\qquad (2.5\text{-}16)
$$

In these equations it is assumed that $y' = f(y)$. $A(y) = f_y$ is a scalar or the Jacobian matrix and if $a_1 = a_2 = a_3 = 0$, (2.5-16) would be explicit. However, we have a semi-implicit type of formula in which each k_i has the same unknown on the right-hand side but no higher k_i, i.e., no $k_{i+1}, k_{i+2}, \ldots, k_v$.

The parameters in this formulation are $w_1, \ldots, w_3, a_1, b_1, a_2, \ldots, e_1$ and these can be evaluated as in the usual explicit Runge–Kutta formulation via the Taylor series comparison. Of importance, however, is that each k_i equation can be solved for the k_i to yield a class of explicit forms. Thus

$$k_1 = hf(y_n) + ha_1 A(y_n)k_1$$

or

$$\mathbf{k}_1 = h[\mathbf{I} - ha_1 \mathbf{A}(\mathbf{y}_n)]^{-1}\mathbf{f}(\mathbf{y}_n)$$

using a matrix symbolism. Equations (2.5-15) and (2.5-16), for k_1 and k_2 only, may now be written as

$$y_{n+1} = y_n + w_1 k_1 + w_2 k_2$$

with

$$
\begin{aligned}
\mathbf{k}_1 &= h[\mathbf{I} - ha_1 \mathbf{A}(\mathbf{y}_n)]^{-1}\mathbf{f}(\mathbf{y}_n) \\
\mathbf{k}_2 &= h[\mathbf{I} - ha_2 \mathbf{A}(\mathbf{y}_n + c_1 \mathbf{k}_1)]^{-1}\mathbf{f}(\mathbf{y}_n + b_1 \mathbf{k}_1)
\end{aligned}
\qquad (2.5\text{-}17)
$$

Two sets of parameters have been given by Rosenbrock:

Rosenbrock semi-implicit $v = 2, p = 3$

$$a_1 = 1 + \sqrt{6}/6 = 1.40824829, \qquad a_2 = 1 - \sqrt{6}/6 = 0.59175171$$

$$b_1 = c_1 = \{-6 - \sqrt{6} + \sqrt{58 + 20\sqrt{6}}\}/(6 + 2\sqrt{6}) = 0.17378667 \quad (2.5\text{-}18)$$

$$w_1 = -0.41315432, \qquad w_2 = 1.41315432$$

and

Rosenbrock semi-implicit $v = 2, p = 2$

$$a_1 = a_2 = 1 - \sqrt{2}/2, \qquad b_1 = (\sqrt{2} - 1)/2$$
$$c_1 = 0, \qquad w_1 = 0, \qquad w_2 = 1 \tag{2.5-19}$$

Haines has presented the parameters in an extension of (2.5-16) as

Haines semi-implicit $v = 4, p = 3$

$$a_1 = 1, \quad a_2 = 1, \quad a_3 = 1, \quad a_4 = \tfrac{2}{3}$$
$$b_1 = 1, \quad b_2 = \tfrac{1}{2}, \quad b_3 = \tfrac{2}{99}$$
$$c_1 = 1, \quad c_2 = \tfrac{1}{2}$$
$$d_1 = \tfrac{1}{2}, \quad d_2 = \tfrac{95}{99} \tag{2.5-20}$$
$$e_1 = \tfrac{1}{2}, \quad f_1 = \tfrac{2}{99}$$
$$c_3 = e_2 = g_1 = 0$$
$$w_1 = \tfrac{19}{9}, \quad w_2 = -\tfrac{43}{18}, \quad w_3 = \tfrac{28}{9}, \quad w_4 = -\tfrac{11}{6}$$

In addition, Calahan [13] has pointed out that the following values

$$a_1 = a_2 = 0.788675134, \qquad c_1 = 0,$$
$$b_1 = -1.15470054, \qquad w_1 = \tfrac{3}{4}, \quad w_2 = \tfrac{1}{4} \tag{2.5-21}$$

yield a $v = 2$ and $p = 3$ form.

Advantages of the implicit Runge–Kutta methods are high orders for the number of stages and desirable stability characteristics. This latter point will be discussed thoroughly in Chapter 3.

A disadvantage of the implicit methods is that an iterative procedure is required to solve the equations for the k_i. This follows, of course, because the a_{ij} matrix is full. The semi-implicit methods retain the desirable stability characteristics of the implicit methods but do not require iteration.

2.6. RUNGE–KUTTA FORMULAS FOR VECTOR DIFFERENTIAL EQUATIONS

We have already mentioned that while most of the Runge–Kutta and other single-step formulas were derived on the basis of a scalar differential equation, the extension to a set of m first-order equations follows in a direct manner. Here we shall briefly show this explicit formulation.

Consider the second-order equation

$$y'' = g(x, y, y') \tag{2.6-1}$$

with initial conditions

$$y(x_0) = y_0, \qquad y'(x_0) = y_0' \qquad (2.6\text{-}2)$$

Let $w = y'$ and define the vectors

$$\mathbf{z} = \begin{pmatrix} y \\ w \end{pmatrix}, \qquad \mathbf{f}(x, \mathbf{z}) = \begin{pmatrix} w \\ g(x, y, y') \end{pmatrix}, \qquad \mathbf{z}_0 = \begin{pmatrix} y_0 \\ w_0 \end{pmatrix}$$

to yield

$$\mathbf{z}' = \mathbf{f}(x, \mathbf{z}), \qquad \mathbf{z}(x_0) = \mathbf{z}_0 \qquad (2.6\text{-}3)$$

This is merely the vector form for two simultaneous ODEs and any of the previous Runge–Kutta formulas can be used directly on (2.6-3). Thus if

$$\begin{aligned} y' &= f(x, y, w) \\ w' &= b(x, y, w) \end{aligned} \qquad (2.6\text{-}4)$$

then for $f(x, y, w) = w$ or $y' = w$ this reduces to the form of (2.6-3). Applying (2.6-4) to the classic fourth-order Runge–Kutta formula we get

$$\begin{aligned} y_{n+1} &= y_n + \tfrac{1}{6}[k_1 + 2k_2 + 2k_3 + k_4] \\ w_{n+1} &= w_n + \tfrac{1}{6}[l_1 + 2l_2 + 2l_3 + l_4] \end{aligned}$$

where

$$\begin{aligned} k_1 &= hf(x_n, y_n, w_n) \\ k_2 &= hf(x_n + (h/2), y_n + (k_1/2), w_n + (l_1/2)) \\ k_3 &= hf(x_n + (h/2), y_n + (k_2/2), w_n + (l_2/2)) \\ k_4 &= hf(x_n + h, y_n + k_3, w_n + l_3) \end{aligned} \qquad (2.6\text{-}5)$$

The l_i are identical to k_i except that b replaces f. It is apparent on this basis that we could extend any of our previous formulas to the case $m = 2$.

The application of these ideas to m ($m \geq 2$) first-order ODEs is obvious. If we define

$$\begin{aligned} y_1' &= f_1(x, y_1, y_2, \ldots, y_m) \\ y_2' &= f_2(x, y_1, y_2, \ldots, y_m) \\ &\;\;\vdots \\ y_m' &= f_m(x, y_1, y_2, \ldots, y_m) \end{aligned} \qquad (2.6\text{-}6)$$

where

$$\mathbf{y} = \begin{pmatrix} y_1 \\ \vdots \\ y_m \end{pmatrix}, \qquad \mathbf{f} = \begin{pmatrix} f_1 \\ \vdots \\ f_m \end{pmatrix}$$

then any of the methods holds directly in terms of $\mathbf{y}$ and $\mathbf{f}$.

Since the present writers prefer to solve m first-order equations rather than one mth-order equation, we shall not pursue this material further. For the interested reader, however, the works of Scraton [67] and Collatz [17] are recommended.

2.7. TWO-STEP RUNGE–KUTTA FORMULAS

In Section 2.5 we observed that the implicit as compared to the explicit formulation was able to lower the number of stages v for a fixed order p Runge–Kutta process. The penalty paid for this advantage was the need to iterate the equations. In this section we shall look at another way to decrease the stages as compared to the explicit forms; now the penalty paid is a different one, namely, that one state or function evaluation is required from outside the interval (x_n, x_{n+1}). As such the method begins to resemble a multistep procedure although we only consider $k = 2$ or two steps. The work we shall discuss is due to Byrne and Lambert [11, 12] although the material by Rosen [58] and by Ceschino and Kuntzmann [15] bears directly on the problem.

Based upon the usual scalar ODE Byrne and Lambert defined a 2-step Runge–Kutta system as

$$y_{n+1} = y_n + \sum_{i=1}^{v} s_i l_i(y_{n-1}) + \sum_{i=1}^{n} w_i k_i(y_n) \tag{2.7-1}$$

where

$$
\begin{aligned}
l_1(y_{n-1}) &= hf(x_{n-1}, y_{n-1}) \\
l_2(y_{n-1}) &= hf(x_{n-1} + u_1 h, y_{n-1} + u_1 l_1) \\
l_3(y_{n-1}) &= hf(x_{n-1} + u_2 h, y_{n-1} + (u_2 - u_3)l_1 + u_3 l_2)
\end{aligned}
\tag{2.7-2}
$$

$$
\begin{aligned}
k_1(y_n) &= hf(x_n, y_n) \\
k_2(y_n) &= hf(x_n + u_1 h, y_n + u_1 k_1) \\
k_3(y_n) &= hf(x_n + u_2 h, y_n + (u_2 - u_3)k_1 + u_3 k_2)
\end{aligned}
\tag{2.7-3}
$$

If $s_i = 0$ in (2.7-1), the normal explicit Runge–Kutta formulas result.

For $v = 3$, the parameters u_1 and u_2 are left free while u_3, the s_i, and the w_i are evaluated as functions of u_1 and u_2. The result is a fourth-order ($p = 4$) process requiring three stages. If $v = 2$, the terms l_3 and k_3 are not needed and a third-order process ($p = 3$) with two stages can be obtained with u_1 as the free parameter.

By expanding in Taylor series and comparing with the exact expansion, agreement term by term can be achieved through $O(h^3)$ or $O(h^4)$. The result of such a comparison is

$$s_1 = (5 - 6u_1)/12u_1$$
$$s_2 = -w_2,$$
$$w_1 = 1 - s_1 \qquad\qquad v = 2, \quad p = 3 \qquad (2.7\text{-}4)$$
$$w_2 = 5/12u_1, \qquad u_1 \text{ free}$$

$$s_1 = [5(u_1 + u_2) - 6u_1u_2 - 4]/12u_1u_2$$
$$s_2 = -w_2, \qquad s_3 = -w_1;$$
$$w_1 = 1 - s_1$$
$$w_2 = [4 - 5u_2]/[12u_1(u_1 - u_2)] \qquad v = 3, \quad p = 4 \qquad (2.7\text{-}5)$$
$$w_3 = [5u_1 - 4]/[12u_2(u_1 - u_2)]$$
$$u_3 = [2u_2(u_2 - u_1)]/[u_1(4 - 5u_1)], \qquad u_2, u_1 \text{ free}$$

where it has been assumed that $u_1 \neq 0$, $u_2 \neq 0$, $u_1 \neq u_2$.

A major problem is to select u_1 and u_1, u_2 in the third- and fourth-order cases. An obvious approach is to select these parameters on the basis of the bound of (2.4-2) and (2.4-3) to minimize the local truncation error. The resulting choice of the free parameters would yield "optimum" third- and fourth-order 2-step Runge–Kutta formulas.

Byrne and Lambert show that the choices $u_1 = \frac{4}{5}$ in the $p = 3$ case and $u_1 = 0.541$, $u_2 = 0.722779927$ in the $p = 4$ case yields optimum results.

The generation of higher-order formulas with more steps $(k > 2)$ is possible. However, since Byrne and Lambert evidence a loss in accuracy over the classic Runge–Kutta formulas as indicated by some numerical examples, there seems little motivation to extend the present results.

2.8. LOCAL TRUNCATION ERROR ESTIMATES IN A SINGLE STEP

Any attempt to explicitly estimate the local truncation error in the Runge–Kutta formulas by the equations already given is extremely difficult, i.e., to use (2.3-12) for (2.3-13). In fact this is one of the serious defects in the whole Runge–Kutta framework. This estimate for $T(x, h)$ is needed to provide a reasonable explicit value of h to use in a single-step calculation.

Since this feature is sufficiently important, we wish to examine the available methods for estimating $T(x, h)$ or the accuracy of a single-step in a calculation. Thus in this section we shall detail some methods which can be used as the computation proceeds from x_n to x_{n+1}; in Sections 2.9 and 2.10 we will develop methods which require two steps or more, i.e., x_n to x_{n+1} to x_{n+2}, to evaluate $T(x, h)$.

2.8.1. Merson and Scraton Forms

Perhaps the first significant attempt to estimate $T(x, h)$ was due to Merson [51] using the formula

Merson formula

$$
\begin{array}{c|ccccc}
0 & & & & & \\
\frac{1}{3} & \frac{1}{3} & & & & \\
\frac{1}{3} & \frac{1}{6} & \frac{1}{6} & & & \\
\frac{1}{2} & \frac{1}{8} & 0 & \frac{3}{8} & & \\
1 & \frac{1}{2} & 0 & -\frac{3}{2} & 2 & \\
\hline
& \frac{1}{6} & 0 & 0 & \frac{2}{3} & \frac{1}{6}
\end{array}
\qquad v = 5, \quad p = 4 \qquad (2.8\text{-}1)
$$

This formula is fourth order ($p = 4$) and yet it uses five ($v = 5$) substitutions. As a result it has an extra degree of freedom built into it. Let us call the result of the calculation using the c_i and a_{ij} above the dashed line and the w_i just below the dashed line $\bar{y}_{n+1}$; let us also call y_{n+1} the result of the full normal calculation indicated by (2.8-1). Merson has shown that if the differential equation is linear or h is small enough to approximate $f(x, y)$ with a linear function, a good estimate of the error in y_{n+1} is

$$
T(x, h) = (1/30)[2k_1 - 9k_3 + 8k_4 - k_5] \qquad (2.8\text{-}2)
$$

It is important to emphasize that the method is valid only when $f(x, y)$ is linear. In the case of a nonlinear $f(x, y)$ the formulas should be used only for a rough guide. Nevertheless, the method does provide an estimate (conservative) for the local truncation error in a single step of the computation. The penalty is of course the extra computation required because of the fifth substitution as compared to four substitutions in the standard fourth-order Runge–Kutta.

Scraton [68] has also proposed another fourth-order process using five substitutions. The formula is given by

Scraton formula

$$
\begin{array}{c|ccccc}
0 & & & & & \\
\frac{2}{9} & \frac{2}{9} & & & & \\
\frac{1}{3} & \frac{1}{12} & \frac{1}{4} & & & \\
\frac{3}{4} & \frac{69}{128} & \frac{243}{128} & \frac{270}{128} & & \\
\hline
\frac{9}{10} & -345 & 2025 & -1224 & 544 & \left(\times\frac{9}{10000}\right) \\
\hline
& \frac{17}{162} & 0 & \frac{81}{170} & \frac{32}{135} & \frac{250}{1377}
\end{array}
\qquad v = 5, \quad p = 4 \quad (2.8\text{-}3)
$$

and the local truncation error is

$$
T(x, h) = -(k_4 - k_1)^{-1}\left(-\tfrac{1}{18}k_1 + \tfrac{27}{170}k_3 - \tfrac{4}{15}k_4 + \tfrac{25}{153}k_5\right)
$$
$$
\times \left(\tfrac{19}{24}k_1 - \tfrac{27}{8}k_2 + \tfrac{57}{20}k_3 - \tfrac{4}{15}k_4\right) \qquad (2.8\text{-}4)
$$

2.8.2. Collatz Rule of Thumb

Collatz [17] has outlined a rule-of-thumb method for indirectly measuring $T(x, h)$ by specifying the correct magnitude of h. He suggested that in the fourth-order classic Runge–Kutta calculation k_2 and k_3 must be approximately equal. To be more specific, it was suggested that the identity

$$\left| \frac{k_3 - k_2}{k_2 - k_1} \right| = 1.0 \qquad (2.8\text{-}5)$$

hold when the proper value of h is used in a calculation. If the ratio is much greater than 1, the local truncation error is too large in this step and h must be decreased. If the ratio is much smaller than 1, the h should be increased.

The use of (2.8-5) to specify h is, of course, subject to considerable interpretation and should be used only with care. As an example, if $f(x, y) = f(x)$ then $k_3 = k_2$ in the Runge–Kutta formulas and (2.8-5) breaks down. Warten [72] has, however, used the idea with some success. He suggested that one try to estimate the local truncation error by means of a weighted linear combination of the k_i. The weighting factors are then adjusted to fit the local truncation error for the linear differential equation $y' = ay + b$ or its vector equivalent.

Obviously, this idea of Warten has a connection with the work of Merson. As pointed out by Scraton, for the special linear case, it is possible to get a single-step error estimate directly for the classic Runge–Kutta process (and others as well) in the form

$$T(x, h) = - \frac{8(k_3 - k_2)^3}{15(k_4 - k_1)^2} \qquad (2.8\text{-}6)$$

However, this approach must be considered as empirical and subject to analysis before use.

2.8.3. Sarafyan Embedding Forms

Since the previous methods all depend on the use of a linear system, we turn to the important work of Sarafyan [62–64] which like Merson involves an embedding of Runge–Kutta formulas. At the same time, however, the method holds for any differential equation and is suggested by the present authors as being the most promising procedure for step-size (h) control.

The embedding principle is simple to understand. Let us call $y_{v,n+1}$ the value of y_{n+1} calculated using a v-stage process; then one attempts to build up the sequence, $y_{1,n+1}$, $y_{2,n+1}$, $y_{3,n+1}$, $y_{4,n+1}$, and $y_{6,n+1}$ using the same set of k_i. Thus, $y_{1,n+1}$ uses a specific k_1, $y_{2,n+1}$ uses the same k_1 and a specific $k_2, \ldots, y_{6,n+1}$ uses the same k_1, k_2, k_3, k_4 and a specific k_5 and k_6. Further, $y_{1,n+1}$ yields a $p = 1$ result, $y_{2,n+1}$ a $p = 2$ result, $\ldots, y_{4,n+1}$ a $p = 4$ result and $y_{6,n+1}$ a $p = 5$ result. Thus a fifth-order Runge–Kutta formula with six stages ($p = 5, v = 6$) has embedded within its formula some of the lower-order and fewer-stages formulas. If $y_{4,n+1}$ and $y_{6,n+1}$, for example, agree with each other up to j decimals then $y_{6,n+1}$ is correct up to the j decimals. This specifies a conservative error for $y_{6,n+1}$. The value of h may then be adjusted up or down based upon this error.

While Sarafyan presents many fifth-order formulas one of the best is (2.3-27) which we write as

Sarafyan form

0							
							$y_{1,n+1}$
$\frac{1}{2}$	$\frac{1}{2}$						
							$y_{2,n+1}$
$\frac{1}{2}$	$\frac{1}{4}$	$\frac{1}{4}$					
1	0	-1	2				
							$y_{4,n+1}$
$\frac{2}{3}$	$\frac{7}{27}$	$\frac{10}{27}$	0	$\frac{1}{27}$			
$\frac{2}{10}$	$\frac{28}{625}$	$-\frac{125}{625}$	$\frac{546}{625}$	$\frac{54}{625}$	$-\frac{378}{625}$		
							$y_{6,n+1}$
	$\frac{14}{336}$	0	0	$\frac{35}{336}$	$\frac{162}{336}$	$\frac{125}{336}$	

$$v = 6, \quad p = 5 \qquad (2.8\text{-}7)$$

to yield $y_{6,n+1}$. Along the way, however (see dashed lines), it is possible to calculate

$$
\begin{aligned}
y_{1,n+1} &= y_n + k_1 \\
y_{2,n+1} &= y_n + k_2 \\
y_{4,n+1} &= y_n + \tfrac{1}{6}[k_1 + 4k_3 + k_4]
\end{aligned}
\qquad (2.8\text{-}8)
$$

where (2.8-8) turns out to be a fourth-order Runge–Kutta process which we have not previously given but which may be found in Kopal [40, p. 211].

Thus if one wishes to use a fifth-order Runge–Kutta formula, (2.8-7) is a natural since concurrently a fourth-order result can be obtained with almost no extra computation. By comparing the two results, an estimate of the accuracy can be obtained and then a judgment can be made on the value of h being used. If a fourth-order formula is being used in the calculation then two extra stages are required to obtain this comparison, approximately a 50% increase in computation time for six stages versus four stages. In this case the method should be used only occasionally to monitor the calculation rather than every step.

Actually Sarafyan has also developed some further points of interest in this area. For the classic fourth-order Runge–Kutta formula of (2.3-13) he has shown that

$$
\begin{aligned}
y_{1,n+1} &= y_n + k_1 \\
y_{2,n+1} &= y_n + k_2 \\
y_{4,n+1} &= y_n + \tfrac{1}{6}[k_1 + 2k_2 + 2k_3 + k_4]
\end{aligned}
\qquad (2.8\text{-}9)
$$

and that

$$y_{1,\,n+3/4} = y_n + \tfrac{3}{4}k_1$$
$$y_{2,\,n+3/4} = y_n + \tfrac{3}{16}(k_1 + 3k_2) \tag{2.8-10}$$
$$y_{3,\,n+3/4} = y_n + \tfrac{3}{32}(2k_1 + 3k_2 + 3k_3)$$

Thus from the same set of k_i it is possible to generate a collection of Runge–Kutta formulas of order 1 to order 4. Based upon these formulas he then develops a scalar measure Θ such that

$$\Theta = \frac{2}{3}\left|\frac{k_3 - k_2}{k_2 - k_1}\right| \tag{2.8-11}$$

fulfills the equivalent role as (2.8-5) previously. When $\Theta < 1$, the step size is too small, etc.

Finally, however, Sarafyan has developed what seems to be the best estimate procedure of all. It is based upon the fifth-order formula of Butcher (2.3-20), which we rewrite here

Butcher form

$$
\begin{array}{c|cccccc}
0 \\
\frac{1}{4} & \frac{1}{4} \\
\frac{1}{4} & \frac{1}{8} & \frac{1}{8} \\
\frac{1}{2} & 0 & -\frac{1}{2} & 1 \\
\frac{3}{4} & \frac{3}{16} & 0 & 0 & \frac{9}{16} \\
1 & -\frac{3}{7} & \frac{2}{7} & \frac{12}{7} & -\frac{12}{7} & \frac{8}{7} \\
\hline
 & \frac{7}{90} & 0 & \frac{32}{90} & \frac{12}{90} & \frac{32}{90} & \frac{7}{90}
\end{array}
\qquad v = 6, \quad p = 5 \tag{2.8-12}
$$

But with these k_i used for $y_{6,n+1}$, he shows that

$$y_{1,\,n+1} = y_n + k_1$$
$$y_{2,\,n+1} = y_n - k_1 + 2k_2 \tag{2.8-13}$$

and

$$y_{1,\,n+1/2} = y_n + \tfrac{1}{2}k_1$$
$$y_{2,\,n+1/2} = y_n + \tfrac{1}{2}k_2$$
$$y_{4,\,n+1/2} = y_n + \tfrac{1}{12}(k_1 + 4k_3 + k_4)$$

Also when a step size $2h$ is used (2.8-12) and (2.8-13) become

$$
\begin{array}{c|cccccc}
0 \\
\frac{1}{2} & \frac{1}{2} \\
\frac{1}{2} & \frac{1}{4} & \frac{1}{4} \\
1 & 0 & -1 & 2 \\
\frac{3}{2} & \frac{3}{8} & 0 & 0 & \frac{9}{8} \\
2 & -\frac{6}{7} & \frac{4}{7} & \frac{24}{7} & -\frac{24}{7} & \frac{16}{7} \\
\hline
 & \frac{7}{45} & 0 & \frac{32}{45} & \frac{12}{45} & \frac{32}{45} & \frac{7}{45}
\end{array}
\tag{2.8-14}
$$

$$y_{1,n+2} = y_n + 2k_1$$
$$y_{2,n+2} = y_n - 2k_1 + 4k_2 \qquad (2.8\text{-}15)$$
$$y_{6,n+2} = y_n + \tfrac{1}{45}[7k_1 + 32k_3 + 12k_4 + 32k_5 + 7k_6]$$

and

$$y_{1,n+1} = y_n + k_1$$
$$y_{2,n+1} = y_n + k_2$$
$$y_{4,n+1} = y_n + \tfrac{1}{6}[k_1 + 4k_3 + k_4]$$

In other words using the fifth-order formula of (2.8-14) and the relations of (2.8-15) one obtains a fifth-order Runge–Kutta form for $2h$ and embedded within this a fourth-order Runge–Kutta form for h. The importance of this feature cannot be overstressed since it means that starting from x_n it is possible to calculate y_{n+1} via a fourth-order formula and y_{n+2} via a fifth-order formula requiring a total of $v = 6$. On a per-step basis only three stages are required as compared to four stages with any other fourth-order formula. Thus this procedure is the most economical that we have encountered which yields at least fourth-order results. At the same time $y_{4,n+1}$ fits within the previous pattern of (2.8-7) which can be used to estimate the proper values of h in the calculation. Obviously this entire formulation is highly recommended for the user.

The work of Rosser [60] is also of interest along the lines of embedding lower-order within higher-order formulas. He proceeds by the use of blocks of calculations, with each block containing a certain number of steps in the overall computations. Fewer stages per step are needed as compared to the usual Runge–Kutta formulas and estimates of the truncation error can be made. This method would seem to warrant further investigation.

2.8.4. Fehlberg Embedding Forms

At roughly the same time Fehlberg [24] developed a fifth-order $(v = 6)$ formula embedded within a sixth-order $(v = 8)$ formula, a sixth-order $(v = 8)$ formula embedded within a seventh-order $(v = 10)$ formula, etc. The length of the formulas precludes our writing them here; however, the fifth- and sixth-order forms can be specified. In fact the fifth-order formula is merely (2.3-28) and thus we write

Fehlberg form

	Equation (2.3-28)							
0	$-\frac{11}{640}$	0	$\frac{11}{256}$	$-\frac{11}{160}$	$\frac{11}{256}$	0		$v = 6,\ p = 5$
1	$\frac{93}{640}$	$-\frac{18}{5}$	$\frac{803}{256}$	$-\frac{11}{160}$	$\frac{99}{256}$	0	1	and
	$\frac{7}{1408}$	0	$\frac{1125}{2816}$	$\frac{9}{32}$	$\frac{125}{768}$	0	$\frac{5}{66}$	$\frac{5}{66}$	$v = 8,\ p = 6$

$$(2.8\text{-}16)$$

Obviously these embedding formulas (and the higher-order ones) can be used to adjust the step size h.

2.9. LOCAL TRUNCTION ERROR ESTIMATE IN TWO STEPS

While the previous methods allow an estimate of the truncation error as the normal computation proceeds, other methods involve the use of two-step formulas. These two steps may involve the integration over a total interval of $2h$ or a forward and backwards pass over a single h interval. In either case we classify the methods as involving two steps of calculation.

2.9.1. Extrapolation

The first method of interest is termed extrapolation and is probably the most popular procedure for estimating $T(x, h)$. While further details will be presented in Chapter 5, we outline here a simple version of the overall procedure. Consider that we start a single-step calculation at (x_{n-1}, y_{n-1}) and proceed to (x_{n+1}, y_{n+1}) along two paths. The first path uses two steps of length h so that we apply the single-step method to go from (x_{n-1}, y_{n-1}) to (x_n, y_n) and then to (x_{n+1}, y_{n+1}) where $h = x_{n+1} - x_n = x_n - x_{n-1}$. The second path is to go from (x_{n-1}, y_{n-1}) to (x_{n+1}, y_{n+1}) in one step of length $2h = h' = x_{n+1} - x_{n-1}$. If we assume that the local truncation error has the form

$$T(x, h) = \phi h^{p+1} \tag{2.9-1}$$

where ϕ is the principal error function (assumed essentially constant), we may write the approximate formula for two steps of length h:

$$y(x_{n+1}) - y_{n+1}^{(h)} = 2\phi h^{p+1} \tag{2.9-2}$$

and one step of length h' or $2h$:

$$y(x_{n+1}) - y_{n+1}^{(2h)} = 2^{p+1}\phi h^{p+1} \tag{2.9-3}$$

Here we assume no round-off error and thus the difference between the exact value $y(x_{n+1})$ and the calculated (by the single-step method) value is merely the local truncation error. In (2.9-2) we assume that two steps of length h each yield twice the truncation error of one step. The use of $h' = 2h$ is arbitrary since we could have used $h' = kh$ where k is any fraction or integer except 0 or 1.

By subtracting (2.9-2) and (2.9-3) to evaluate ϕ and then substituting into (2.9-1), we can obtain the formula

$$T(x, h) = \phi h^{p+1} = \frac{y_{n+1}^{(h)} - y_{n+1}^{(2h)}}{2^{p+1} - 2} \tag{2.9-4}$$

which specifies the local truncation error for one step of length h. For the various Runge–Kutta formulas the local truncation error may be tabulated as

Order of formula p:	1	2	3	4	5	6
Local truncation error times $(y_{n+1}^{(h)} - y_{n+1}^{2h})$:	$\frac{1}{2}$	$\frac{1}{6}$	$\frac{1}{14}$	$\frac{1}{30}$	$\frac{1}{62}$	$\frac{1}{126}$

Since this procedure is general it may be used to obtain an asymptotic type of bound on the local truncation error for any single-step method. The defect associated with the procedure is that an extra step of length $h' = 2h$ is required above and beyond the normal portion of the computation. Thus it probably is best to use only occasionally to monitor the behavior of the calculation.

2.9.2. Gorbunov–Shakhov Method

An alternative procedure might be to proceed from y_n to y_{n+1} using a single-step method whose local truncation error is written as $T(x, h) = \phi h^{p+1}$ and then reverse the calculation by going from the now known y_{n+1} to recalculate y_n, call this y_n^*. In the first case a step of h is used while in the reverse path a step of $(-h)$ is used. Assuming that p is an even integer it is simple to show that

$$y_n^* - y_n = 2\phi h^{p+1} \tag{2.9-5}$$

Thus the principal error function or the local truncation error can be estimated by one-half the difference in the two calculated values y_n^* and y_n. Butcher has made extensive use of the idea of this procedure for the Case 1, 2, and 3 implicit formulas of Section 2.5. He is able to estimate the local truncation error in the 1 and 2 processes by going from y_n to y_{n+1} to y_{n+2} and then passing backward. Such procedures would seem worthy of further investigation.

Obviously this forward–backward procedure could be replaced by two forward integrations from (x_n, y_n) to (x_{n+1}, y_{n+1}) where the two integrations are carried out using one single-step method whose principal error function is ϕ and then another single-step method (of the same order) whose principal error function is $-\phi$. The work of Gorbunov and Shakhov [29] is indicative of this approach. They write the truncation error as

$$T(x, h) = \alpha\phi h^{p+1} \tag{2.9-6}$$

where α is a free parameter and derive two sets of Runge–Kutta formulas which yield $\bar{y}_{n+1}$ with $+\alpha$ and $\bar{\bar{y}}_{n+1}$ with $-\alpha$. The final value of y_{n+1} is then taken as

$$y_{n+1} = \tfrac{1}{2}[\bar{y}_{n+1} + \bar{\bar{y}}_{n+1}] \tag{2.9-7}$$

Typical explicit formulas are given by

$$
\begin{array}{c|cc}
0 & & \\
\frac{1}{4} & \frac{1}{4} & \\
\frac{2}{3} & 0 & \frac{2}{3} \\
\hline
& \frac{1}{4} & 0 & \frac{3}{4}
\end{array}
\qquad
\begin{array}{c|cc}
0 & & \\
\frac{5}{12} & \frac{5}{12} & \\
\frac{2}{3} & 0 & \frac{2}{3} \\
\hline
& \frac{1}{4} & 0 & \frac{3}{4}
\end{array}
\qquad v = 3, \quad p = 2 \qquad (2.9\text{-}8)
$$

$$
\alpha = \tfrac{1}{24} \qquad\qquad \alpha = -\tfrac{1}{24}
$$

$$
T(x, h) = \pm\tfrac{1}{24}h^3(f_x f_y + ff_y^2)
$$

$$
\begin{array}{c|cccc}
0 & & & & \\
\frac{2}{3} & \frac{2}{3} & & & \\
1 & \frac{1}{4} & \frac{3}{4} & & \\
\frac{1}{2} & \frac{1}{4} & \frac{3}{8} & -\frac{1}{8} & \\
\hline
& \frac{1}{6} & 0 & \frac{1}{6} & \frac{2}{3}
\end{array}
\qquad
\begin{array}{c|cccc}
0 & & & & \\
\frac{2}{3} & \frac{2}{3} & & & \\
1 & \frac{13}{4} & -\frac{9}{4} & & \\
\frac{3}{2} & \frac{27}{8} & 0 & -\frac{15}{8} & \\
\hline
& \frac{2}{9} & \frac{9}{10} & -\frac{1}{6} & \frac{2}{45}
\end{array}
\qquad v = 4, \quad p = 3 \qquad (2.9\text{-}9)
$$

$$
\alpha = +\tfrac{1}{12} \qquad\qquad \alpha = -\tfrac{1}{12}
$$

$$
T(x, h) = \pm\tfrac{1}{12}h^4(f_x f_y^2 + ff_y^3)
$$

Note that in order to obtain the $\pm\alpha$ value it is necessary to sacrifice one stage per order, i.e., the third-order method requires four stages. Nevertheless, the method is a most interesting one.

2.10. LOCAL TRUNCATION ERROR ESTIMATES IN MORE THAN TWO STEPS

Finally we turn to a method to estimate the local truncation error which requires more than two steps in a Runge–Kutta or single-step calculation. In the particular case we choose to illustrate the procedure, three steps are necessary for a fourth-order process but the extension to other cases will be obvious.

The basic idea is detailed by Ceschino and Kuntzmann [15] and Chai [16] and lies in the use of the Hermite interpolation polynomials. We have previously written the formula (1.3-7)

$$
y_{n+1} = 10y_{n-2} + 9y_{n-1} - 18y_n + 3h[y'_{n-2} + 6y'_{n-1} + 3y_n'] \qquad (2.10\text{-}1)
$$

which has a truncation error of $O(h^6)$. An alternative formula with the same order truncation error is given by

$$
hy'_{n-2} = 57y_{n-1} - 24y_n - 33y_{n+1} + h(24y'_{n-1} + 57y_n' + 10'_{n+1}) \qquad (2.10\text{-}2)
$$

By using the first of these Hermite formula as an indication of the correct values of $y_{n+1}, \ldots$, Ceschino and Kuntzmann derive the local truncation error estimate. [The indices in (2.10-1) and (2.10-2) need to be raised by 1.]

$$
30T(x, h) = y_{n+2} + 18y_{n+1} - 9y_n - 10y_{n-1} - 3h[3f_{n+1} + 6f_n + f_{n-1}] \qquad (2.10\text{-}3)
$$

This value of $T(x, h)$ corresponds, however, to the local truncation error at approximately x_{n+1} and not at x_{n+2}. In other words we use the fourth-order Runge–Kutta formula to take three steps, from x_{n-1} to x_n, from x_n to x_{n+1} and from x_{n+1} to x_{n+2}. When this third step is completed, we can estimate the local truncation error at the end of the second step. Actually because of the method of derivation this is really an average type of error over all three steps and we can probably use it for the x_{n+2} point. Note that this estimate is picked up without any extra calculations and thus there is considerable merit to its use.

Alternatively by using (2.10-2) it is possible to obtain the local truncation error estimate

$$90T(x, h) = 33y_{n+2} + 24y_{n+1} - 57y_n - h[10f_{n+2} + 57f_{n+1} + 24f_n - f_{n-1}]$$

$$(2.10\text{-}4)$$

By contrast to the previous estimate, this formula gives an estimate of the local truncation error at approximately x_{n+2}.

Other formulas of this type can be derived for any order single-step method. The one defect is that if the calculated truncation error is too large for a particular calculation, it may be necessary to go back three steps to restart with a new value of h.

2.11. LOCAL ROUND-OFF ERROR IN SINGLE-STEP METHODS

In this section we shall briefly discuss certain aspects of the local round-off error in single-step (including Runge–Kutta) methods. The general equation of interest is

$$\tilde{y}_{n+1} = \tilde{y}_n + h\Phi(x_n, \tilde{y}_n; h) + e_n \qquad (2.11\text{-}1)$$

Here e_n is the local round-off error associated with the direct evaluation of the normal single-step algorithm. $\tilde{y}_n$ is taken to mean a rounded or otherwise approximate value of y_n.

It is quite possible to work with (2.11-1) to obtain strict bounds for the local round-off error. However, these are usually much too conservative. The interested reader should consult Henrici [32] for these details. Instead we turn to a discussion based upon computational experience (recall the numerical results discussed in Chapter 1).

To illustrate this approach we select the fourth-order classic Runge–Kutta formula for which

$$\Phi(x_n, y_n; h) = \tfrac{1}{6}[k_1 + 2k_2 + 2k_3 + k_4] \qquad (2.11\text{-}2)$$

This particular increment function has an excellent form for minimizing the local round-off error. All of the k_i have the same sign, the weights on the

k_i are approximately equal and thus each term in Φ is of approximately equal magnitude and the values are all added together. This criteria that all the weights be of equal sign and magnitude is probably as good a rule of thumb as one can devise for minimizing the local round-off error.

It is also probably true that in most cases the round-off error is concentrated in the last step of the algorithm, i.e., in the equation

$$y_{n+1} = y_n + (h/6)[k_1 + 2k_2 + 2k_3 + k_4]$$

This follows because it is assumed that each k_i is calculated with a high degree of accuracy and the value of $h/6$ is small. The round-off associated with the sum of the k_is, especially after multiplying by $h/6$, will not affect the final sum.

One obvious procedure for minimizing the round-off error in each step is to use double-precision arithmetic. This means that any errors which occur are in the digits well beyond those actually needed in the calculation. In other words we use double length numbers with the extra length used to protect or buffer the calculation. Obviously this has a drawback in the sense that the overall computation time may rise. Nevertheless, this is frequently the path chosen to produce a valid calculation.

Henrici has pointed out an alternative way of achieving many of the advantages of double precision while minimizing the overall computing time. This method involves a partial double precision in which $\tilde{y}_n$ is carried in double precision (obviously $\tilde{y}_{n+1}$ is also in double precision when calculated) but $h\Phi(x_n, \tilde{y}_n; h)$ is carried in single precision. This algorithm requires only a minor amount of computing time over a single precision calculation. As described, it represents one of the best approaches to using any Runge–Kutta formulas.

2.12. THE EXPLICIT USE OF SINGLE-STEP FORMULAS

The explicit computer implementation of the single-step and Runge–Kutta formulas can be summarized by the algorithm shown below. Basically one programs the loop shown for $n = 1, 2, \ldots$, with $h = h_n$ as a feature of the loop. The algorithm is self-starting in the sense that only the initial conditions $\mathbf{y} = \mathbf{y}_0$, $x = x_0$ are needed.

SINGLE-STEP COMPUTER ALGORITHM

$$n = 0 \qquad \mathbf{y} = \mathbf{y}_0, \quad x = x_0$$
$$\Downarrow$$
$$n = 1, 2, \ldots \qquad \begin{array}{l} \mathbf{y}_n{}' = \mathbf{f}(x_n, \mathbf{y}_n) \\ \mathbf{y}_{n+1} = \mathbf{y}_n + h_n \Phi(x_n, \mathbf{y}_n; h_n) \end{array}$$

The advantages of the single-step algorithm are: (1) the algorithm is self-starting; (2) the step size h_n can be changed without affecting the use of the algorithm, and (3) high-order accuracy can be achieved and for a small enough h the result is always stable (see Chapter 3). The disadvantages of the algorithm are: (1) how to select the proper value of h_n, and (2) the number of stages v required per step may result in excessive computing time.

We have already discussed in Sections 2.8–2.10 a variety of ways to estimate the local truncation error as the single-step algorithm is used. The desire is then to use this information to monitor h_n in the algorithm so that the overall calculation maintains a certain accuracy. Thus if $T(x, h)$ is smaller than seems necessary to obtain an answer with the desired accuracy the value of h_n can be increased; obviously the converse feature also holds. Here we see that by monitoring the step-by-step accuracy we hope to maintain the overall calculation accuracy below a predetermined bound; this is achieved by varying h_n. Unfortunately the methods for evaluating the per-step accuracy are complex to use, are probably quite conservative, and largely ignore round-off error. As such they must be used with caution.

The second major disadvantage of single-step methods is the number of stages per step. However, we will see later that this is often not as disadvantageous as first thought. This is because in assessing the total desirability of a method we must consider both the accuracy and stability of the method. Often more stages per step result in better stability characteristics than methods of comparable order with a smaller number of stages per step.

2.13. MODERN TAYLOR SERIES EXPANSIONS

One of the oldest and best known ways to evaluate y_{n+1} in terms of y_n is by a Taylor series expansion (see Section 1.2). Unfortunately, while this method has the desirable features of convergence and accuracy, it has been only in the last few years that certain of the disadvantages associated with the explicit use of this method have become overcome to some degree. Here we wish to examine the features of this method and point out some of the more recent work.

Assuming that $f(x, y)$ has sufficient derivatives we write a finite Taylor series as

$$y_{n+1} = y_n + hf(x_n, y_n) + \frac{h}{2!} f'(x_n, y_n) + \cdots + \frac{h^p}{p!} f^{[p-1]}(x_n, y_n) \quad (2.13\text{-}1)$$

where the error associated with truncating the true infinite series is given by

$$T(x, h) = \frac{h^{p+1}}{(p+1)!} f^{[p]}(\zeta), \qquad x_n \le \zeta \le x_{n+1} \quad (2.13\text{-}2)$$

We can write the standard form of a single-step method as

$$y_{n+1} = y_n + h\Phi[x_n, y_n; h] \qquad (2.13\text{-}3)$$

where, for the finite Taylor series of order p,

$$\Phi(x_n, y_n; h) = \sum_{i=0}^{p-1} \frac{h^i}{(i+1)!} f^{[i]}(x_n, y_n) \qquad (2.13\text{-}4)$$

equivalent to (2.13-1). Note that for $p = 1$, (2.13-3) becomes Euler's method.

The main difficulty associated with the use of (2.13-3) and (2.13-4) is the need to generate the successive derivatives $f^{[1]}, f^{[2]}, \ldots$. Even for a moderately complex $f(x, y)$ these derivatives are extremely time-consuming to evaluate. In fact, this is the main reason for the Runge–Kutta formulation, since it bypasses the need for evaluating the higher-order derivatives.

2.13.1. Error Bounds

Let us now briefly consider some of the error bounds of the Taylor series expansion. We may carry this through in the same manner as we previously did in Section 1.11.2 for Euler's method. If we assume a Lipschitz condition

$$|\Phi(x_n, y_n; h) - \Phi(x_n, y(x_n); h)| \leq L[y_n - y(x_n)]$$

or, equivalently if $f(x, y)$ has continuous derivatives of order $p + 1$,

$$|f^{[j-1]}(x_n, y_n) - f^{[j-1]}(x_n, y(x_n))| \leq L[y_n - y(x_n)] \qquad (2.13\text{-}5)$$

and a bound on the exact derivative

$$|y^{[p+1]}(x)| \leq M_{p+1} \qquad (2.13\text{-}6)$$

then as before

$$|\varepsilon_n| \leq \exp[L(x_n - x_0)]|\varepsilon_0| + \frac{h^p M_{p+1}}{(p+1)!}\left[\frac{\exp[L(x_n - x_0)] - 1}{L}\right] \qquad (2.13\text{-}7)$$

Thus as $h \to 0$, ε_n tends to zero at least like h^p at the point x_n.

It is interesting to note that if the special system

$$y' = y, \qquad y(0) = 1$$

is used the bound of (2.13-7) can be written as (for $\varepsilon_0 = 0$)

$$|\varepsilon_n| \leq \frac{h^p}{(p+1)!} x_n \exp x_n \qquad (2.13\text{-}8)$$

Thus the absolute error ε_n, for a fixed h, is bounded by an increasing exponential whereas if we considered the error relative to the true solution, $y(x_n) = \exp x_n$, the error is bounded by a linear growth rate.

2.13.2. Nonarithmetic Computer Operations

What has made the Taylor series method attractive as a single-step computer algorithm is the availability of nonarithmetic operations on a digital computer. In fact, there currently exist many nonarithmetic computer routines for a variety of digital computers which, given the explicit form of $f(x, y)$, will automatically generate f', f'', ... (See Blum [4] and Howard and Tashjian [34] for a discussion on this and comparative points.) As such these routines remove one of the main objections to the use of Taylor series in the sense that higher-order derivatives are generated by the machine itself. A typical such routine is FORMAC [61] as devised for many of the IBM computers and a typical example of the use of this routine is the work of Eisenpress [22] in which the second and third derivatives were generated for a set of 20 simultaneous equations. Ball and Berns [2] have used alternative routines to generate the same type of results.

Unfortunately, there is at least one defect to the use of these derivative-generating routines. Because of the manner of generating the derivatives, computer storage is rapidly used up. As a result, it is necessary to turn to some very clever programming to bypass this storage problem. The work of Eisenpress illustrates this feature quite nicely.

Actually it is not too difficult to illustrate a convenient method for generating the higher-order derivatives for many explicit forms of $f(x, y)$. Of particular importance is the ability to generate $f^{[p]}$ in terms of $f^{[p-1]}$, $f^{[p-2]}$, ... Since we always have available a certain number of the $f^{[i]}$, depending on the boundary conditions and the order of the differential equations, such recurrence relations are quite powerful. For the interested reader we recommend the work of Gibbons [28], Miller [52], Moore [53], Nikolaev [54], and Zee [74].

2.14. PUBLISHED NUMERICAL RESULTS

In this section we present some of the numerical results which have been published in the literature dealing with the analysis of single-step methods. Because of the voluminous amount of such material which is available, i.e., almost every publication presents certain numerical results, only a small fraction will actually be mentioned. The main desire is to give the reader some information on where to find certain results.

The first necessity in testing the performance of different integration formulas is to have a collection of test ODEs with many unusual and interesting features, e.g., singularities, discontinuities, infinite second derivatives, widely separated eigenvalues, etc. Several of such systems are presented by Rosser [60].

Many authors have compared different Runge–Kutta methods on test ODEs in order to compare accuracy and computation time. Ceschino and Kuntzmann [15] present many short calculations on a number of test ODEs using a wide variety of Runge–Kutta formulas. They use (2.3-13), (2.3-15), and (2.4-8), the classic, Gill, and optimum fourth-order Runge–Kutta methods on

$$y' = y - 1.5e^{-0.5x}, \qquad y(0) = 1 \qquad (2.14\text{-}1)$$

and

$$y' = -2xy^2, \qquad y(0) = 1 \qquad (2.14\text{-}2)$$

They find that with $h = 0.1$ the results obtained on (2.14-1) and (2.14-2) with the three methods are almost identical, indicating the difficulty in choosing one Runge–Kutta method over another of the same order. Fehlberg [24, 25] and Benyon [3] have both also come to roughly the same conclusion, namely, that it is difficult to distinguish between neighboring order Runge–Kutta methods when essentially identical accuracy is the goal of a calculation.

By contrast, Scraton [68] and Shanks [69] show the increased accuracy associated with increased order formulas. Scraton uses the equation

$$y'' = -xy, \qquad y(0) = 1, \qquad y'(0) = 0 \qquad (2.14\text{-}3)$$

with a variety of different Runge–Kutta formulas. Shanks compares the classic Runge–Kutta $(4, 4)$ with the $(5, 5)$ form of (2.3-29) and the $(6, 6)$ form of (2.3-33) on $y' = y$, $y(0) = 1$. They find, as expected, that the higher-order formulas are more accurate and that smaller values of h lead to improved results.

Hull used a number of test ODEs ranging from simple linear systems to fifty nonlinear equations, different bounds on the accuracy required in the integrations and Runge–Kutta formulas of all orders from two to six as well as an equation equivalent to (2.4-8) and to that of Lawson given by (2.3-30). The results of a large amount of computation showed that higher-order methods were generally best, i.e., the fourth-order method was always better than any of the lower-order methods, no matter what error tolerances are used. These results are reinforced by Babuska [1] on the system

$$y' = x(x + 2)y^3 + (x + 3)y^2, \qquad y(\tfrac{1}{2}) = \tfrac{8}{5} \qquad (2.14\text{-}4)$$

By using Runge–Kutta formulas from orders two to six, he showed that as long as only the truncation error was involved, the higher-order formulas gave lower errors in single precision. When h was reduced to a point where round-off error became important this distinction between orders tended to disappear and all the formulas gave about equal error. These results are equivalent to those already detailed in Section 1.11.3.

Next we mention the results of Sarafyan applied to the embedding formula (2.8-7). Using the test system

$$(1 - x^2)y'' - 2xy' + 6y = 0, \qquad y(0) = -\tfrac{1}{2}, \quad y'(0) = 0 \qquad (2.14\text{-}5)$$

and an h of 0.0125 he showed that $y_{4,n+1}$ and $y_{6,n+1}$ agree to within 11 decimal places (at a certain x). Thus the calculation is correct to that many decimal places. As such, the values $y_{4,n+1}$ and $y_{6,n+1}$ can be used to adjust the magnitude of h as the calculation proceeds. Other, and more detailed, calculations also confirm experimentally the use of the embedding method to adjust h. In addition, Sarafyan shows how the rule of (2.8-11) may be used to specify a usable h in a calculation. This is done by choosing the system

$$y' = 10y^2, \qquad y(0) = 1 \qquad (2.14\text{-}6)$$

which has a finite solution only for $x < 0.1$. By selecting $h = 0.2$, and the classic fourth-order Runge–Kutta method, (2.8-11) yields $\Theta = 4.66 > 1$; for $h = 0.1$, $\Theta = 1.2083 > 1$ and for $h = 0.05$, $\Theta = 0.4401 < 1$. Thus the rule specified by (2.8-11) predicts that an $h < 0.1$ must be used to obtain a valid calculation.

Finally, we point out the work quoted by Waters [73] using a test system of 64 first-order differential equations in which $f(x, y)$ was easy (or trivial) to calculate. His conclusion, upon comparing many single-step and multiple-step formulas, was that the Butcher fifth-order formula of (2.3-20) was the best in terms of minimum computation time and integration errors.

From these results cited it is possible to develop certain conclusions regarding the use of single-step integration methods:

1. It is often difficult to distinguish among Runge–Kutta methods of the same order, and, in fact, among methods of adjacent orders, e.g., third and fourth order or fourth and fifth order. However, in general, a higher-order formula yields a higher accuracy result than a lower-order formula.

2. As h is decreased, a point is reached at which round-off error becomes important. Further decreases in h will lead to decreased accuracy and may tend to obscure the differences between formulas of different orders.

3. Fifth-order methods appear to be highly desirable.

2.15. NUMERICAL EXPERIMENTS

In order to test many of the methods presented in this chapter we wish to present numerical results on a variety of different ODEs. We shall be concerned with the questions of the truncation errors and the round-off errors in the calculations as well as the cost of computation, i.e., perhaps the computer time involved. Further results dealing with stability bounds associated with the use of the various single-step methods will be given in Chapter 3.

TABLE 2.3

System Number	System	Boundary condition	Analytical solution	
I.	$y' = -y$	$y_0 = 1$	$y(x) = e^{-x}$	(2.15-1)
II.	$y' = +y$	$y_0 = 1$	$y(x) = e^{+x}$	(2.15-2)
III.	$y' = a + by$	$y_0 = 0$	$y(x) = (1/b)(e^{bx} - 1)$ (for $a = 1$)	(2.15-3)
IV.	$y_1' = 1/y_2$ $y_2' = -1/y_1$	$y_{10} = y_{20} = 1$	$y_1(x) = e^x$ $y_2(x) = e^{-x}$	(2.15-4)
V.	$y'' = -y$ or $y_1' = y_2$ $y_2' = -y_1$	$y_{10} = y_{20} = 1$	$y_1(x) = \sin x$ $y_2(x) = \cos x$	(2.15-5)
VI.	$\mathbf{y}' = \mathbf{A}\mathbf{y}$ $\mathbf{A} = \begin{bmatrix} -0.1 & -49.9 & 0 \\ 0 & -50 & 0 \\ 0 & 70 & -120 \end{bmatrix}$	$\mathbf{y}_0 = \begin{bmatrix} 2 \\ 1 \\ 2 \end{bmatrix}$	$y_1(x) = e^{-0.1x} + e^{-50x}$ $y_2(x) = e^{-50x}$ $y_3(x) = e^{-50x} + e^{-120x}$	(2.15-6)
VII.	$y' = 1 - y^2$	$y_0 = 0$	$y(x) = (e^{2x} - 1)/(e^{2x} + 1)$	(2.15-7)

VIII. A six variable nonlinear system with no analytical solution. The six equations are given below [42] and represent the dynamics of a gas absorber:

$$y_1' = \{-[40.8 + 66.7(M_1 + 0.08y_1)]y_1 + 66.7(M_2 + 0.08y_2)y_2\}/z_1$$
$$+ 40.8v_1/z_1$$

$$y_i' = \{40.8y_{i-1} - [40.8 + 66.7(M_i + 0.08y_i)]y_i$$
$$+ 66.7(M_{i+1} + 0.08y_{i+1})y_{i+1}\}/z_i, \quad i = 2, 3, 4, 5 \quad (2.15\text{-}8)$$

$$y_6' = \{40.8y_5 - [40.8 + 66.7(M_6 + 0.08y_6)y_6\}/z_6$$
$$+ 66.7(M_7 + 0.08v_2)v_2/z_6$$

with $z_i = (M_i + 0.16y_i) + 75, \quad i = 1, 2, \dots, 6$

The initial conditions for this system and the specified values of the parameters M_i are given by:

$$\mathbf{y}_0 = \begin{bmatrix} -0.03424992 \\ -0.06192031 \\ -0.08368619 \\ -0.10042889 \\ -0.11306320 \\ -0.12243691 \end{bmatrix}; \quad \mathbf{M} = \begin{bmatrix} 0.73576500 \\ 0.74875687 \\ 0.75929635 \\ 0.76774008 \\ 0.77443837 \\ 0.77971110 \\ 0.78383672 \end{bmatrix}$$

For the case $v_1 = v_2 = 0$, the result is a trajectory starting at $\mathbf{y}_0$ and going to the origin ($\mathbf{y} = \mathbf{0}$).

IX. A two variable problem representing the mass and temperature balance in a chemical reactor [42]:

$$y_1' = -(1 + \exp E)y_1 - 0.5(\exp E - 1), \quad y_{10} = -0.1111889$$
$$y_2' = (\exp E)y_1 - 8.9y_2^2 - (2 + 2.225)y_2 \quad y_{20} = 0.0323358 \quad (2.15\text{-}9)$$
$$+ 0.5(\exp E - 1)$$

where $E = 25y_2/(y_2 + 2)$

As we shall see, the distinction between single precision and double precision calculations will be extremely important. Thus we point out that almost all the results were obtained on an IBM 360/50-67 computer for which single precision yields seven significant digits and double precision yields fourteen–fifteen digits.

The systems of ODEs which we shall analyze are given in Table 2.3. The rationale in using these systems is interesting. Systems I–VII all have known analytical solutions and thus numerical calculations can be checked for accuracy. By contrast, Systems VIII and IX have no analytical solutions which can be used. System IV, when compared to I and II, may illustrate an interaction effect. System V has two oscillating solutions while System VI is a "stiff" problem in the sense of widely differing constants in the exponential solutions (see Chapter 6).

TABLE 2.4

EXACT SOLUTIONS FOR DIFFERENT SYSTEMS

System I		System II		System VII	
x	$y(x)$	x	$y(x)$	x	$y(x)$
0	1.0	0	1.0	0	0
0.2	0.818731	0.2	$0.122140_{10}{}^{+1}$	0.125	0.124352
0.4	0.670320	0.4	$0.149182_{10}{}^{+1}$	0.250	0.244919
0.6	0.548812	0.6	$0.182212_{10}{}^{+1}$	0.375	0.358357
0.8	0.449329	0.8	$0.222554_{10}{}^{+1}$	0.500	0.462117
1.0	0.367879	1.0	$0.271828_{10}{}^{+1}$	1.0	0.761594
2.5	$0.820850_{10}{}^{-1}$	2.5	$0.121825_{10}{}^{+2}$	2.0	0.964027
5.0	$0.673795_{10}{}^{-2}$	5.0	$0.148413_{10}{}^{+3}$	3.0	0.995054
7.5	$0.553084_{10}{}^{-3}$	7.5	$0.180804_{10}{}^{+4}$	4.0	0.999329
10.0	$0.453999_{10}{}^{-4}$	10.0	$0.220265_{10}{}^{+5}$	5.0	0.999909
15.0	$0.305902_{10}{}^{-6}$	15.0	$0.326902_{10}{}^{+7}$	6.0	0.999987
20.0	$0.206115_{10}{}^{-8}$	20.0	$0.485165_{10}{}^{+9}$		
25.0	$0.138879_{10}{}^{-10}$	25.0	$0.720049_{10}{}^{+11}$		

System VI

x	$y_1(x)$	$y_2(x)$	$y_3(x)$
0	2.0	1.0	2.0
0.2	0.98024	$0.45400_{10}{}^{-4}$	$0.45400_{10}{}^{-4}$
0.4	0.96079	$0.20611_{10}{}^{-8}$	$0.20611_{10}{}^{-8}$
0.6	0.94176	$0.93576_{10}{}^{-13}$	$0.93576_{10}{}^{-13}$
0.8	0.92312	$0.42484_{10}{}^{-17}$	$0.42484_{10}{}^{-17}$
1.0	0.90484	$0.19287_{10}{}^{-21}$	$0.19287_{10}{}^{-21}$
2.0	0.81873	$0.37201_{10}{}^{-43}$	$0.37201_{10}{}^{-43}$

The solution of System VIII should show a simple monotonic trajectory to the origin, but it does have six variables. System IX, by contrast, represents the self-sustained oscillation of an energy–mass system [42]. This is termed a limit cycle and on a y_1-y_2 plane, a closed elliptic curve will be achieved. Starting from the given initial conditions the values of y_1 and y_2 will change and eventually return again to the starting point.

For ease in visualizing the magnitude of many of the numbers to be analyzed, we present in Table 2.4 some exact values of the dependent variables for Systems I, II, VI, and VII (obviously System IV is also included within this format). The following nomenclature will also be of interest in subsequent analysis: ε [or ε_i] is the error or deviation between the exact and the numerically calculated value of $y(x)$ [or $y_i(x)$]; R [or R_i] is the ratio of the exact to the numerically calculated value of $y(x)$ [or $y_i(x)$]; and ε_{iSP} [or ε_{iDP}] is the error of $y_i(x)$ in single precision (SP) [or double precision (DP)].

2.15.1. Use of Explicit Single-Step Methods

We start our discussion of the numerical integration of these systems with the explicit methods of this chapter. Table 2.5 shows selected results for Systems I, II, and IV respectively using a third-, fourth-, and fifth-order Runge–Kutta method. The data were obtained with $h = 0.5$ and with a double precision format. Explicitly the third-order method is (2.3-11), the fourth-order method is (2.3-13) and the fifth-order method is (2.3-20). As can be seen, with this value of h, each method yields excellent approximations to the exact values for Systems I and II. Of interest is that at $x = 25.0$ a deviation of $\varepsilon = 0.231_{10+7}$ resulted with System II and the fifth-order method. While it appears that the algorithm is seriously in error, comparison with the exact value (Table 2.4) at $x = 25.0$, $y_1(x) = 0.72_{10+11}$ shows that the computation is correct to 4 decimal places.

System IV is also integrated quite well, but the interaction of the two variables yields results which are not as good as for the single elements given by the System I and II results. This behavior will be observed throughout many of the numerical calculations in this chapter.

Other items of interest here are that a higher-order formula, as expected, always yields better (more accurate) results and that other data not shown for $h = 0.1$ generates ratios for Systems I and II with the fifth-order method which are identically 1.0 to seven significant digits. Thus the integrations can be carried out to almost any accuracy desired with the use of $h \leq 0.1$. When h is raised to 1.0 or greater, instabilities, inaccuracies, and even oscillations occur; the computations become worthless at these large values of h. Explicit details on this point will be presented in Section 3.10.

TABLE 2.5

DOUBLE PRECISION $h = 0.5$

	Third-order $R–K$ (2.3-11)	Fourth-order $R–K$ (2.3-13)	Fifth-order $R–K$ (2.3-20)	
		System I		
x	R	R	R	
0	1.0	1.0	1.0	
1.0	1.007841	0.9992085	0.9999862	
2.5	1.019718	0.9980225	0.9999656	
5.0	1.039824	0.9960488	0.9999312	
10.0	1.081235	0.9921132	0.9998625	
15.0	1.124294	0.9881932	0.9997937	
20.0	1.169068	0.9842887	0.9997250	
25.0	1.215626	0.9803996	0.9996562	
		System II		
0	1.0	1.0	1.0	
1.0	1.003512	1.000344	0.9999987	
2.5	1.008804	1.000861	0.9999968	
5.0	1.017686	1.001723	0.9999936	
10.0	1.035685	1.003449	0.9999871	
15.0	1.054002	1.005177	0.9999807	
20.0	1.072644	1.006909	0.9999743	
25.0	1.091615	1.008644	0.9999679	
		System IV		
x	R_2	R_2	R_1	R_2
0	1.0	1.0	1.0	1.0
1.0	1.00888	0.998311	1.00001	1.00005
2.5	1.04048	0.995974	0.999919	1.00022
5.0	1.15216	0.992600	0.999490	1.00079
10.0	1.75982	0.987787	0.997586	1.00298
15.0	3.89496	0.985528	0.994293	1.00658
20.0	14.5637	0.985812	0.989622	1.01162
25.0	126.009	0.988649	0.983590	1.01811

In addition to these results we have also used other Runge–Kutta forms and spanned a wide sequence of different h. Included are the forms; third-order Heun (2.3-11) and Ralston (2.4-6); fourth-order classic (2.3-13) and Ralston (2.4-8); fifth-order Nystrom (2.3-16), Butcher (2.3-20), and Luther (2.3-24); sixth-order Lawson [44]. Figure 2.1 shows some typical single precision results for most of the forms with the error or deviation plotted versus x for System I.

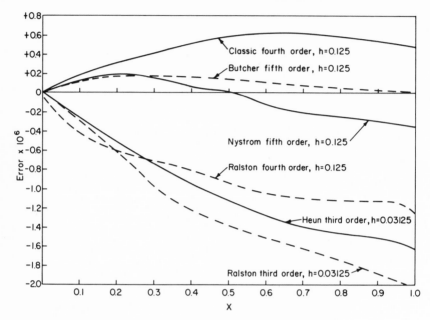

Figure 2.1. Error versus x. System I. Single precision.

Of real interest however are the single-precision results shown in Table 2.6 for System I. Here the absolute error is given as a function of $1/h$ for the different methods at the selected value of $x = 1.0$. In essentially all cases the results show a minimum in the error versus $1/h$ data. Qualitatively this may be interpreted as due to a competition between decreasing truncation error and increasing round-off error as mentioned in Chapter 1. The rise in the error for decreasing $1/h$ values (increasing h) is due to increasing truncation error; the rise for larger $1/h$ values (smaller h) is due to the intrusion of round-off error. This represents the classic pattern in which an increased accuracy associated with a decreasing value of h is stopped because of round off. Further details to be presented shortly will completely confirm these points.

Other data, not presented here, on System IV and System VII also show equivalent results to those presented in Table 2.6.

To test this feature of the importance of round-off in causing the error curve to increase beyond a certain small value of h, we present the data in Tables 2.7 and 2.8. Table 2.8 shows data for System I using the fourth-order classic Runge–Kutta form in single and double precision. Table 2.7 is equivalent but relates to System V with its oscillatory rather than monotonic behavior. In both cases it can be seen that double precision (with an increased computer time) does effectively extend the region of decreasing truncation

TABLE 2.6

SYSTEM I. SINGLE PRECISION ABSOLUTE ERROR VERSUS $1/h$ FOR $x = 1.0$

$1/h$	Third-order Heun	Third-order Ralston	Fourth-order Classic	Fourth-order Ralston	Fifth-order Nystrom	Fifth-order Butcher	Sixth-order Lawson
1	$0.345_{10^{-1}}$	$0.345_{10^{-1}}$	$0.712_{10^{-2}}$	$0.712_{10^{-2}}$	$0.121_{10^{-2}}$	$0.349_{10^{-3}}$	$0.229_{10^{-2}}$
2	$0.286_{10^{-3}}$	$0.286_{10^{-3}}$	$0.291_{10^{-3}}$	$0.289_{10^{-3}}$	$0.244_{10^{-4}}$	$0.506_{10^{-5}}$	$0.732_{10^{-3}}$
4	$0.292_{10^{-3}}$	$0.292_{10^{-3}}$	$0.147_{10^{-4}}$	$0.130_{10^{-4}}$	$0.476_{10^{-6}}$	$0.100_{10^{-8}}$	$0.275_{10^{-3}}$
8	$0.332_{10^{-4}}$	$0.333_{10^{-4}}$	$0.477_{10^{-6}}$	$0.124_{10^{-5}}$	$0.357_{10^{-6}}$	$0.476_{10^{-6}}$	$0.123_{10^{-3}}$
16	$0.447_{10^{-5}}$	$0.477_{10^{-5}}$	$0.101_{10^{-5}}$	$0.262_{10^{-5}}$	$0.107_{10^{-5}}$	$0.137_{10^{-5}}$	$0.591_{10^{-4}}$
32	$0.161_{10^{-5}}$	$0.202_{10^{-5}}$	$0.190_{10^{-5}}$	$0.381_{10^{-5}}$	$0.232_{10^{-5}}$	$0.256_{10^{-5}}$	$0.304_{10^{-4}}$
64	$0.232_{10^{-5}}$	$0.316_{10^{-5}}$	$0.411_{10^{-5}}$	$0.620_{10^{-5}}$	$0.453_{10^{-5}}$	$0.494_{10^{-5}}$	$0.190_{10^{-4}}$
128	$0.423_{10^{-5}}$	$0.638_{10^{-5}}$	$0.834_{10^{-5}}$	$0.106_{10^{-4}}$	$0.834_{10^{-5}}$	$0.989_{10^{-5}}$	$0.175_{10^{-4}}$

TABLE 2.7

SYSTEM V. SINGLE AND DOUBLE PRECISION WITH FOURTH-ORDER R–K

	$x = 0.5$		$x = 2.0$		$x = 6.0$	
h	$(\sin x - y_{1DP})/\sin x$	$(\sin x - y_{1SP})/\sin x$	$(\sin x - y_{1DP})/\sin x$	$(\sin x - y_{1SP})/\sin x$	$(\sin x - y_{1DP})/\sin x$	$(\sin x - y_{1SP})/\sin x$
0.5	$0.5400_{10^{-3}}$	$0.5399_{10^{-3}}$	$-0.1384_{10^{-4}}$	$-0.1370_{10^{-4}}$	$-0.8519_{10^{-2}}$	$-0.8520_{10^{-2}}$
0.25	$0.3250_{10^{-4}}$	$0.3270_{10^{-4}}$	$-0.1567_{10^{-4}}$	$0.1526_{10^{-3}}$	$-0.6158_{10^{-3}}$	$-0.6201_{10^{-3}}$
0.125	$0.1957_{10^{-5}}$	$0.2362_{10^{-5}}$	$-0.1429_{10^{-5}}$	$-0.1245_{10^{-4}}$	$-0.4045_{10^{-4}}$	$-0.4458_{10^{-4}}$
0.10	$0.7947_{10^{-6}}$	$0.1368_{10^{-5}}$	$-0.6213_{10^{-6}}$	$-0.2622_{10^{-6}}$	$-0.1670_{10^{-4}}$	$-0.2112_{10^{-4}}$
0.05	$0.4872_{10^{-7}}$	$0.8703_{10^{-6}}$	$-0.4330_{10^{-6}}$	$0.1180_{10^{-5}}$	$-0.1060_{10^{-5}}$	$-0.1707_{10^{-5}}$
0.01	$0.7688_{10^{-10}}$	$0.3232_{10^{-5}}$	$-0.7513_{10^{-10}}$	$0.6686_{10^{-5}}$	$-0.1719_{10^{-8}}$	$0.1323_{10^{-4}}$

TABLE 2.8

SYSTEM I. SINGLE AND DOUBLE PRECISION WITH FOURTH-ORDER R–K

h	$x = 5.0$		$x = 10.0$	
	$(e^{-x} - y_{DP})/e^{-x}$	$(e^{-x} - y_{SP})/e^{-x}$	$(e^{-x} - y_{DP})/e^{-x}$	$(e^{-x} - y_{SP})/e^{-x}$
0.5	-0.396685_{10}^{-2}	-0.396693_{10}^{-2}	$-.0794944_{10}^{-2}$	-0.794940_{10}^{-2}
0.25	-0.200600_{10}^{-3}	-0.199591_{10}^{-3}	-0.401241_{10}^{-3}	-0.398415_{10}^{-3}
0.125	-0.112909_{10}^{-4}	-0.718747_{10}^{-5}	-0.225819_{10}^{-4}	-0.134621_{10}^{-4}
0.10	-0.452907_{10}^{-5}	0.221153_{10}^{-5}	-0.905817_{10}^{-5}	0.448738_{10}^{-5}
0.05	-0.271467_{10}^{-6}	0.939900_{10}^{-5}	-0.542934_{10}^{-6}	0.237190_{10}^{-4}
0.01	-0.413607_{10}^{-9}	0.707689_{10}^{-4}	-0.837189_{10}^{-9}	0.145519_{10}^{-3}

error. In fact, the changes are startling especially at the larger values of x.

Next we consider two further sets of calculations. The first applies to System VI, the "stiff" system, and the second to System VIII, the nonlinear equations of (2.15-8). The stiff system is interesting because at small values of x the three solutions should be quite different; as x increases, however, y_2 and y_3 should be approximately equal but still different from y_1. Further, y_1 should decrease much more slowly than either y_2 or y_3.

Table 2.9 shows the deviation in y_1, y_2, and y_3 for selected values of h using the fifth-order method of Butcher in double precision. As can be seen, the behavior mentioned above as well as the influences of double precision are fully confirmed. Data with Lawson's fifth-order method behave almost identically at these values of h. In Chapter 3 we shall detail further results for values of $h > 0.02$. At that point we will be able to distinguish between the two fifth-order forms.

The analysis of nonlinear System VIII follows in much the same way as previously except for one difficulty. Since an analytical solution is not available, the deviation or error cannot be calculated. We can however use the fifth-order Butcher method with $1/h = 8$ and compare all other results with this value. This choice of $1/h$ is somewhat arbitrary, but the results to be quoted do not seem to be a strong function of this selection. Table 2.10 presents some single-precision results for y_6 at $x = 5.0$ using a variety of different single-step methods. Only y_6 is shown to conserve space, because it varies more at $x = 5.0$ than do $y_1, \ldots, y_5$. The data have all significant figures in italic which agree with the fifth-order Butcher method, $1/h = 8$.

It is obvious that the behavior in this case is analogous to those single precision data previously presented for System I. Most of the methods show a minimum point in the error as a function of $1/h$, and the methods tend to get better as the order increases except that the sixth-order Lawson is worst.

TABLE 2.9

SYSTEM VI. DOUBLE PRECISION

Number of steps in x		Fifth-order Butcher		
		ε_1	ε_2	ε_3
$h = 0.001$	25	$8.70_{10^{-10}}$	$-2.16_{10^{-11}}$	$-8.47_{10^{-10}}$
	50	$1.77_{10^{-9}}$	$-1.24_{10^{-11}}$	$-9.46_{10^{-10}}$
	75	$2.66_{10^{-9}}$	$-5.32_{10^{-12}}$	$-1.15_{10^{-11}}$
	100	$3.54_{10^{-9}}$	$-2.03_{10^{-12}}$	$-2.44_{10^{-12}}$
$h = 0.005$	25	$1.06_{10^{-9}}$	$-3.35_{10^{-9}}$	$-3.54_{10^{-9}}$
	50	$8.71_{10^{-9}}$	$-1.30_{10^{-11}}$	$-1.30_{10^{-11}}$
	75	$1.29_{10^{-8}}$	$-3.75_{10^{-14}}$	$-3.75_{10^{-14}}$
	100	$1.70_{10^{-8}}$	$-9.65_{10^{-17}}$	$-9.65_{10^{-17}}$
$h = 0.01$	25	$8.08_{10^{-9}}$	$-6.41_{10^{-10}}$	$-6.41_{10^{-10}}$
	50	$1.70_{10^{-8}}$	$-4.78_{10^{-15}}$	$-4.78_{10^{-15}}$
	75	$2.49_{10^{-8}}$	$-2.67_{10^{-20}}$	$-2.67_{10^{-20}}$
	100	$3.24_{10^{-8}}$	$-1.33_{10^{-25}}$	$-1.33_{10^{-25}}$
$h = 0.02$	25	$1.70_{10^{-8}}$	$-3.34_{10^{-13}}$	$-3.34_{10^{-13}}$
	50	$3.24_{10^{-8}}$	$-9.38_{10^{-24}}$	$-9.38_{10^{-24}}$
	75	$4.62_{10^{-8}}$	$-1.98_{10^{-34}}$	$-1.98_{10^{-34}}$
	100	$5.86_{10^{-8}}$	$-3.71_{10^{-45}}$	$-3.71_{10^{-45}}$

There does not seem to be any obvious reason for the poor performance of the sixth-order method, but Hull [37] confirms the present results.

In addition we have also indicated the computation time required for each method using $1/h = 128$. The increased time as the order increases is, of course, due to the number of function evaluations v. Thus if we select the fourth-order classical, *one function evaluation* can be suggested as requiring 2.07 sec. On this rough basis one could predict that:

	Predicted
Third-order case ($v = 3$):	6.21 sec
Fourth-order case ($v = 4$):	8.28 sec
Fifth-order case ($v = 6$):	12.42 sec
Sixth-order case ($v = 7$):	14.49 sec

These numbers agree closely enough with values in Table 2.10 to suggest that most of the computation time is involved in the function evaluations.

The results thus far tend to point out qualitatively that in terms of accuracy the fifth-order method is the best. In an effort to be more quantitative regarding this point, we have analyzed further data in double precision equivalent to Table 2.5, but with a variety of different h. Without actually

TABLE 2.10

SYSTEM VIII. SINGLE PRECISION VALUES OF y_6 AT $x = 5.0$

$1/h$	Third-order Heun	Third-order Ralston	Fourth-order classic	Fourth-order Ralston	Fifth-order Nystrom	Fifth-order Butcher	Sixth-order Lawson
2	-0.02373383	-0.02373045	-0.02374224	-0.02374239	-0.02373378	-0.02373712	-0.02404642
4	-0.02374451	-0.02374220	-0.02374085	-0.02374105	-0.02373864	-0.02373675	-0.02404120
8	-0.02374502	-0.02374309	-0.02374044	-0.02374067	-0.02373835	-0.02373661 6^a	-0.02404099
16	-0.02374454	-0.02374259	-0.02373964	-0.02373990	-0.02373745	-0.02373520	-0.02404054
32	-0.02374359	-0.02374147	-0.02373811	-0.02373829	-0.02373595	-0.02373324	-0.02403918
64	-0.02374192	-0.02373922	-0.02373504	-0.02373516	-0.02373270	-0.02372938	-0.02403608
128	-0.02373887	-0.02373456	-0.02372899	-0.02372888	-0.02372647	-0.02372188	-0.02402993
Computation time, in seconds for $1/h = 128$	7.03	7.02	8.28	8.47	11.16	11.11	12.78

^{a} Taken as correct value.

presenting the numerical calculations we can still make a number of interesting statements.

1. When we plot the deviation or error versus $1/h$ for the third-, fourth-, and fifth-order methods, the fifth-order method yields a curve which is *always below* the third- and fourth-order curves.

2. If we normalize the value of h to include the number of function evaluations v, then we can define

$$H = 2h/v \qquad (2.15\text{-}10)$$

Here H is equivalent to the distance advanced in two function evaluations and it will be $H = 2h/3$, $= 2h/4$, and $= 2h/6$ for the third-, fourth-, and fifth-order methods respectively. The use of two function evaluations allows a comparison with the multiple-step methods of Chapter 4.

On the basis of this normalized H a plot of the error versus $1/H$ shows once again that the fifth-order method is best. Its curve is always below the other curves, although they approach each other at low values of $1/H$ (high values of h). At high values of $1/H$ the fifth-order method is far superior to the other methods.

3. Finally, we turn to a comparison on the basis of actual computer time. The times listed below in sec/100 steps in the calculation are indicative of the results:

	System I	System II	System IV
Third-order	0.50	0.525	0.45
Fourth-order	0.552	0.564	1.08
Fifth-order	0.651	0.651	1.31

The increased accuracy achieved in going from the third- to the fourth- to the fifth-order methods is almost always about 10 to 100 and the corresponding computation times are about the same. Thus these results would seem to indicate that the added computation time is much less important than the decreasing error.

On the basis of these results, the conclusion seems to be that *the fifth-order method is to be preferred as long as high accuracy is desired.* Thus we next turn to an analysis to distinguish between some of the fifth-order methods and also to more carefully define the question of the round-off error.

To do this we have selected Systems I, IV, and VI, the fifth-order methods of Luther (2.3-17), Butcher (2.3-20), Luther (2.3-25), Sarafyan (2.3-27), Fehlberg (2.3-28) and the sixth-order method of Butcher (2.3-32). Numerical calculations in single and double precision were carried out with each method as well as one additional calculation bearing upon the round-off error. As we shall detail in Chapter 3, if we select System I, we may get the exact difference equation associated with any of the above methods. As an illustration, for the Butcher fifth-order method

$$y_{n+1} = \left[1 - h + \frac{h^2}{2} - \frac{h^3}{6} + \frac{h^4}{24} - \frac{h^5}{120} + \frac{h^6}{640} \right] y_n \qquad (2.15\text{-}11)$$

represents the numerical solution of System I. Given the step-by-step solution of (2.15-11) a comparison with $y(x) = \exp(-x)$, the exact solution of System I, will yield the true local truncation error. When this truncation error is compared to the normally computed value (which includes both truncation and round-off error) an estimate of the local round-off error can be obtained.

In order to preserve space we present only a limited number of these results here. Nevertheless, these are sufficient to illustrate certain significant points. Table 2.11 illustrates the behavior for all six methods using System IV and double precision. The use of single precision and System I and VI all reveal essentially the same behavior, namely *that* (2.3-17), (2.3-20), *and* (2.3-27) *are by far the best methods to use.* A further distinction between these three can be made by suggesting that Butcher's fifth-order method (2.3-20) is the best, but this conclusion is not as definite. The computation time for (2.3-28) is lower than the other fifth-order methods, 4.3 sec/100 steps versus 4.8 sec/100 steps, but this difference is not important as compared to the accuracy results. The Butcher sixth-order method (2.3-32) takes longer, about 5.1 sec/100 steps, and thus this method cannot be compared on any basis with the fifth-order methods.

Table 2.12 shows single- and double-precision results on System I using the six different methods. Also shown is the cumulative round-off error after 100 steps, $RO_{\text{cum}/100 \text{ steps}}$, and the local round-off error for (2.3-17), RO/step; these values being calculated for equations analogous to (2.15-11). The significant influence of the round-off in the single and double precision cases is obvious. It is also apparent that the three fifth-order methods mentioned above are orders of magnitude better than the other methods.

We can summarize the results of this section:

1. The three explicit fifth-order methods (2.3-17), (2.3-20), and (2.3-27) and the fifth-order method of Lawson (2.3-30) discussed earlier are excellent algorithms for solving the systems considered here.

2. If high accuracy results with no round-off difficulties are desired, the obvious procedure is to use double precision. In fact, double precision seems to be a natural mode of operation if one is using a computer for which this operation is not heavily time consuming, e.g., for the IBM 360 series used here, double precision only requires about a maximum of 35% more computing time than single precision.

TABLE 2.11

SYSTEM IV. DOUBLE PRECISION R_1 AND R_2 FOR DIFFERENT METHODS

			(2.3-17)	(2.3-20)	(2.3-25)	(2.3-27)	(2.3-28)	(2.3-32)
$h = 0.01$	$x = 0.5$	R_1	1.0	1.0	1.0	1.0	1.0046	0.99944
		R_2	1.0	1.0	1.0	1.0	0.99793	1.0033
$h = 0.1$	$x = 0.5$	R_1	1.0	1.0	1.0	1.0	1.0090	0.99980
		R_2	1.0	1.0	1.0	1.0	1.0173	1.0330
	$x = 10.0$	R_1	1.0	1.0	1.0	1.0	0.41171	0.31795
		R_2	1.0	1.0	1.0	0.99985	3.0080	4.0619
$h = 0.5$	$x = 0.5$	R_1	1.0003	0.99999	0.99990	0.99988		
		R_2	0.99955	0.99995	0.99976	0.99749		
	$x = 10.0$	R_1	1.0082	0.99758	1.0137	1.1313		
		R_2	0.99073	1.0029	0.98295	0.85916		

TABLE 2.12

System I. $h = 0.1$. R and RO$_{\text{cum}/100 \text{ steps}}$ for Different Methods

		(2.3-17)	(2.3-20)	(2.3-25)	(2.3-27)	(2.3-28)	(2.3-32)
$x = 1.0$	SP	0.99999 $4.696_{10^{-5}}$	0.99999 $5.096_{10^{-5}}$	0.99999 $4.595_{10^{-5}}$	0.99999 $4.679_{10^{-5}}$	0.99672 $3.714_{10^{-2}}$	1.0016 $4.464_{10^{-2}}$
	DP	1.0 $8.319_{10^{-14}}$	1.0 $7.537_{10^{-14}}$	1.0 $9.360_{10^{-6}}$	1.0 $8.398_{10^{-14}}$	0.99672 $8.719_{10^{-2}}$	1.0016 $4.459_{10^{-2}}$
$x = 10.0$	SP	0.99993 $8.081_{10^{-6}}$	0.99994 $1.174_{10^{-5}}$	0.99993 $1.177_{10^{-5}}$	0.99993 $8.963_{10^{-6}}$	1.0962 $9.267_{10^{-2}}$	1.19152 $1.782_{10^{-1}}$
	DP	1.0 $5.931_{10^{-15}}$	1.0 $7.065_{10^{-15}}$	1.0 $3.747_{10^{-6}}$	1.0 $4.646_{10^{-15}}$	1.0962 $9.266_{10^{-2}}$	1.19159 $1.781_{10^{-1}}$

For (2.3-17), $h = 0.1$

x	RO/step (SP)	RO/step (DP)
0.2	$-3.576_{10^{-7}}$	$-4.996_{10^{-16}}$
0.4	$-4.768_{10^{-7}}$	$-8.743_{10^{-16}}$
0.6	$-5.364_{10^{-7}}$	$-1.054_{10^{-15}}$
0.8	$-5.960_{10^{-7}}$	$-1.110_{10^{-15}}$
1.0	$-5.960_{10^{-7}}$	$-1.110_{10^{-15}}$

2.15.2. Use of Implicit Single-Step Methods

Next we shall investigate the behavior of some of the implicit single-step methods of Section 2.5. In particular we shall use the semi-implicit third-order Rosenbrock form (2.5-18), and two of the Butcher fifth order fully implicit forms (2.5-8) and (2.5-10).

To start our analysis, we repeat certain of the previous calculations using the third-order Rosenbrock form in double precision. Table 2.13, as illustra-

TABLE 2.13

DOUBLE-PRECISION. ROSENBROCK THIRD-ORDER METHOD, $h = 0.5$

	R		
x	System I	System II	System IV
0	1.0	1.0	1.0
1.0	1.0091	1.0618	1.0089
2.5	1.0230	1.1618	1.1015
5.0	1.0466	1.3498	1.6048
10.0	1.0954	1.8219	8.6738
15.0	1.1464	2.4593	$9.4440_{10}{}^{+1}$
20.0	1.1999	3.3196	$1.1478_{10}{}^{+3}$
25.0	1.2558	4.4808	$1.4066_{10}{}^{+4}$

tion, presents results for Systems I, II, and IV using $h = 0.5$. When these are compared to the explicit results of Table 2.5 it is apparent that the present method is never as good as, and frequently is worse than, the equivalent explicit third-order results.

Table 2.14 shows the magnitude of the round-off error in single and double precision calculations on System I. This round-off was calculated in the same manner as previously, i.e., by obtaining the difference equation for System I and the Rosenbrock method.

Obviously it is dangerous to extrapolate the present results to large nonlinear systems, but our conclusion is that there is no clear advantage for the semi-implicit method over the normal explicit methods when the sole criterion is accuracy. However, we will see later that implicit methods have desirable *stability* characteristics, which in some cases is the primary consideration.

Next we turn to the use of Butcher's fully implicit forms on Systems I, II, and V. Since the calculated differences between the two fifth-order equations (2.5-8) and (2.5-10) are quite small, we shall report only on the results of using (2.5-8). First we should point out that since each system investigated

TABLE 2.14

SYSTEM I

ROSENBROCK THIRD-ORDER METHOD, $h = 0.1$

x	R	RO/step
	Double precision	
0.2	1.00002	$6.3092_{10}-11$
0.4	1.00004	$1.0330_{10}-10$
0.6	1.00006	$1.2687_{10}-10$
0.8	1.00008	$1.3849_{10}-10$
1.0	1.00010	$1.4173_{10}-10$
	Single precision	
0.2	1.00002	$7.7480_{10}-7$
0.4	1.00004	$1.7285_{10}-6$
0.6	1.00006	$2.2649_{10}-6$
0.8	1.00008	$2.5033_{10}-6$
1.0	1.00010	$2.4437_{10}-6$

was linear, it was merely necessary to solve a set of linear algebraic equations at each step to calculate the various k_i. Since (2.5-8) uses three substitutions, there are three linear equations to solve. In all cases the equation are extremely well-conditioned with a typical inverse of the coefficient matrix times the coefficient matrix yielding off-diagonal elements of at least 0.5_{10-19}. Of course, if we had solved nonlinear differential equations, the evaluation of the k_i's would have required some type of iterative procedure.

Table 2.15 shows some single and double precision results for Systems I and II. Note, as expected, that single precision is as good as double precision until small values of h are used. When these are compared to the fifth-order data of Table 2.5, it can be seen that the results are all about equivalent. However, the implicit method only requires three function evaluations to achieve these results whereas the explicit method requires six evaluations. Thus there may be a considerable saving in actual computer time in the implicit path. Unfortunately, the solution of the simultaneous algebraic equations compensates for this. As an example we previously quoted the computation time/100 steps for the explicit double precision fifth-order solution of Systems I and II as 0.651 sec. By contrast the present implicit method requires 1.16 sec/100 steps. Thus even for linear algebraic equations the implicit form is not superior. Of course, in other problems the function evaluations may be more important and thus the implicit form may show up better.

While we are at this point, we also note that the double precision calculations take 1.16 sec/100 steps while the single-precision results take 0.86 sec/100

TABLE 2.15

SYSTEM I AND II. BUTCHER IMPLICIT METHOD

h = 0.5

	System I				System II			
	x = 1.0		x = 10.0		x = 1.0		x = 10.0	
$1/h$	ε_{SP}	ε_{DP}	ε_{SP}	ε_{DP}	ε_{SP}	ε_{DP}	ε_{SP}	ε_{DP}
1.0	$0.6335_{10^{-4}}$	$0.6334_{10^{-4}}$	$0.7769_{10^{-7}}$	$0.7811_{10^{-7}}$	$0.3356_{10^{-3}}$	$0.3331_{10^{-3}}$	$0.2719_{10^{+2}}$	$0.2697_{10^{+2}}$
2.0	$0.1430_{10^{-5}}$	$0.1755_{10^{-5}}$	$0.1658_{10^{-8}}$	$0.2165_{10^{-8}}$	$0.1335_{10^{-4}}$	$0.1096_{10^{-4}}$	$0.1042_{10^{+1}}$	0.8884
10.0	$-0.4172_{10^{-6}}$	$-0.2173_{10^{-8}}$	$0.2124_{10^{-8}}$	$-0.2681_{10^{-11}}$	$0.1239_{10^{-4}}$	$0.2000_{10^{-7}}$	-0.4453	$0.1620_{10^{-2}}$
100.0	$0.2741_{10^{-5}}$	$-0.2487_{10^{-8}}$	$0.3507_{10^{-7}}$	$-0.2903_{10^{-11}}$	$0.7534_{10^{-4}}$	$0.1967_{10^{-7}}$	$-0.1008_{10^{+2}}$	$0.1525_{10^{-2}}$
1000.0	$0.2211_{10^{-4}}$	$-0.2487_{10^{-8}}$	$0.2957_{10^{-6}}$	$-0.3134_{10^{-11}}$	$0.7839_{10^{-3}}$	$0.1816_{10^{-7}}$	$-0.6739_{10^{+2}}$	$0.1386_{10^{-2}}$

h = 0.5

	System I		System II	
x	ε_{SP}	ε_{DP}	ε_{SP}	ε_{DP}
1.0	$0.1430_{10^{-5}}$	$0.1755_{10^{-5}}$	$0.1335_{10^{-4}}$	$0.1096_{10^{-4}}$
2.5	$0.7748_{10^{-6}}$	$0.9789_{10^{-6}}$	$0.1440_{10^{-3}}$	$0.1228_{10^{-3}}$
5.0	$0.1192_{10^{-6}}$	$0.1607_{10^{-6}}$	$0.3585_{10^{-2}}$	$0.2993_{10^{-2}}$
10.0	$0.1658_{10^{-8}}$	$0.2165_{10^{-8}}$	$0.1042_{10^{+1}}$	0.8884
15.0	$0.1739_{10^{-10}}$	$0.2188_{10^{-10}}$	$0.2310_{10^{+3}}$	$0.1977_{10^{+3}}$
20.0	$0.1560_{10^{-12}}$	$0.1966_{10^{-12}}$	$0.4633_{10^{+5}}$	$0.3913_{10^{+5}}$

steps. There is about a 35% increase in computation time for the double precision as compared to the single precision.

2.15.3. Truncation Error Estimates

Next we wish to assess the usefulness of some of the methods of Sections 2.8 and 2.9 for estimating the truncation error, and in turn the allowable h, in the single-step computations. All the data were obtained in double precision to minimize the influence of round-off error. Three methods have been analyzed:

1. Merson form, (2.8-1) and (2.8-2). This is a fourth-order Runge–Kutta form using five substitutions (4, 5). The one extra substitution is used to estimate $T(x, h)$.
2. The extrapolation using steps of h and $2h$ as given in Section 2.9.1. The equation of interest, with the fourth-order classic Runge–Kutta form, is then $T(x, h) = [y_{n+1}^{(h)} - y_{n+1}^{(2h)}]/30$.
3. The embedding method due to Sarafyan which uses Butcher's fifth-order form adapted to a step size $2h$ (2.8-14) and the embedded formulas given in (2.8-15).

Systems IV and VII have been used in the calculations.

Table 2.16 presents some results for the Merson and the $(h–2h)$ method (1 and 2 above) on the two systems. The ε_i are the actual deviations or errors while the T_i are the predicted truncation errors. In all cases the truncation estimation using $(h–2h)$ is orders better than that in the Merson approach. As a general rule the $(h–2h)$ prediction is too small by about one decimal place whereas the Merson method is too large by as many as eight decimal places. It would seem apparent that the $(h–2h)$ calculation could be used to estimate the magnitude of the local truncation error and to then adjust the value of h to meet any desired accuracy demands.

Table 2-17 presents results on these same two systems using the embedding technique of Sarafyan. The terminology is $y = $ exact value, $y_4 = y_{4,n+1}$ and $y_6 = y_{6,n+1}$ as given in (2.8-15). In each column all the elements in the dependent variables are given. Obviously, y_6 is calculated with an actual h which is twice that listed in the table, whereas y_4 is calculated with the actual h given. Both y_1 and y_2 are shown for System IV.

Turning first to the data for System IV, we see that for $h = 0.01$ and $x = 0.5$ the value of y is accurate to at least 10^{-8}. From the data used to construct Table 2.11 we can state that the actual deviation is about 10^{-14}. Because we have not tabulated enough digits, we cannot really predict the true accuracy of y_6. For the cases $h = 0.1$, $x = 0.5$ and $h = 0.1$, $x = 10.0$, the results are respectively that y_6 is accurate to 1.3_{10-6}, 4.7_{10-8} and 1.6_{10-2}, 3.8_{10-11}. When

TABLE 2.16
DOUBLE PRECISION. ESTIMATE OF TRUNCATION ERROR

System IV

		Merson				(h–2h)		
x	ε_1	ε_2	T_1	T_2	ε_1	ε_2	T_1	T_2
$h=0.01$								
0.2	$-5.13_{10^{-12}}$	$3.37_{10^{-12}}$	$4.07_{10^{-4}}$	$-2.73_{10^{-4}}$	$7.13_{10^{-11}}$	$-4.77_{10^{-11}}$	$3.57_{10^{-12}}$	$-2.38_{10^{-12}}$
0.5	$-1.73_{10^{-11}}$	$6.21_{10^{-12}}$	$5.50_{10^{-4}}$	$-2.02_{10^{-4}}$	$2.41_{10^{-11}}$	$-8.83_{10^{-11}}$	$4.82_{10^{-12}}$	$-1.76_{10^{-12}}$
1.0	$-5.66_{10^{-11}}$	$7.50_{10^{-12}}$	$9.06_{10^{-4}}$	$-1.23_{10^{-4}}$	$7.93_{10^{-10}}$	$-1.07_{10^{-10}}$	$7.95_{10^{-12}}$	$-1.07_{10^{-12}}$
$h=0.1$								
0.5	$-1.84_{10^{-7}}$	$5.36_{10^{-8}}$	$5.49_{10^{-3}}$	$-2.02_{10^{-3}}$	$2.42_{10^{-6}}$	$-8.69_{10^{-7}}$	$4.90_{10^{-7}}$	$-1.72_{10^{-7}}$
1.0	$-5.75_{10^{-7}}$	$6.08_{10^{-8}}$	$9.06_{10^{-3}}$	$-1.23_{10^{-3}}$	$7.93_{10^{-6}}$	$-1.05_{10^{-6}}$	$8.08_{10^{-7}}$	$-1.05_{10^{-7}}$
3.0	$-9.96_{10^{-6}}$	$1.78_{10^{-8}}$	$6.69_{10^{-2}}$	$-1.66_{10^{-4}}$	$1.72_{10^{-4}}$	$-4.15_{10^{-7}}$	$5.97_{10^{-6}}$	$-1.41_{10^{-8}}$
5.0	$-8.82_{10^{-5}}$	$2.44_{10^{-9}}$	$4.95_{10^{-1}}$	$-2.25_{10^{-5}}$	$2.06_{10^{-3}}$	$-9.14_{10^{-8}}$	$4.41_{10^{-5}}$	$-1.91_{10^{-9}}$

System VII

	Merson		(h–2h)	
x	ε	T	ε	T
$h=0.01$				
0.2	$-1.11_{10^{-11}}$	$3.20_{10^{-4}}$	$1.81_{10^{-11}}$	$1.08_{10^{-12}}$
0.5	$-2.46_{10^{-11}}$	$2.62_{10^{-4}}$	$5.92_{10^{-11}}$	$2.16_{10^{-12}}$
1.0	$-2.17_{10^{-11}}$	$1.40_{10^{-4}}$	$1.35_{10^{-11}}$	$3.01_{10^{-12}}$
$h=0.1$				
0.2	$-1.15_{10^{-7}}$	$3.20_{10^{-3}}$	$1.76_{10^{-7}}$	$8.45_{10^{-8}}$
0.5	$-2.69_{10^{-7}}$	$2.62_{10^{-3}}$	$5.90_{10^{-7}}$	$1.70_{10^{-7}}$
1.0	$-2.59_{10^{-7}}$	$1.40_{10^{-3}}$	$1.45_{10^{-6}}$	$3.31_{10^{-7}}$
3.0	$2.23_{10^{-8}}$	$3.29_{10^{-5}}$	$3.00_{10^{-7}}$	$1.83_{10^{-8}}$
5.0	$1.20_{10^{-9}}$	$6.07_{10^{-7}}$	$1.12_{10^{-8}}$	$3.42_{10^{-10}}$

2.   Runge–Kutta and Allied Single-Step Methods

TABLE 2.17

DOUBLE PRECISION. SARAFYAN EMBEDDING

System IV

	x	y	y_4	y_6
$h = 0.01$	0.1	1.1051709	1.1051709	1.1051709
		$9.0483742_{10^{-1}}$	$9.0483742_{10^{-1}}$	$9.0483742_{10^{-1}}$
	0.5	1.6487213	1.6487213	1.6487213
		$6.0653066_{10^{-1}}$	$6.0653066_{10^{-1}}$	$6.0653066_{10^{-1}}$
	1.0	2.7182818	2.7182818	2.7182818
		$3.6787944_{10^{-1}}$	$3.6787944_{10^{-1}}$	$3.6787944_{10^{-1}}$
	1.5	4.4816891	4.4816891	4.4816891
		$2.2313016_{10^{-1}}$	$2.2313016_{10^{-1}}$	$2.2313016_{10^{-1}}$
	2.0	7.3890561	7.3890561	7.3890561
		$1.3533528_{10^{-1}}$	$1.3533528_{10^{-1}}$	$1.3533528_{10^{-1}}$
$h = 0.1$	0.6	1.8221188	1.8221173	1.8221186
		$5.4881164_{10^{-1}}$	$5.4881197_{10^{-1}}$	$5.4881150_{10^{-1}}$
	1.0	2.7182818	2.7182797	2.7182817
		$3.6787944_{10^{-1}}$	$3.6787956_{10^{-1}}$	$3.6787925_{10^{-1}}$
	2.0	7.3890561	7.3890541	7.3890596
		$1.3533528_{10^{-1}}$	$1.3533518_{10^{-1}}$	$1.35133507_{10^{-1}}$
	5.0	$1.4841316_{10^{+2}}$	$1.4841385_{10^{+2}}$	$1.4841396_{10^{+2}}$
		$6.7379470_{10^{-3}}$	$6.7378976_{10^{-3}}$	$6.7378919_{10^{-3}}$
	10.0	$2.2026466_{10^{+4}}$	$2.2026994_{10^{+4}}$	$2.2027010_{10^{+4}}$
		$4.5399930_{10^{-5}}$	$4.5398593_{10^{-5}}$	$4.5398555_{10^{-5}}$

System VII

	x	y	y_4	y_6
$h = 0.01$	0.2	$1.9737532_{10^{-1}}$	$1.9737532_{10^{-1}}$	$1.9737532_{10^{-1}}$
	0.6	$5.3704957_{10^{-1}}$	$5.3704957_{10^{-1}}$	$5.3704957_{10^{-1}}$
	1.0	$7.6159416_{10^{-1}}$	$7.6159416_{10^{-1}}$	$7.6159416_{10^{-1}}$
	1.6	$9.2166855_{10^{-1}}$	$9.2166855_{10^{-1}}$	$9.2166855_{10^{-1}}$
	2.0	$9.6402758_{10^{-1}}$	$9.6402758_{10^{-1}}$	$9.6402758_{10^{-1}}$
$h = 0.1$	0.2	$1.9737532_{10^{-1}}$	$1.9737515_{10^{-1}}$	$1.9737529_{10^{-1}}$
	0.6	$5.3704957_{10^{-1}}$	$5.3704913_{10^{-1}}$	$5.3704928_{10^{-1}}$
	1.0	$7.6159416_{10^{-1}}$	$7.6159335_{10^{-1}}$	$7.6159366_{10^{-1}}$
	2.0	$9.6402758_{10^{-1}}$	$9.6402713_{10^{-1}}$	$9.6402728_{10^{-1}}$
	3.0	$9.9505475_{10^{-1}}$	$9.9505465_{10^{-1}}$	$9.9505467_{10^{-1}}$

compared to the closest values used for Table 2.11, we can determine that the correct deviations for y_6 are 1.7_{10-9}, 6.1_{10-9} and 1.7_{10-2}, 4.3_{10-11}. Especially in the $h = 0.1$, $x = 10.0$ case, the method is predicting the deviation in a superb fashion. Turning to System VII we can calculate a predicted accuracy in y_6 at $h = 0.1$, $x = 3.0$ of 2_{10-8} and an actual deviation of 8_{10-8}.

Thus one must suggest that this embedding method yields the best estimates of the local truncation error. The method of calculation and the fact that it does use the fifth-order Butcher form (2.3-20), which is among the best we have found, further points out the extreme usefulness of this method.

2.15.4. Recommendations

The single-step methods of this chapter are highly recommended as suitable integration procedures. They are self-starting, easy to program for a computer and require only a moderate amount of computer storage. The fifth-order forms, and in particular the Butcher equation used in an embedded form to estimate the truncation error and adjust the step size h, are extremely accurate especially when double precision is used. The main disadvantage of single-step methods is the number of function evaluations per step which may lead to large computer times. This point will be discussed further in Chapter 4 and compared with multiple-step methods.

REFERENCES

1. Babuska, I., Prager, M., and Vitosek, E., "Numerical Processes in Differential Equations." Wiley (Interscience), New York, 1966.
2. Ball, W. E. and Berns, R. I., Automast, *Comm. ACM* **9**, 626 (1966).
3. Benyon, P. R., A review of numerical methods for digital simulation, *Simulation* **11**, 219 (1968).
4. Blum, E. K., A formal system for differentiation, *J. Assoc. Comput. Mach.* **13**, 495 (1966).
5. Butcher, J. C., Coefficients for the study of Runge–Kutta integration processes, *J. Austral. Math. Soc.* **3**, 185 (1963).
6. Butcher, J. C., On the integration processes of A. Huta, *J. Austral. Math. Soc.* **3**, 203 (1963).
7. Butcher, J. C., On Runge–Kutta processes of high order, *J. Austral. Math. Soc.* **4**, 179 (1964).
8. Butcher, J. C., Implicit Runge–Kutta processes, *Math. Comp.* **18**, 50 (1964).
9. Butcher, J. C., Integration processes based on Radau quadrature formulas, *Math. Comp.* **18**, 233 (1964).
10. Butcher, J. C., On the attainable order of Runge–Kutta methods, *Math. Comp.* **19**, 408 (1965).
11. Byrne, G. D. and Lambert, R. J., Pseudo-Runge–Kutta methods involving two points, *J. Assoc. Comput. Mach.* **13**, 114 (1966).

12. Byrne, G. D., Parameters for pseudo-Runge–Kutta methods, *Comm. ACM* **10**, 102 (1967).
13. Calahan, D. A., A stable, accurate method of numerical integration for nonlinear systems, *Proc. IEEE* **56**, 744 (1968).
14. Cassity, C. R., Solutions of the fifth-order Runge–Kutta equations, *SIAM J. Numer. Anal.* **3**, 598 (1966).
15. Ceschino, F. and Kuntzmann, J., "Numerical Solution of Initial Value Problems." Prentice-Hall, Englewood Cliffs, New Jersey, 1966.
16. Chai, A. S., Error estimate of a fourth-order Runge–Kutta method with only one initial derivative evaluation, *AFIPS Conf. Proc.* **32**, 467 (1968).
17. Collatz, L., "The Numerical Treatment of Differential Equations," 3rd ed. Springer, Berlin, 1960.
18. Davison, E. J., A high order Crank–Nicholson technique for solving differential equations," *Comput. J.* **10**, 195 (1967).
19. Day, J. T., A one-step method for the numerical solution of second-order linear ordinary differential equations," *Math. Comp.* **18**, 664 (1964).
20. Day, J. T., A one-step method for the numerical integration of the differential equation $y'' = f(x)y + g(x)$, *Comput. J.* **7**, 314 (1965).
21. Day, J. T., Quadrature methods of arbitrary order for solving linear ordinary differential equations, *BIT* **6**, 181 (1966).
22. Eisenpress, H. and Bomberault, A., Efficient symbolic differentiation using PL/1-FORMAC, IBM New York Scientific Center Report 320–29561 (Sept. 1968).
23. El-Sherif, H. H., "Implicit Implementation of the Weighted Backward Euler Formula", *IBM J. Res. Develop.* **6**, 336 (1968).
24. Fehlberg, E., Classical fifth-, sixth-, seventh-, and eighth-order Runge–Kutta formulas with stepsize control, *NASA Technical Report*, NASA TR R-287 (Oct. 1968).
25. Fehlberg, E., Low-order classical Runge–Kutta formulas with stepsize control, *NASA Technical Report*, M-256 (1969).
26. Fyfe, D. J., Economical evaluation of Runge–Kutta formulae, *Math. Comput.* **20**, 392 (1966).
27. Gates, L. D., Numerical solution of differential equations by repeated quadratures, *SIAM Rev.* **6**, 134 (1964).
28. Gibbons, A., A program for the automatic integration of differential equations using the method of Taylor series, *Comput. J.* **3**, 108 (1960).
29. Gorbunov, A. D. and Shakhov, Y. A., On the approximate solution of Cauchy's problem with ordinary differential equations to a given number of correct figures, *U.S.S.R. Comput. Math. and Math. Phys.* **3**, 239 (1963); *ibid.*, **4**, 427 (1964).
30. Haines, C. F., Implicit integration processes with error estimation for the numerical solution of differential equations, *Comput. J.* **12**, 183 (1969).
31. Hamming, R. W., "Numerical Methods for Scientists and Engineers." McGraw-Hill, New York, 1962.
32. Henrici, P. "Discrete Variable Methods in Ordinary Differential Equations." Wiley, New York, 1962.
33. Hildebrand, F. B., "Introduction to Numerical Analysis," McGraw-Hill, New York, 1956.
34. Howard, J. C. and Tashjian, H., An algorithm for deriving the equations of mathematical physics by symbolic manipulation," *Comm. ACM* **11**, 814 (1968).
35. Hull, T. E. and Johnston, R. L., Optimum Runge–Kutta methods, *Math. Comp.* **18**, 306 (1964).
36. Hull, T. E. and Swenson, J. R., Test of probabilistic models for propagation of round-off errors, *Comm. ACM* **9**, 108 (1966).

37. Hull, R. E., A search for optimum methods for the numerical integration of ordinary differential equations, *SIAM Rev.* **9**, 647 (1967).
38. King, R., Runge–Kutta methods with constrained minimum error bounds, *Math. Comp.* **20**, 386 (1966).
39. Konen, H. P. and Luther, H. A., Some singular explicit fifth-order Runge–Kutta solutions, *SIAM J. Numer. Anal.* **4**, 607 (1967).
40. Kopal, Z., " Numerical Analysis." Wiley, New York, 1955.
41. Kuntzmann, J., Deux Formules Optimales du Fyce de Runge–Kutta, *Chiffres* **2**, 21 (1959).
42. Lapidus, L. and Luus, R., "Optimal Control of Engineering Processes." Random House (Blaisdell), New York, 1967.
43. Lawson, J. D., An order five Runge–Kutta process with extended region of stability, *SIAM J. Numer. Anal.* **3**, 593 (1966).
44. Lawson, J. D., An order six Runge–Kutta process with extended region of stability, *SIAM J. Numer. Anal.* **4**, 620 (1967).
45. Lawson, J. D., Generalized Runge–Kutta processes with stable systems with large Lipschitz constants, *SIAM J. Numer. Anal.* **4**, 372 (1967).
46. Luther, H. A., Further explicit fifth-order Runge–Kutta formulas, *SIAM Rev.* **8**, 374 (1966).
47. Luther, H. A., An explicit sixth-order Runge–Kutta formula, *Math. Comp.* **22**, 434 (1968).
48. Luther, H. A., and Konen, H. P., Some fifth-order classical Runge–Kutta formulas, *SIAM Rev.* **7**, 551 (1965).
49. Makinson, G. J., High order implicit methods for the numerical solution of systems differential equations, *Comput. J.* **11**, 305 (1968).
50. Martens, H. R., A comparative study of digital integration methods, *Simulation* **12**, 87 (1969).
51. Merson, R. H., An operational method for study of integration processes, *Proceedings of Symposium on Data Processing*, Weapons Research Establishment, Salisbury, Australia (1957).
52. Miller, J. C. P., The numerical solution of ordinary differential equations. In " Numerical Analysis," (J. Walsh, ed.), Chapter 4. Thompson Book, 1967.
53. Moore, R. E., "Interval Analysis," Prentice-Hall, Englewood Cliffs, New Jersey, 1966.
54. Nikolaev, V. S., The solution of systems of ordinary differential equations by expansion in power series on high speed computers, *U.S.S.R. Comp. Math. and Math. Phys.* **5**, 608 (1965).
55. Pope, D. A. An exponential method of numerical integration of ordinary differential equations, *Comm. ACM* **6**, 491 (1963).
56. Ralston, A., Runge–Kutta methods with minimum error bounds, *Math. Comp.* **16**, 431 (1962).
57. Rosen, J. S., The Runge–Kutta equations by quadrature methods, *NASA Technical Report*, NASA TR R-275 (Nov. 1967).
58. Rosen, J. S., Multi-step Runge–Kutta methods, *NASA Technical Note*, NASA TN D-4400 (April, 1968).
59. Rosenbrock, H. H., Some general implicit processes for the numerical solution of differential equations, *Comput. J.* **5**, 329 (1963).
60. Rosser, J. B., A Runge–Kutta for all seasons, *SIAM Rev.* **9**, 417 (1967).
61. Sammet, J. E., Formula manipulation compiler, *Datamation*, **12**, 32 (July, 1966).
62. Sarafyan, D., Error estimation for Runge–Kutta methods through Pseudo-iterative formulas, Louisiana State University, Technical Report No. 14 (May, 1966)

63. Sarafyan, D., Estimation of errors for the approximate solution of differential equations and their systems, Louisiana State University, Technical Report No. 15 (August, 1966).

64. Sarafyan, D. Composite and multi-step Runge–Kutta formulas, Louisiana State University, Technical Report No. 18 (November, 1966).

65. Sarafyan, D. Multi-order property of Runge–Kutta formulas and error estimation, Louisiana State University, Technical Report No. 29 (November, 1967).

66. Sarafyan, D. and Brown R., Computer deriviations of algebraic equations associated with Runge-Kutta formulas, *BIT* 7, 156 (1967).

67. Scraton, R. E., The numerical solution of second-order differential equations not containing the first derivative explicitly, *Comput. J.* 6, 368 (1964).

68. Scraton, R. E., Estimation of the truncation error in Runge–Kutta and allied processes, *Comput. J.* 7, 246 (1965).

69. Shanks, E. B., Solutions of differential equations by evaluations of functions, *Math. Comp.* 20, 21 (1966).

70. Stoller, L. and Morrison, D., A method for the numerical integration of ordinary differential equations, *MTAC* 12, 269 (1958).

71. Thompson, W. E., Solution of linear differential equations, *Comput. J.* 10, 417 (1967).

72. Warten, R. M., Automatic step-size control for Runge–Kutta integration, *IBM J. Res. Develop.* 2, 340 (1963).

73. Waters, J. Methods of numerical integration applied to a system having trivial function evaluations, *Comm. ACM* 9, 293 (1966).

74. Zee, C., On solving second-order nonlinear differential equations, *Quart. Appl. Math.* 22, 71 (1964).

3

Stability of Multistep and Runge–Kutta Methods

At various points in the first two chapters we have referred in a qualitative sense to the concept of stability of numerical methods for the integration of ODEs as related to the propagation of errors committed in a single step through n steps In this chapter the stability of numerical methods for ODEs will be treated in detail The two key considerations in the selection of an algorithm for a particular problem are accuracy and stability, and we will develop the relationship between these two aspects in this chapter.

3.1. LINEAR MULTISTEP METHODS

Linear multistep methods (1.7-1) encompass a large number of formulas used in practice and provide a convenient framework for an introduction to numerical stability. In this section we will develop the general properties of this class of methods. Two restrictions are inherent in (1.7-1). Only values of y and y' are used; higher derivatives do not appear Also, it is required that (1.7-1) be applied with equispaced points. These restrictions may limit somewhat the performance of methods based on (1.7-1), however (1.7-1) represents a very important and extensive class of formulas, the general properties of which are instrumental to study. In later sections we will consider other methods in which the restrictions of (1.7-1) will be relaxed. Most important for our purposes, it is convenient to develop the concepts of numerical stability with reference to this class of methods,

$$y_{n+1} = \alpha_1 y_n + \cdots + \alpha_k y_{n-k+1} + h[\beta_0 y'_{n+1} + \cdots + \beta_k y'_{n-k+1}]$$

$$(3.1\text{-}1)$$

In this chapter we will be concerned with those methods which involve only a single application of (3.1-1) to advance the solution one step. Often it is desirable to use an explicit form $(\beta_0 = 0)$ of (3.1-1) to get a first approximation to y_{n+1} (the *predicted* value) and then use an implicit form $(\beta_0 \neq 0)$ of (3.1-1) to improve on this value with y'_{n+1} evaluated using the predicted value of y_{n+1} (the *corrected* value). Such methods, which involve two or more applications of (3.1-1) to advance the solution one step will be the subject of Chapter 4.

Let us now outline briefly two essential features of (3.1-1), convergence and consistency, as a prelude to a study of numerical stability. The results in this section are due to Dahlquist [4,5]. The detailed proofs will be omitted. The interested reader may consult the original references for these.

If we introduce the generating polynomials,

$$\rho(\xi) = -\xi^k + \alpha_1 \xi^{k-1} + \cdots + \alpha_k \qquad (3.1\text{-}2)$$

$$\sigma(\xi) = \beta_0 \xi^k + \beta_1 \xi^{k-1} + \cdots + \beta_k \qquad (3.1\text{-}3)$$

and the shift operator E, defined by $E^k y_n = y_{n+k}$, (3.1-1) can be written compactly as

$$\rho(E) y_{n-k+1} + h\sigma(E) y'_{n-k+1} = 0 \qquad (3.1\text{-}4)$$

In order that (3.1-1) be feasible for numerical integration it is necessary that when $y_{n+1}, y_n, \ldots, y'_{n+1}, y_n', \ldots$ are replaced by the exact solution, the discrepancy between the left- and right-hand sides be small. In particular, we want (3.1-4) to be small when the exact solution is used for $y_{n+1}, y_n, \ldots$ and $y'_{n+1}, y_n', \ldots$ for all sufficiently regular functions $y(x)$. This imposes restrictions on the α_i and β_i in (3.1-1). In order that for arbitrary $y(x)$, the right-hand side of (3.1-4) be $O(h^{p+1})$ it is necessary that

$$\sum_{i=0}^{k} \alpha_i = 0 \qquad (3.1\text{-}5)$$

and

$$\sum_{i=0}^{k} \left[\frac{\alpha_{k-i} i^j}{j!} - \frac{\beta_{k-i} i^{j-1}}{(j-1)!} \right] = 0, \qquad j = 1, 2, \ldots, p \qquad (3.1\text{-}6)$$

We will make the following assumptions:

1. The coefficient β_0 is a real constant (α_0 is taken as -1).
2. The polynomials $\rho(\xi)$ and $\sigma(\xi)$ have no common factor.
3. p is at least equal to one.

Assumption 2 is made for convenience, since if $\rho(\xi)$ and $\sigma(\xi)$ possessed a common factor, then (3.1-1) effectively reduces to an equation of lower order.

The method defined by (3.1-1) is *convergent* if for all ODEs that satisfy a Lipschitz condition and for all y_0

$$\lim_{\substack{h \to 0 \\ nh \to x}} y_n = y(x),$$

holds for all $x \in [a, b]$ and for all solutions y_n of (3.1-1) having initial values satisfying

$$\lim_{h \to 0} y_i(h) = y_0, \qquad i = 0, 1, 2, \ldots, k - 1$$

It can be shown that (3.1-5), or $\rho(1) = 0$, is a necessary condition for convergence.

The second important concept is that of *consistency*, that is, if y_n converges uniformly to a continuous function $y(x)$, $y(x)$ satisfies the correct differential equation. Consistency can be shown equivalent to the condition that

$$\rho'(1) = \sigma(1) \qquad (3.1\text{-}7)$$

In this section we have in a very brief manner tried to point out some of the theoretical considerations in the construction of linear multistep methods. The full theoretical treatment of these methods was developed by Dahlquist [4,5] in a series of classic papers. For our purposes, this material is most relevant as a preface to a study of numerical stability and for this reason is presented in an abridged form.

3.2. NUMERICAL STABILITY OF LINEAR MULTISTEP METHODS

In the numerical solution of an ODE a sequence of approximations y_n to the true solution $y(x_n)$ is generated. Roughly speaking, the stability of a numerical method refers to the behavior of the difference or error, $y_n - y(x_n)$, as n becomes large. In order to begin our discussion let us consider the various types of errors which are incurred in numerical integration with (3.1-1).

We will place the errors incurred in a single step in two categories:

1. The local truncation error—the error introduced by the approximation of the differential equation by a difference equation.

2. Errors due to the deviation of the numerical solution from the exact theoretical solution of the difference equation (3.1-1). Included in this class are round-off errors and errors which are incurred if (3.1-1) is implicit and not solved exactly at each step.

An additional source of error results from the use of an auxiliary method, usually a single-step method, to develop the $k - 1$ starting values for (3.1-1).

Let the accumulated error at step n, ε_n, be given by (1.11-18),

$$\varepsilon_n = y_n - y(x_n) \tag{3.2-1}$$

If we substitute the true solution in (3.1-4), we obtain

$$\rho(E)y(x_n) + h\sigma(E)f(x_n, y(x_n)) - T_n = 0 \tag{3.2-2}$$

where we have changed the index on y from $n - k + 1$ to n for convenience, and where $T_n = T(x_n, h)$ is the local truncation error. The actual numerical solution obeys the relation

$$\rho(E)y_n + h\sigma(E)f(x_n, y_n) + \eta_n = 0 \tag{3.2-3}$$

where η_n represents the second category of errors described above, the errors which occur in the calculation of y_{n+1} because the difference equation is not solved exactly. Thus, y_n is the estimate obtained in a numerical solution of the theoretical solution of the difference equation (3.1-1).

Subtracting (3.2-2) from (3.2-3) we obtain

$$\rho(E)[y_n - y(x_n)] + h\sigma(E)[f(x_n, y_n) - f(x_n, y(x_n))] + T_n + \eta_n = 0 \tag{3.2-4}$$

Using the mean-value theorem,

$$f(x_n, y_n) - f(x_n, y(x_n)) = f_{\bar{y}}(x_n, \bar{y})(y_n - y(x_n)) \tag{3.2-5}$$

where $y_n \le \bar{y} \le y(x_n)$ and if we let $f_{\bar{y}}(x_n, \bar{y})$ be denoted by λ_n, (3.2-4) and (3.2-5) can be combined to give

$$\rho(E)\varepsilon_n + h\sigma(E)\lambda_n \varepsilon_n + T_n + \eta_n = 0 \tag{3.2-6}$$

The reader will recall a specific application of (3.2-6) in (1.11-25) for Euler's method. Equation (3.2-6) is the k-step difference equation for the accumulated error ε_n for a method given by (3.1-1).

If estimates of $\varepsilon_0, \ldots, \varepsilon_{k-1}$ are available and λ_n, T_n, and η_n are known, the inhomogeneous k-step difference equation (3.2-6) can be solved for all n. The stability of the method refers explicitly to the behavior of ε_n as n gets large, so that the solution of (3.2-6) will yield all the stability information desired, since (3.2-6) determines the accumulated error at each step. It is now clear why the accumulated error is not simply equal to the sum of the local errors. It must be computed as the solution of (3.2-6), a linear k-step difference equation.

The solution of linear difference equations parallels that for linear differential equations. As we know, it is much simpler to solve linear ODEs with constant coefficients rather than with coefficients which are functions of x. By the same token, it is much simpler to solve linear difference equations with constant coefficients rather than with coefficients which are functions of

n. Thus, as a first approximation, let us assume that λ_n, T_n, and η_n are constants (independent of n) and equal to λ, T, and η. Actually, knowledge of λ_n, T_n, and η_n as a function of n is often quite difficult to obtain so that we may be forced to make this assumption anyway. The accumulated error now obeys the simplified form of (3.2-6)

$$\rho(E)\varepsilon_n + h\lambda\sigma(E)\varepsilon_n + T + \eta = 0 \qquad (3.2\text{-}7)$$

a linear, inhomogeneous, k-step difference equation with constant coefficients.

A brief review of linear ordinary difference equations with constant coefficients is necessary to solve (3.2-7). Let us rewrite (3.2-7) in the general form

$$[E^k + a_{k-1}E^{k-1} + \cdots + a_0]\varepsilon_n = b \qquad (3.2\text{-}8)$$

where the coefficients $a_0, \ldots, a_{k-1}$ are real and independent of n. The theory of linear difference equations is similar to that of linear ODEs. The characteristic roots of (3.2-8) are the k roots of the characteristic equation

$$\mu^k + a_{k-1}\mu^{k-1} + \cdots + a_0 = 0 \qquad (3.2\text{-}9)$$

As with ODEs, (3.2-8) has homogeneous and particular solutions. If the k roots are distinct the homogeneous solution of (3.2-8) is

$$\varepsilon_{n_h} = c_1\mu_1{}^n + c_2\mu_2{}^n + \cdots + c_k\mu_k{}^n \qquad (3.2\text{-}10)$$

where $c_1, \ldots, c_k$ are constants determined by the initial conditions. If one of the roots, say μ_1, has multiplicity m, the homogeneous solution is

$$\varepsilon_{n_h} = (c_1 + c_2 n + \cdots + c_{m-1}n^{m-1})\mu_1{}^n + c_m\mu_m{}^n + \cdots + c_k\mu_k{}^n \qquad (3.2\text{-}11)$$

If $a_0, \ldots, a_{k-1}$ are real, any complex root of (3.2-9) must have a conjugate which is also a root. For the roots $\mu_1 = v_R + iv_I$ and $\mu_2 = v_R - iv_I$ we let

$$r^2 = v_R{}^2 + v_I{}^2 \qquad (3.2\text{-}12)$$

$$\theta = \tan(v_R/v_I)$$

and the homogeneous solution of (3.2-8) is

$$\varepsilon_{n_h} = c_1 r^n \cos n\theta + c_2 r^n \sin n\theta + c_3\mu_3{}^n + \cdots + c_k\mu_k{}^n \qquad (3.2\text{-}13)$$

If the characteristic roots are distinct and unequal to one, the particular solution of (3.2-8) is given by

$$\varepsilon_{n_p} = b/(1 + a_{k-1} + \cdots + a_0) \qquad (3.2\text{-}14)$$

These ideas may readily be extended to the case in which we have m linear ordinary difference equations,

$$\varepsilon_{n+k} + A_{k-1}\varepsilon_{n+k-1} + \cdots + A_0\varepsilon_n = b \qquad (3.2\text{-}15)$$

Equation (3.2-15) can be written $G(E)\varepsilon_n = b$, where G is an $m \times m$ matrix. The characteristic equation for (3.2-15) is

$$\det G(\mu) = 0 \qquad (3.2\text{-}16)$$

Since the dimension of G is $m \times m$ and μ appears to the kth power in $G(\mu)$, there are km characteristic roots of (3.2-16). If they are all distinct the solution of (3.2-15) is

$$\varepsilon_{i_n} = \sum_{j=1}^{m} \sum_{l=1}^{m} c_{ij}(\mu_{jl})^n + \{b[I + A_{k-1} + \cdots + A_0]^{-1}\}_i \qquad (3.2\text{-}17)$$

for $i = 1, 2, \ldots, m$.

Returning to the error difference equation (3.2-7), the characteristic equation is

$$\rho(\mu) + h\lambda\sigma(\mu) = 0 \qquad (3.2\text{-}18)$$

a kth order polynomial, the k solutions of which are the characteristic roots. The solution of (3.2-7) is

$$\varepsilon_n = \varepsilon_{n_h} + \varepsilon_{n_p} \qquad (3.2\text{-}19)$$

where ε_{n_h} and ε_{n_p} are given by (3.2-10) and (3.2-14), respectively, with $b = -(T + \eta)$ and $(1 + a_{k-1} + \cdots + a_0)$ replaced by $(-1 + \sum_1^k \alpha_i + h\lambda \sum_0^k \beta_i)$.

We are interested in the behavior of ε_n as n increases—in particular, if ε_n remains bounded as n gets large. Since the particular solution ε_{n_p} is a constant, only the homogeneous solution ε_{n_h} need be considered. The c_i in (3.2-10) depend on the initial k errors, $\varepsilon_0, \ldots, \varepsilon_{k-1}$. The k roots have been assumed to be distinct. However, if we go back to (3.1-4), we see that the characteristic equation (3.2-18) of the difference equation for the accumulated error ε_n, (3.2-7), is exactly the same as the characteristic equation for the original difference equation in y_n. Thus, we can write

$$y_n = d_1\mu_1^n + d_2\mu_2^n + \cdots + d_k\mu_k^n \qquad (3.2\text{-}20)$$

as the solution to (3.1-4). The constants d_i depend on the initial starting values used for (3.1-4) and are, of course, different than the c_i in (3.2-9). The d_i would be determined from knowledge of $y_0, \ldots, y_{k-1}$, usually computed by a single step method.

Consider for a moment the integration of $y' = \lambda y$, $y(x_0) = 1$, the solution of which is $y(x) = \exp(\lambda x)$, where, in general, λ may be a complex constant. One of the roots of the characteristic equation (3.2-18) approximates the Taylor series expansion of the true solution with a truncation error corresponding to the order of the method. If we let this root be μ_1, then

$$\mu_1 = e^{h\lambda} + O(h^{p+1}) \qquad (3.2\text{-}21)$$

as $h \to 0$. This root, called the *principal root*, is the root which we wish to be represented in the numerical solution, since μ_1^n approximates $\exp(nh\lambda)$. The other $k - 1$ roots are called *spurious, parasitic*, or *extraneous roots* and are a result of the use of a difference equation of order k to represent a first-order differential equation. The extraneous roots have no relation to the exact solution of the differential equation but, nevertheless, are unavoidable.

We can now outline the problem of numerical instability in general terms. Comparing (3.2-20) to the exact solution of our system $y' = \lambda y$,

$$y(x_n) = \exp(nh\lambda) \tag{3.2-22}$$

we see that as long as $|\mu_1| > |\mu_i|$, $i = 2, 3, \ldots, k$, as n increases the extraneous solutions, $d_2 \mu_2^n + \cdots + d_k \mu_k^n$, will become small when compared to the principal solution, $d_1 \mu_1^n$. In this situation we may expect that our numerical solution will be stable, although we will see that the requirements are some-what more stringent. However, if $|\mu_i| > |\mu_1|$ for any i, μ_i^n will become large with respect to μ_1^n, and the numerical solution component corresponding to μ_i will predominate. Since this root has no relation to the exact solution, the numerical solution will bear no relation to the exact solution. This situation is referred to as numerical instability.

We can now be more precise about the requirements for numerical stability. We have already seen that intuitively our solution should be valid if $|\mu_1| > |\mu_i|$, $i = 2, 3, \ldots, k$. However, recall from (3.2-18) that the characteristic roots of the difference equation (3.1-4) are the same as those of the difference equation for the error ε_n, i.e. (3.2-10). For a valid numerical solution we require that ε_n not grow with n. From (3.2-10) we see that this is equivalent to the condition, $|\mu_i| \leq 1$, $i = 1, 2, \ldots, k$. It is thus possible to define the stability characteristics of a method in the following way. A linear multistep method (3.1-1) is called

Absolutely stable if $|\mu_i| \leq 1$, $i = 1, 2, \ldots, k$
Relatively stable if $|\mu_i| \leq |\mu_1|$, $i = 2, 3, \ldots, k$

Absolute stability does not imply relative stability. In other words, a numerical solution may have $|\mu_i| \leq 1$, $i = 1, 2, \ldots, k$, but $|\mu_1| < |\mu_i| \leq 1$, $i = 2, 3, \ldots, k$.

The single ODE $y' = \lambda y$ will be called *inherently stable* if $\mathrm{Re}(\lambda) < 0$. In this case the exact solution is decreasing with x_n, and the important condition is absolute stability, since the numerical solution must also decrease with x_n. If, however, $\mathrm{Re}(\lambda) \geq 0$, the exact solution is growing with x_n, and we do not want $|\mu_i| \leq 1$, rather it is relative stability that is the important consideration. In other words, we will have a valid solution as long as no component of the numerical solution, μ_i^n, increases faster than the one corresponding to the principal root. The condition of inherent stability can be extended to the vector ODE $\mathbf{y}' = \mathbf{f}(x, \mathbf{y})$, which is the subject of the theory of stability of ODEs. For the moment we will confine our discussion to the case in which $\lambda_n = \lambda = \text{constant}$.

Since the characteristic roots μ_n are obtained from (3.2-18), they depend on the product $h\lambda$ and the generating polynomials $\rho(E)$ and $\sigma(E)$. For fixed λ in a particular method, h can be increased. The value of $h\lambda$ for which $|\mu_i| = 1$ and for which a small increase in $|h\lambda|$ makes $|\mu_i| > 1$ is called the *general stability boundary*. Since in general λ is complex, we can let $\lambda = \bar{\lambda} \exp(i\theta)$ where $\bar{\lambda}$ and θ are real. Then we can make similar definitions of the *real* and *imaginary stability boundaries*, corresponding to the values of $\bar{\lambda}h$ and $i(\bar{\lambda}h)$ where the root condition is obeyed. Any method with a finite general stability boundary can be called *conditionally* stable, whereas any method with an infinite general stability boundary can be called *unconditionally* stable, or *A-stable*. Thus, a linear multistep method is *A*-stable if all solutions of (3.1-1) tend to zero, as $n \to \infty$, when the method is applied with fixed $h > 0$ to $y' = \lambda y$, where λ is a complex constant with $\text{Re}(\lambda) < 0$.

It is worthwhile for several reasons to consider the stability of linear multistep methods as $h \to 0$. First, if a numerical method is unstable in the limit of $h \to 0$, then there is probably little chance that the algorithm will be stable for some finite h. Also, while the elements of a stability analysis for finite h have already been presented, rigorous theoretical results of a general nature for methods (3.1-1) can only be obtained for $h \to 0$. It is of interest to examine briefly these classic results.

For numerical stability in the limit of $h \to 0$ we require that the accumulated error remain bounded as $n \to \infty$. This requires that solutions of the difference equation

$$\rho(E)\varepsilon_n + T + \eta = 0 \qquad (3.2\text{-}23)$$

remain bounded as $n \to \infty$. The characteristic equation of (3.2-23) is $\rho(\mu) = 0$ and the homogeneous solution of (3.2-23) can be expressed

$$\varepsilon_n = c_1'\mu_1{}^n + c_2'\mu_2{}^n + \cdots + c_k'\mu_k{}^n \qquad (3.2\text{-}24)$$

where the k roots of $\rho(\mu) = 0$ have been assumed to be distinct. Again if $|\mu_i| \leq 1$, $i = 1, 2, \ldots, k$, the numerical solution will be valid, and if $|\mu_i| > 1$ for any i, the numerical solution will become invalid as $n \to \infty$.

Dahlquist [4] has shown that the condition of asymptotic stability for $h \to 0$ is obeyed if: (1) The roots of $\rho(\mu) = 0$ are located within or on the unit circle. (2) The roots on the unit circle are distinct.

Thus, for asymptotic stability, which Dahlquist has called *strong stability*, it is necessary that the roots of $\rho(\mu) = 0$ obey the condition, $|\mu_i| \leq 1$. *Consistency* and *strong stability* can be shown to be necessary and sufficient conditions for *convergence*.

Some methods may have more than one root on the unit circle. These methods are usually called *weakly stable*. The stability of such methods can be measured by means of certain growth parameters introduced by Dahlquist [5]. For each root on the unit circle, there exists a growth parameter v_i such that

$$\mu_i(h\lambda) = \mu_i(h \to 0)e^{v_i h\lambda} + O(h^2) \qquad (3.2\text{-}25)$$

where $\mu_i(h\lambda)$ and $\mu_i(h \to 0)$ are the ith characteristic roots of (3.2-18) and $\rho(\mu) = 0$ respectively. For $i = 1$, $v_i = 1$. The more that a particular v_i differs from one for roots on the unit circle, the more undesirable is the component of the numerical solution corresponding to that root.

Let us consider an example of numerical instability. Henrici [17] has considered the integration of $y' = \lambda y$, $y_0 = 1$, by the midpoint rule (1.6-31)

$$y_{n+1} = y_{n-1} + 2hy_n' \qquad (3.2\text{-}26)$$

The characteristic equation of (3.2-26) is

$$\mu^2 - 2h\lambda\mu - 1 = 0 \qquad (3.2\text{-}27)$$

which has the two roots,

$$\mu_1 = h\lambda + [1 + (h\lambda)^2]^{1/2} \qquad (3.2\text{-}28)$$

$$\mu_2 = -1/\mu_1 \qquad (3.2\text{-}29)$$

If we expand μ_1 using the binomial formula, we obtain

$$\mu_1 = 1 + h\lambda + \tfrac{1}{2}(h\lambda)^2 + O(h^4) \qquad (3.2\text{-}30)$$

By comparison to the series expansion of $\exp(h\lambda)$,

$$\mu_1 = e^{h\lambda} - \tfrac{1}{6}(h\lambda)^3 + O(h^4) \qquad (3.2\text{-}31)$$

The numerical solution is given by

$$y_n = d_1\mu_1{}^n + d_2\mu_2{}^n \qquad (3.2\text{-}32)$$

where

$$\mu_1{}^n = e^{\lambda n h}[1 - \tfrac{1}{6}nh^3\lambda^3 + \cdots] \qquad (3.2\text{-}33)$$

and

$$\mu_2{}^n = (-1)^n e^{-\lambda n h}[1 + \tfrac{1}{6}nh^3\lambda^3 + \cdots] \qquad (3.2\text{-}34)$$

Using (3.2-32) we obtain with $d_1 = 1 - d_2$ (since $y_0 = 1$)

$$y_n = e^{\lambda x_n} - \tfrac{1}{6}\lambda^3 x_n e^{\lambda x_n}h^2 + d_2 e^{\lambda x_n} + d_2(-1)^n e^{-\lambda x_n} \qquad (3.2\text{-}35)$$

The first term is the desired solution of the ODE. The second term is the truncation error, due to the approximation of the ODE by a discrete formula.

The last two terms arise because of the extraneous root μ_2. From (3.2-32) we see that

$$d_2 = (y_1 - \mu_1)/(\mu_2 - \mu_1) \qquad (3.2\text{-}36)$$

In general $y_1 \neq \mu_1$ so that $d_2 \neq 0$. The disasterous effect of these extra components is particularly evident if $\text{Re}(\lambda) < 0$. Then the last term increases exponentially, quickly invalidating the numerical solution y_n.

In addition, if we consider the limit of $h \to 0$, the two roots (3.2-28) and (3.2-29) are ± 1. Thus, the midpoint rule is a weakly stable method. Referring to (3.2-25) we see that the growth parameter $v_2 = -1$, so that even in the limit of $h = 0$, for a large enough n, stability problems will arise. The conclusion to be drawn is that the midpoint rule is a poor method for numerical integration.

We can summarize the results of this section as applied to the case $y' = \lambda y$. Numerical instability results from extraneous solutions of the difference equation which bear no connection to the exact solution. The conditions of asymptotic (strong) and absolute stability can be seen clearly with reference to the extraneous solutions. If, in the limit $h \to 0$, the extraneous solutions vanish as $n \to \infty$, then the method is strongly stable and convergent. If, for values of h less than some h_0, the extraneous solutions vanish as $n \to \infty$, the method is absolutely stable. Relative stability is a significant concept for inherently unstable ODEs. The condition provides that extraneous solutions will not grow more rapidly or decay more slowly than the true solution.

3.3. DAHLQUIST STABILITY THEOREMS

Dahlquist [5] has proved several important theorems relating p and k for stable multistep methods. In the following we will present the most important of these theorems with a discussion of their implications in selecting possible multistep methods.

Theorem 3.3.1.   Although for any positive k there exists a consistent method of order $p = 2k$, the order of a strongly stable k-step method cannot exceed $k + 2$. If k is odd, it cannot exceed $k + 1$.

Methods for which p is large have desirable accuracy, but unfortunately have an associated undesirable error growth resulting in instability.

Theorem 3.3.2.   An explicit k-step method cannot be A-stable.

Theorem 3.3.3.   The order p of an A-stable linear multistep method cannot exceed two. The smallest truncation error in such a case is obtained for the trapezoidal rule for which $p = 2$, $k = 1$.

The concept of A-stability can be modified to allow for methods of higher accuracy. Widlund [31] has introduced the following definition. A k-step method is $A(\alpha)$-*stable*, $\alpha \in (0, \pi/2)$, if all solutions of (3.1-1) tend to zero as $n \to \infty$ when the method is applied with fixed $h > 0$ to $y' = \lambda y$ where λ is a complex constant which lies in the set

$$S_\alpha = \{z : |\arg(-z)| < \alpha, \quad z \neq 0\}$$

The region S_α is shown in Figure 3.1. A method is $A(\pi/2)$ stable if it is $A(\alpha)$-stable for all $\alpha \in (0, \pi/2)$. Theorem 3.3.2 states that no explicit method can be $A(\pi/2)$-stable and Theorem 3.3.3 states that the trapezoidal rule is the most accurate $A(\pi/2)$-stable method.

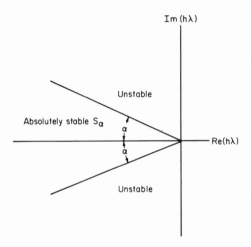

Figure 3.1. Stability region in complex $h\lambda$ plane for Theorem 3.3.6.

Assume we know that the complex λ lies in some region S_α. Then it is only necessary to require $A(\alpha)$-stability for a numerical method. Based on this reasoning, Widlund has proved the following theorems.

Theorem 3.3.4. No explicit method can be $A(0)$-stable.

Theorem 3.3.5. There is only one $A(0)$-stable method with $p \geq k + 1$, the trapezoidal rule

Theorem 3.3.6. For all $\alpha \in [0, \pi/2)$ there exist $A(\alpha)$-stable methods for which $k = p = 3$ and $k = p = 4$.

Theorem 3.3.6 is an encouraging result if an unconditionally stable method is desired. Of particular importance is the case when all the eigenvalues are real, so that α is known to be zero.

The theorems in this section are, however, basically discouraging in that it is impossible to construct arbitrarily accurate A-stable linear multistep methods. In fact, the trapezoidal rule is the most accurate A-stable linear multistep method. We will see (Section 3.4) that the trapezoidal rule will play a key role in situations in which stability is a primary consideration. A very important point, however, is that the results of this section are confined to linear multistep methods of the class (3.1-1). The theorems in no way govern the construction of more accurate A-stable methods which are not of the linear multistep type; such methods can indeed be constructed.

3.4. STABILITY OF MULTISTEP METHODS IN INTEGRATING COUPLED ODEs

The question we wish to answer in this section is: How do we determine the characteristic roots of (3.1-1) when integrating coupled ODEs? The development in the last section was made in relation to the integration of a single ODE. In fact, the assumption that we could replace λ_n by λ, a constant, really restricted the analysis to a single linear ODE, even though the stability definitions and concepts are completely general. For the moment, let us still confine our attention to linear ODEs. In the next section we will discuss the stability of nonlinear coupled ODEs.

First, let us analyze the stability of single-step methods which are special cases of the general linear multistep method (3.1-1). Of particular interest will be the application of the methods to both scalar and vector ODEs. Since single-step methods have only one characteristic root, absolute rather than relative stability will be the important consideration. Our discussion will center on three simple methods: forward Euler, backward Euler, and the trapezoidal rule (modified Euler). Let us first consider the application of each of these methods to the scalar ODE $y' = \lambda y$.

When applied to $y' = \lambda y$, Euler's forward method (1.6-7) yields

$$y_{n+1} = (1 + h\lambda)y_n \qquad (3.4\text{-}1)$$

The characteristic root $\mu_1 = 1 + h\lambda$ is plotted in Figure 3.2. The condition for absolute stability is

$$|1 + h\lambda| \leq 1 \qquad (3.4\text{-}2)$$

In general, λ may be complex, as we noted, and thus can be expressed as $\lambda = \lambda_R + i\lambda_I$. An alternative way of expressing λ is by letting $\lambda = \bar{\lambda}e^{i\theta}$, where $\bar{\lambda}$ is real and θ is the angle measured from the positive real axis if $\lambda_R > 0$ [or $\text{Re}(\lambda) > 0$] or the angle measured from the negative real axis if $\lambda_R < 0$ [or $\text{Re}(\lambda) < 0$]. If $\lambda_R = 0$, then $\theta = \pi/2$ in either case. Note that $\bar{\lambda}$ is not the real part of λ, λ_R, but rather $\lambda_R = \bar{\lambda}\cos\theta$ and $\lambda_I = \bar{\lambda}\sin\theta$. Thus,

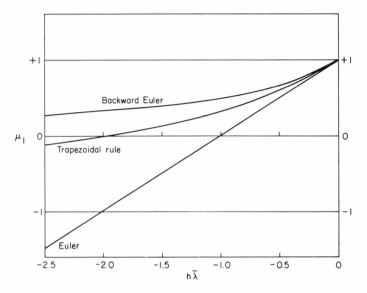

Figure 3.2. Characteristic root for Euler, backward Euler, and trapezoidal rule methods.

$\lambda = \lambda_R + i\lambda_I$ and $\lambda = \bar{\lambda}e^{i\theta}$ represent alternate descriptions of the complex constant λ. In this case (3.4-2) can be written as

$$|1 + h\bar{\lambda}e^{i\theta}| \le 1 \qquad (3.4\text{-}3)$$

or

$$|1 + h\lambda_R + ih\lambda_I| \le 1 \qquad (3.4\text{-}4)$$

If λ is real, i.e., $\lambda_I = 0$ or $\theta = 0$, the stability bound for Euler's forward method is $h\lambda \ge -2$. It will be convenient to be able to condense (3.4-3) and (3.4-4) into plots representing regions of stability. For various values of θ, the value of $h\bar{\lambda}$ for which the equality in (3.4-3) holds can be determined. This has been carried out and the results are shown in Figure 3.3, in which values of $h\bar{\lambda}$ below the boundary are stable combinations. The alternative to Figure 3.3 is to plot the stability region in the complex plane, corresponding to the equality in (3.4-4). This is the most common representation and is shown in Figure 3.4 in which the curve labeled $p = 1$ corresponds to Euler's forward method. Complex values of $h\lambda$ located within the region shown represent stable combinations.

Early work by Gray, Gurk, and Rubinoff [10–14, 27, 28] was focused on numerical stability as part of studies of the feasibility of using digital computers in aircraft flight simulators. They introduced a "stability chart" which is more comprehensive than the types represented by Figures 3.3 and 3.4. Just as there is a complex exponential $e^{\lambda x}$ corresponding to the eigenvalue

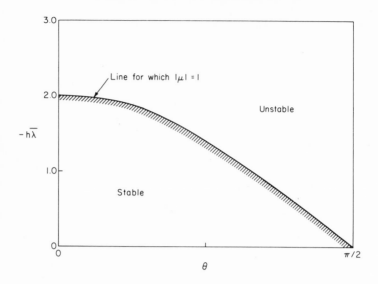

Figure 3.3. Stability boundary for Euler's method [22].

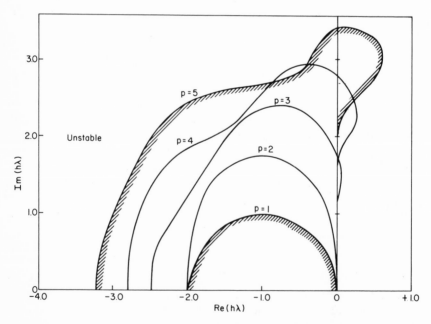

Figure 3.4. Stability regions in the complex plane for classic explicit Runge–Kutta methods of orders 1–5.

λ of the **ODE**, so there is a complex exponential $\mu_i{}^n$ corresponding to each root μ_i of the characteristic equation of the numerical method. We can let $\mu_i{}^n = e^{w_i h n} = e^{w_i x}$, where w_i is defined as $(\ln \mu_i)/h$. For numerical stability we consider the properties of μ_i, or, equivalently w_i. The "stability chart" presents the root w_i with the most positive real part, call it w, as a function of λ for a given method. Normally hw is plotted as a function of $h\lambda$. Since both hw and $h\lambda$ are complex, i.e., $hw = hw_R + ihw_I$ and $h\lambda = h\lambda_R + ih\lambda_I$, the chart takes the form of contours of constant hw_R and hw_I in the $h\lambda_R$ and $h\lambda_I$ plane. Figure 3.5 shows the stability chart for a multiple-step method proposed by Gurk [14], the specifics of which need not concern us here. In this figure $U = hw_R$ and $V = hw_I$. The stability boundary is given by the curve on which $U = 0$, since positive values of U will result in an exponentially increasing numerical solution. In addition, the stability chart has also been used as a rough guide to the accuracy of a method. A method is assumed accurate for that choice of h for which the U-V grid closely matches the $h\lambda_R - h\lambda_I$ grid. The stability region plot of the type of Figure 3.4 is simply the $U = 0$ line from the stability chart of the same method.

For example, suppose $\lambda = -4 + 4i$ and consider the numerical integration using the method of Figure 3.5 [28]. The stability chart shown in Figure 3.5 is square out to about -0.25. Thus, choice of $h = 0.05$ will yield accurate simulation, choice of $h = 0.10$ leads to slight inaccuracy, since $hw = -0.39 + 0.38i$ for $h\lambda = -0.4 + 0.4i$ and choice of $h = 0.15$ leads to $hw = +0.07 +$

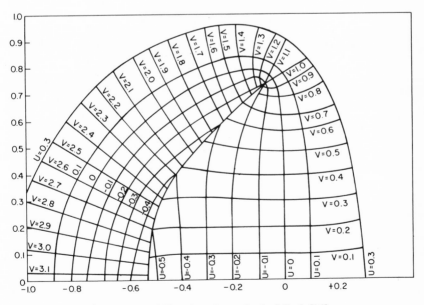

Figure 3.5. Stability chart for method of Gurk [14].

2.19i and hence an exponentially increasing, unstable response. In general, close agreement between the U-V and $h\lambda_R - h\lambda_I$ grids is only a rough guide to the accuracy of the method. Hence, we will not rely on full stability charts in the remainder of the book; only the $U = 0$ stability limit will be plotted. The question of accuracy will be discussed subsequently.

Next we consider the backward Euler method, which comes from a backward difference equation in the same manner as Euler's forward method,

$$y_{n+1} = y_n + hy'_{n+1} \qquad (3.4\text{-}5)$$

which when applied to $y' = \lambda y$ yields

$$y_{n+1} = y_n/(1 - h\lambda) \qquad (3.4\text{-}6)$$

where the characteristic root is $(1 - h\lambda)^{-1}$. This root is plotted also in Figure 3.2 and we see that for $\mathrm{Re}(\lambda) < 0$, $|\mu_1|$ is always less than 1. Thus, the backward Euler method is A-stable.

Finally, let us consider the trapezoidal rule (the modified Euler method) (1.6-20). When applied to $y' = \lambda y$ we obtain

$$y_{n+1} = \frac{1 + \frac{1}{2}h\lambda}{1 - \frac{1}{2}h\lambda}\, y_n \qquad (3.4\text{-}7)$$

the characteristic root of which is

$$\mu_1 = \frac{1 + \frac{1}{2}h\lambda}{1 - \frac{1}{2}h\lambda} \qquad (3.4\text{-}8)$$

From Figure 3.2 we see that $|\mu_1|$ is always less than 1 as long as $\mathrm{Re}(\lambda) < 0$. (We already knew from Theorem 3.3.3 that the trapezoidal rule is A-stable.)

Seldom are we interested in numerically integrating a single ODE, but rather a system of coupled ODEs. For convenience, let us begin our discussion by considering the two-dimensional linear ODE with constant coefficients

$$\mathbf{y}' = \mathbf{A}\mathbf{y}, \qquad \mathbf{y}(0) = \mathbf{y}_0 \qquad (3.4\text{-}9)$$

where the elements of $\mathbf{A}$ are denoted as a_{ij}. In vector notation the exact solution of (3.4-9) is

$$\mathbf{y}(x) = \exp(\mathbf{A}x)\mathbf{y}_0 \qquad (3.4\text{-}10)$$

It will be convenient for us to determine the individual solutions y_1 and y_2. If the eigenvalues of $\mathbf{A}$ are λ_1 and λ_2, then the general solution of (3.4-9) is

$$y_1(x) = c_1 u_1 e^{\lambda_1 x} + c_2 v_1 e^{\lambda_2 x}$$
$$y_2(x) = c_1 u_2 e^{\lambda_1 x} + c_2 v_2 e^{\lambda_2 x} \qquad (3.4\text{-}11)$$

where $\mathbf{u} = (u_1, u_2)$ and $\mathbf{v} = (v_2, v_2)$ are the eigenvectors of (3.4-9). c_1 and c_2 can be determined from the initial conditions,

$$c_1 u_1 + c_2 v_1 = y_{1_0}$$
$$c_1 u_2 + c_2 v_2 = y_{2_0}$$

(3.4-12)

The exact solution will change over a step h by the factor $\exp(\mathbf{A}h)$, i.e.

$$\mathbf{y}(x_{n+1}) = \exp(\mathbf{A}h)\mathbf{y}(x_n)$$

(3.4-13)

so that any single-step method which can be expressed as

$$\mathbf{y}_{n+1} = \mathbf{M}(\mathbf{A}h)\mathbf{y}_n$$

(3.4-14)

has properties which are essentially determined by how well $\mathbf{M}(\mathbf{A}h)$ approximates $\exp(\mathbf{A}h)$. It is this approximation that we now wish to study in detail for single-step integration methods.

In order to begin our discussion, let us consider the application of Euler's method to (3.4-9). This yields

$$\mathbf{y}_{n+1} = (\mathbf{I} + h\mathbf{A})\mathbf{y}_n$$

(3.4-15)

It is obvious that $(\mathbf{I} + h\mathbf{A}) = \mathbf{M}(h\mathbf{A})$ represents an approximation to $\exp(\mathbf{A}h)$ by the first two terms in an infinite series expansion of $\exp(\mathbf{A}h)$ in the form,

$$\exp(\mathbf{A}h) = \mathbf{I} + h\mathbf{A} + \frac{h^2\mathbf{A}^2}{2!} + \frac{h^3\mathbf{A}^3}{3!} + \cdots$$

(3.4-16)

In the case when $\mathbf{A} = \lambda$, a scalar, (3.4-15) reduces to (3.4-1) as expected. It will be somewhat clearer to rewrite (3.4-15) in its component form,

$$y_{1_{n+1}} = y_{1_n} + h(a_{11}y_{1_n} + a_{12}y_{2_n})$$
$$y_{2_{n+1}} = y_{2_n} + h(a_{21}y_{1_n} + a_{22}y_{2_n})$$

(3.4-17)

This system of two coupled first-order difference equations can be solved by assuming a solution of the form [29]

$$y_{1_n} = u\mu^n, \qquad y_{2_n} = v\mu^n$$

(3.4-18)

Substituting (3.4-18) into (3.4-17) yields

$$a_{11}u + a_{12}v = -(1 - \mu)u/h$$
$$a_{21}u + a_{22}v = -(1 - \mu)v/h$$

(3.4-19)

If we set

$$(1 - \mu)/h = -\lambda, \qquad \mu = 1 + h\lambda$$

(3.4-20)

we obtain the same eigenvalue problem that arose in the exact solution of (3.4-9), where the eigenvectors $\mathbf{u}$ and $\mathbf{v}$ are the same as before. Thus, the solution of (3.4-17) is

$$
\begin{aligned}
y_{1_n} &= c_1 u_1 (1 + \lambda_1 h)^n + c_2 v_1 (1 + \lambda_2 h)^n \\
y_{2_n} &= c_1 u_2 (1 + \lambda_1 h)^n + c_2 v_2 (1 + \lambda_2 h)^n
\end{aligned}
\tag{3.4-21}
$$

where v_1, c_2, u, and v are the same coefficients as in (3.4-11).

Rather than comparing $\mathbf{M}(h\mathbf{A})$ and $(\mathbf{I} + h\mathbf{A})$ directly, we have reduced the problem to the comparison of (3.4-11) and (3.4-21). Obviously $(1 + \lambda_1 h)^n$ and $(1 + \lambda_2 h)^n$ represent approximations to $\exp(\lambda_1 nh)$ and $\exp(\lambda_2 nh)$ respectively. Now since we are interested in inherently stable ODEs, i.e., $\mathrm{Re}(\lambda_1) < 0$, $\mathrm{Re}(\lambda_2) < 0$, it is necessary for stability that

$$
|1 + \lambda_1 h| < 1, \qquad |1 + \lambda_2 h| < 1
\tag{3.4-22}
$$

It is evident that the eigenvalue larger in absolute value will govern the choice of h in maintaining stability. The accuracy of the method is determined by the approximations

$$
\begin{aligned}
(1 + \lambda_1 h) &\cong e^{\lambda_1 h} \\
(1 + \lambda_2 h) &\cong e^{\lambda_2 h}
\end{aligned}
\tag{3.4-23}
$$

Since (3.4-23) are better approximations as $\lambda_i h \to 0$, the poorer approximation will be associated with the larger eigenvalue, say λ_1, $\mathrm{Re}(\lambda_1) < \mathrm{Re}(\lambda_2)$.

Let us now consider the application of the trapezoidal rule to (3.4-9). In vector notation, the trapezoidal rule (1.6-20) yields

$$
\mathbf{y}_{n+1} = (\mathbf{I} - \tfrac{1}{2}h\mathbf{A})^{-1}(\mathbf{I} + \tfrac{1}{2}h\mathbf{A})\mathbf{y}_n
\tag{3.4-24}
$$

where in the notation of our general single-step method

$$
\mathbf{M}(h\mathbf{A}) = (\mathbf{I} - \tfrac{1}{2}h\mathbf{A})^{-1}(\mathbf{I} + \tfrac{1}{2}h\mathbf{A})
\tag{3.4-25}
$$

In component notation (3.4-24) is

$$
y_{1_{n+1}} = y_{1_n} + (h/2)[(a_{11}y_{1_{n+1}} + a_{12}y_{2_{n+1}}) + (a_{11}y_{1_n} + a_{12}y_{2_n})]
\tag{3.4-26}
$$

$$
y_{2_{n+1}} = y_{2_n} + (h/2)[(a_{21}y_{1_{n+1}} + a_{22}y_{2_{n+1}}) + (a_{21}y_{1_n} + a_{22}y_{2_n})]
$$

Again, letting $y_{1_n} = u\mu^n$ and $y_{2_n} = v\mu^n$ we obtain

$$
\begin{aligned}
a_{11}u + a_{12}v &= -\frac{2}{h}\frac{1 - \mu}{1 + \mu}u \\[2mm]
a_{21}u + a_{22}v &= -\frac{2}{h}\frac{1 - \mu}{1 + \mu}v
\end{aligned}
\tag{3.4-27}
$$

If we let $\mu = (1 + \frac{1}{2}h\lambda)/(1 - \frac{1}{2}h\lambda)$ we obtain the difference equation solution

$$y_{1n} = c_1 u_1 \left[\frac{1 + \frac{1}{2}h\lambda_1}{1 - \frac{1}{2}h\lambda_1}\right]^n + c_2 v_1 \left[\frac{1 + \frac{1}{2}h\lambda_2}{1 - \frac{1}{2}h\lambda_2}\right]^n$$

$$y_{2n} = c_1 u_2 \left[\frac{1 + \frac{1}{2}h\lambda_1}{1 - \frac{1}{2}h\lambda_1}\right]^n + c_2 v_2 \left[\frac{1 + \frac{1}{2}h\lambda_2}{1 - \frac{1}{2}h\lambda_2}\right]^n$$

(3.4-28)

We have already pointed out that for $\text{Re}(\lambda_i) < 0$, the trapezoidal rule is A-stable. Accuracy is determined by the approximations,

$$\left(\frac{1 + \frac{1}{2}h\lambda_1}{1 - \frac{1}{2}h\lambda_1}\right) \cong e^{\lambda_1 h}$$

$$\left(\frac{1 + \frac{1}{2}h\lambda_2}{1 - \frac{1}{2}h\lambda_2}\right) \cong e^{\lambda_2 h}$$

(3.4-29)

These approximations improve as $\lambda_i h \to 0$, so, as with Euler's method, the poorer of the two approximations will be for the eigenvalue with the largest real part in absolute value.

A very important result has been obtained for Euler's method and the trapezoidal rule, namely that *when used to integrate a coupled set of linear ODEs the stability of the method depends only on the eigenvalues of* $\mathbf{A}$. This result can be generalized to any single-step method.

Now let us show that for an m-dimensional linear ODE and any single-step method, the stability of the algorithm depends only on the eigenvalues of the ODE. Obviously we could continue as above, writing the difference equation, determining the characteristic root and solving the difference equation. A more concise demonstration can be made based on a similarity transformation. Let us define

$$\mathbf{y} = \mathbf{Pz} \qquad (3.4\text{-}30)$$

so that (3.4-9) becomes

$$\mathbf{z}' = \mathbf{P}^{-1}\mathbf{APz}$$
$$\mathbf{z}(0) = \mathbf{P}^{-1}\mathbf{y}_0 \qquad (3.4\text{-}31)$$

If the eigenvalues λ_i of $\mathbf{A}$ are distinct, (3.4-31) reduces to

$$\mathbf{z}' = \mathbf{\Lambda z} \qquad (3.4\text{-}32)$$

where $\mathbf{\Lambda}$ is the diagonal matrix of eigenvalues λ_i. The solution of (3.4-9) is

$$\mathbf{y}(x) = \mathbf{P}\exp(\mathbf{\Lambda}x)\mathbf{P}^{-1}\mathbf{y}_0 \qquad (3.4\text{-}33)$$

or, in a recursive notation,

$$\mathbf{y}(x_{n+1}) = \mathbf{P}\exp(\mathbf{\Lambda}h)\mathbf{P}^{-1}\mathbf{y}(x_n) \qquad (3.4\text{-}34)$$

A single-step method can be expressed as (3.4-14), so that we want

$$M(P\Lambda P^{-1}h) = P \exp(\Lambda h)P^{-1} \qquad (3.4\text{-}35)$$

If we consider the general rational form,

$$M(B) = \left(\sum_{i=1}^{n} b_i B^i \right)^{-1} \left(\sum_{i=1}^{m} a_i B^i \right) \qquad (3.4\text{-}36)$$

and let

$$B = P\Lambda P^{-1} \qquad (3.4\text{-}37)$$

then

$$M(P\Lambda P^{-1}) = \left(\sum_{i=1}^{n} b_i (P\Lambda P^{-1})^i \right)^{-1} \left(\sum_{i=1}^{m} a_i (P\Lambda P^{-1})^i \right)$$

$$= P\left(\sum_{i=1}^{n} b_i \Lambda^i \right)^{-1} \left(\sum_{i=1}^{m} a_i \Lambda^i \right) P^{-1} \qquad (3.4\text{-}38)$$

$$= PM(\Lambda)P^{-1}$$

Referring to (3.4-35), we see that

$$M(\Lambda h) \cong \exp(\Lambda h) \qquad (3.4\text{-}39)$$

Since each matrix is diagonal, the corresponding diagonal elements of $M(\Lambda h)$ approximate those of $\exp(\Lambda h)$. Each diagonal element of M is simply the characteristic root μ_1, so that (3.4-39) can be written

$$\mu_1(h\lambda_i) \cong \exp(h\lambda_i), \qquad i = 1, 2, \ldots, m \qquad (3.4\text{-}40)$$

Absolute stability requires that

$$\lim_{n \to \infty} [M(\Lambda h)]^n = 0 \qquad (3.4\text{-}41)$$

or, correspondingly, that

$$|\mu_i(h\lambda_i)| < 1, \qquad i = 1, 2, \ldots, m \qquad (3.4\text{-}42)$$

The important point is that in numerically integrating a coupled set of linear ODEs is it sufficient to consider the method as applied to the scalar equation $y' = \lambda_i y$, where λ_i takes on the values of the eigenvalues of the ODE.

So far in this section we have considered single-step methods of the class (3.1-1) in integrating coupled linear ODEs. Now we wish to extend the analysis to general linear multistep methods (3.1-1). As before, we will consider the two-dimensional linear ODE (3.4-9). Applying (3.1-1) to (3.4-9) yields

$$\alpha_k y_{n+k} + \alpha_{k-1} y_{n+k-1} + \cdots + \alpha_0 y_n = hA(\beta_k y_{n+k} + \cdots + \beta_0 y_n)$$
$$(3.4\text{-}43)$$

an m-dimensional vector difference equation of order k. The exact solution of (3.4-9) for $m = 2$ is given by (3.4-11). It is this solution which we wish to approximate as closely as possible by the solution of the difference equation (3.4-43). The difference equation (3.4-43) will have km roots ($2k$ roots in the case of $m = 2$). In component notation (3.4-43) is,

$$\sum_{i=0}^{k} \alpha_i \begin{bmatrix} y_{1_{n+k-i}} \\ y_{2_{n+k-i}} \end{bmatrix} + h \sum_{i=0}^{k} \beta_i \begin{bmatrix} a_{11} y_{1_{n+k-i}} + a_{12} y_{2_{n+k-i}} \\ a_{21} y_{1_{n+k-i}} + a_{22} y_{2_{n+k-i}} \end{bmatrix} = 0 \qquad (3.4\text{-}44)$$

Let us denote the $2k$ roots of (3.4-44) by μ_{ij}, $i = 1, 2; j = 1, 2, \ldots, k$. Thus, there is a set of k roots corresponding to each of the two components of $\mathbf{y}$. Two of these roots, the principal roots, designated by μ_{11} and μ_{21}, will approximate the Taylor series expansions of $e^{\lambda_1 h}$ and $e^{\lambda_2 h}$ respectively with a truncation error corresponding to the degree of the difference equation. The remaining roots μ_{1j} and μ_{2j}, $j = 2, \ldots, k$ are extraneous and arise, as we have previously noted, because a k-order difference equation has been used to approximate a first-order differential equation. The solution of the difference equation can be written

$$\begin{aligned} y_{1_n} &= C_{11} \mu_{11}^n + C_{12} \mu_{12}^n + \cdots + C_{1k} \mu_{1k}^n \\ &\quad + D_{11} \mu_{21}^n + D_{12} \mu_{22}^n + \cdots + D_{1k} \mu_{2k}^n \\ y_{2_n} &= C_{21} \mu_{11}^n + C_{22} \mu_{12}^n + \cdots + C_{2k} \mu_{1k}^n \\ &\quad + D_{21} \mu_{21}^n + D_{22} \mu_{22}^n + \cdots + D_{2k} \mu_{2k}^n \end{aligned} \qquad (3.4\text{-}45)$$

where $C_{11} = c_1 u_1$, $D_{11} = c_2 v_1$, $C_{21} = c_1 u_2$, and $D_{21} = c_2 v_2$.

The condition for absolute stability is that $|\mu_{ij}| \leq 1$, $i = 1, 2; j = 1, 2, \ldots, k$. The condition for relative stability is that $|\mu_{11}| \geq |\mu_{1j}|$ and $|\mu_{21}| \geq |\mu_{2j}|$, $j = 2, 3, \ldots, k$. The principal roots approximate the exact solution components,

$$\mu_{11} \cong \exp(h\lambda_1), \qquad \mu_{21} \cong \exp(h\lambda_2) \qquad (3.4\text{-}46)$$

so that the accuracy of the method is determined by the degree of the error term in (3.4-46).

We now wish to show that the roots of (3.4-44) depend only on h and the eigenvalues of $\mathbf{A}$ and not on the individual elements of $\mathbf{A}$. We have

$$\sum_{i=0}^{k} (\alpha_i \mathbf{I} + h\beta_i \mathbf{A}) \mathbf{y}_{n+i} = 0 \qquad (3.4\text{-}47)$$

Using the similarity transform (3.4-30), (3.4-47) becomes

$$\sum_{i=0}^{k} (\alpha_i \mathbf{I} + h\beta_i \mathbf{\Lambda}) \mathbf{z}_{n+i} = 0 \qquad (3.4\text{-}48)$$

Defining the diagonal shifting matrix $\mathbf{E}$, such that $\mathbf{z}_{n+1} = \mathbf{E}\mathbf{z}_n$, (3.4-48) is written

$$\sum_{i=0}^{k} (\alpha_i \mathbf{I} + h\beta_i \mathbf{\Lambda})\mathbf{E}^i \mathbf{z}_n = 0 \qquad (3.4\text{-}49)$$

The $2k$ characteristic roots are the roots of

$$\sum_{i=0}^{k} (\alpha_i + h\beta_i \lambda_j)\mu_j{}^i = 0, \qquad j = 1, 2 \qquad (3.4\text{-}50)$$

The form of the roots is determined by α_i and β_i, but depend only on $h\lambda_j$, $j = 1, 2$.

3.5. STABILITY OF INTEGRATION OF NONLINEAR ODEs

Our ultimate aim in discussing stability of numerical integration methods is to determine the stability of integration of the nonlinear ODE, $\mathbf{y}' = \mathbf{f}(x, \mathbf{y})$. At any step n we can form a Taylor series about $(\mathbf{y}_n, x_n)$ truncated after the first order terms,

$$\mathbf{y}' = \mathbf{f}_n + [\mathbf{f}_{\bar{y}}]_n (\mathbf{y} - \mathbf{y}_n) + O(h^2) \qquad (3.5\text{-}1)$$

giving a linear, nonhomogeneous ODE at each step. The eigenvalues of (3.5-1), λ_i, $i = 1, 2, \ldots, m$ are the eigenvalues of the Jacobian matrix $[\mathbf{f}_{\bar{y}}]_n$ and will vary from step to step.

In the foregoing we have dealt with the linear ODE (3.4-9) to formulate an exact theory of stability. Hildebrand [18] has shown that the stability characteristics of the linearized ODE (3.5-1) are very similar to those of $\mathbf{y}' = \mathbf{f}(x, \mathbf{y})$ for small h. It can be shown that the same extraneous solutions exist in both cases, even though the growth of these extraneous solutions relative to the true solution will be slightly different in each case [8]. Considering the asymptotic case $(h \to 0)$, Henrici [15] has considered the variability of $\mathbf{f}_{\bar{y}}$. An exact equation for the accumulated truncation error in the asymptotic case will be shown in Section 5.1.1. However, the differential equation presented there for the accumulated truncation error involves knowledge of the Jacobian matrix of the exact solution. Thus, at this time there does not exist a general theory of absolute stability of linear multistep methods, such that the stability behavior of different multistep methods when applied to nonlinear ODE can be determined in a systematic manner.

A recommended procedure is to determine stability bounds on the basis of the eigenvalues of $\mathbf{f}_{\bar{y}}$. The eigenvalues of $\mathbf{f}_{\bar{y}}$ can be periodically computed and used to adjust h. It appears that a fruitful area of further study is the theoretical and computational analysis of stability bounds for nonlinear ODEs.

3.6. STABILITY OF RUNGE–KUTTA METHODS

Up to this point we have developed the general theory of stability of numerical integration methods with reference to linear multistep methods. Although the elements of the stability analysis are the same, it is convenient to treat Runge–Kutta methods separately.

3.6.1. Explicit Runge–Kutta Methods

In this subsection we consider the stability of the explicit Runge–Kutta methods of Section 2.3. In all cases we will treat the scalar ODE $y' = \lambda y$, which, as we have seen, will be sufficient for studying stability of coupled ODEs, provided λ assumes the values of the eigenvalues of the ODE.

As an example of the derivation of the characteristic root for the entire class of explicit methods in Section 2.3, we consider the application of (2.3-7), the improved Euler method, to $y' = \lambda y$. The method is

$$
\begin{aligned}
y_{n+1} &= y_n + \tfrac{1}{2}k_1 + \tfrac{1}{2}k_2 \\
k_1 &= hf_n, \qquad k_2 = hf(x_n + h, y_n + k_1)
\end{aligned}
\tag{3.6-1}
$$

Using $f_n = \lambda y_n$, $k_1 = h\lambda y_n$, and $k_2 = h\lambda(y_n + k_1) = h\lambda(y_n + h\lambda y_n)$, the single-step recurrence relation is

$$
y_{n+1} = \left(1 + h\lambda + \tfrac{1}{2}h^2\lambda^2\right)y_n
\tag{3.6-2}
$$

It is easy to verify that each of the second-order methods (2.3-5)–(2.3-7) yields exactly the same recurrence relation as (3.6-2). The characteristic root of (3.6-1) is thus

$$
\mu_1 = 1 + h\lambda + \tfrac{1}{2}h^2\lambda^2
\tag{3.6-3}
$$

The stability boundary in the complex plane for the $p = 2$ methods is shown in Figure 3.4. Note that the absolute real stability boundary for (2.3-5)–(2.3-7) is $h\lambda > -2$, and that each of the methods is unstable for any purely imaginary λ. In the cases to follow we will not show the algebra required to obtain the single-step recurrence relation, $y_{n+1} = \mu_1(h\lambda)y_n$, but as the order of the methods increases the amount of manipulation increases substantially.

Next we consider the third-order methods (2.3-9)–(2.3-11). It is straightforward to show that each of these methods has the characteristic root

$$
\mu_1 = 1 + h\lambda + \tfrac{1}{2}h^2\lambda^2 + \tfrac{1}{6}h^3\lambda^3
\tag{3.6-4}
$$

The stability boundary of (2.3-9)–(2.3-11) in the complex plane is shown in Figure 3.4. We see that the absolute real stability boundary for these methods is -2.5.

We presented three fourth-order methods, (2.3-13)–(2.3-15), each of which has the characteristic root

$$\mu_1 = 1 + h\lambda + \tfrac{1}{2}h^2\lambda^2 + \tfrac{1}{6}h^3\lambda^3 + \tfrac{1}{24}h^4\lambda^4 \qquad (3.6\text{-}5)$$

Figure 3.4 presents the stability region in the complex plane for (2.3-13)–(2.3-15). The absolute real stability boundary is -2.785.

When $p = v =$ number of stages, i.e., for $p \leq 4$ we always get

$$\mu_1 = \sum_{i=0}^{p=v} \frac{(h\lambda)^i}{i!} \qquad (3.6\text{-}6)$$

But for $p = 5$, $v = 6$, etc., the number of terms is greater than the order of the method. In particular,

$$\mu_1 = \sum_{i=0}^{p} \frac{(h\lambda)^i}{i!} + \sum_{i=p+1}^{v} a_i \frac{(h\lambda)^i}{i!} \qquad (3.6\text{-}7)$$

where the a_i are constants. Different values of a_i will correspond to each of the fifth-order methods in Section 2.3.5. We can thus characterize the fifth-order methods of Section 2.3.5 by the value of a_6 in the characteristic root. Note that for $|h\lambda| < 1$, the added term in (3.6-7) is quite small, e.g., $(h\lambda)^6$ for $p = 5$, but for $|h\lambda| > 1$ this term tends to change the stability bounds considerably; and usually the real region of interest is $|h\lambda| > 1$.

Values of a_6 for some of the fifth-order methods are given in Table 3.1, which also summarizes much of the stability information of this section. Figures 3.4 and 3.6 also show, as examples, the regions of stability for

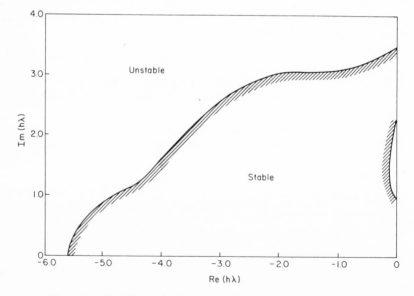

Figure 3.6. Stability region in the complex $h\lambda$ plane for (2.3-30) [20].

Nystrom's form, (2.3-16), and for Lawson's form, (2.3-30). In particular, Nystrom's form [and also (2.3-25)] have values of $a_6 = 0$ and thus correspond to a classic fifth-order method. The Lawson form, (2.3-30), is of interest since here the parameter a_6 is adjusted to maximize the real stability boundary [20]. As seen in Table 3.1, the stability bound is large as compared to any other method; next is the Butcher fifth-order method of (2.3-20).

All explicit Runge–Kutta methods have a finite stability boundary and will be stable for a small enough h. In general, as the order increases, the stability bounds increase.

3.6.2. Implicit Runge–Kutta Methods

Before we go directly into the stability analysis of implicit Runge–Kutta methods, we want to discuss the Padé approximation [19]. Given some function $f(x)$ we may desire to approximate this function by a rational function involving polynomials. If we have the two polynomials,

$$P_m(x) = \sum_{j=0}^{m} a_j x^j \tag{3.6-8}$$

$$Q_n(x) = \sum_{j=0}^{n} b_j x^j, \qquad b_0 = 1 \tag{3.6-9}$$

we could form the rational function

$$F_{m,n}(x) = P_m(x)/Q_n(x) \tag{3.6-10}$$

to approximate the given function $f(x)$. The index of the approximation is $m + n$. To generate the coefficients we require the following:

1. $f(x) = F_{m,n}(x)$ at $x = 0$.
2. The first $m + n$ derivatives of the Maclaurin series expansion of $f(x)$,

$$f(x) = \sum_{j=0}^{\infty} c_j x^j \tag{3.6-11}$$

and $F_{m,n}(x)$ be made equal at $x = 0$.

There are $m + n + 1$ constants in $P_m(x)$ and $Q_n(x)$ and $m + n + 1$ conditions in (1) and (2).

We can write

$$f(x) - F_{m,n}(x) = \left[\left(\sum_0^{\infty} c_j x^j \right) \left(\sum_0^{n} b_j x^j \right) - \sum_0^{m} a_j x^j \right] \bigg/ \sum_0^{n} b_j x^j \tag{3.6-12}$$

In order for conditions (1) and (2) to be met we want the numerator on the right-hand side of (3.6-12) to have the lowest power of x. Thus,

$$\left(\sum_0^{\infty} c_j x^j \right) \left(\sum_0^{n} b_j x^j \right) - \sum_0^{m} a_j x^j = \sum_{m+n+1}^{\infty} d_j x^j \tag{3.6-13}$$

TABLE 3.1

THEORETICAL STABILITY LIMITS FOR DIFFERENT EXPLICIT RUNGE–KUTTA METHODS, $q = h\lambda$

Method	Order	Re(q)	Im(q)	Characteristic root
(2.3-5)	2	−2.0	1.75i	$1 + q + \frac{1}{2}q^2$
(2.3-6)	2	−2.0	1.75i	
(2.3-7)	2	−2.0	1.75i	
(2.3-9)	3	−2.5	2.4i	$1 + q + \frac{1}{2}q^2 + \frac{1}{6}q^3$
(2.3-10)	3	−2.5	2.4i	
(2.3-11)	3	−2.5	2.4i	
(2.3-13)	4	−2.785	2.95i	$1 + q + \frac{1}{2}q^2 + \frac{1}{6}q^3 + \frac{1}{24}q^4$
(2.3-14)	4	−2.785	2.95i	
(2.3-15)	4	−2.785	2.95i	
(2.3-16)	5	−3.217	3.4i	$1 + q + \frac{1}{2}q^2 + \frac{1}{6}q^3 + \frac{1}{24}q^4 + \frac{1}{120}q^5$; $a_6 = 0$
(2.3-17)	5	−2.52	2.7i	(3.6-7); $a_6 = -2.25$
(2.3-18)	5	−2.65	2.85i	(3.6-7); $a_6 = -1.50$
(2.3-20)	5	−3.386	1.5i	(3.6-7); $a_6 = +1.128$
(2.3-24)	5	−2.65	2.85i	same as (2.3-18); $a_6 = -1.50$
(2.3-25)	5	−3.217	3.4i	same as (2.3-16); $a_6 = 0$
(2.3-26)	5	−2.65	2.85i	same as (2.3-18); $a_6 = -1.50$
(2.3-27)	5	−2.65	2.85i	same as (2.3-18); $a_6 = -1.50$
(2.3-28)	5	−3.15	—	
(2.3-30)	5	−5.7	3.5i	(3.6-7); $a_6 = 0.5625$
(2.3-32)	6	−2.15	2.01i	$1 + q + \frac{1}{2}q^2 + \frac{1}{6}q^3 + \frac{1}{24}q^4 + \frac{1}{120}q^5 + \frac{1}{720}q^6$; $a_7 = -2.33$

TABLE 3.2

PADÉ APPROXIMANTS TO e^x

	$m = 0$	$m = 1$	$m = 2$	$m = 3$
$n = 0$	$\dfrac{1}{1}$	$\dfrac{1+x}{1}$	$\dfrac{1+x+\frac{1}{2}x^2}{1}$	$\dfrac{1+x+\frac{1}{2}x^2+\frac{1}{6}x^3}{1}$
$n = 1$	$\dfrac{1}{1-x}$	$\dfrac{1+\frac{1}{2}x}{1-\frac{1}{2}x}$	$\dfrac{1+\frac{2}{3}x+\frac{1}{6}x^2}{1-\frac{1}{3}x}$	$\dfrac{1+\frac{3}{4}x+\frac{1}{4}x^2+\frac{1}{24}x^3}{1-\frac{1}{4}x}$
$n = 2$	$\dfrac{1}{1-x+\frac{1}{2}x^2}$	$\dfrac{1+\frac{1}{3}x}{1-\frac{2}{3}x+\frac{1}{6}x^2}$	$\dfrac{1+\frac{1}{2}x+\frac{1}{12}x^2}{1-\frac{1}{2}x+\frac{1}{12}x^2}$	$\dfrac{1+\frac{3}{5}x+\frac{3}{20}x^2+\frac{1}{60}x^3}{1-\frac{2}{5}x+\frac{1}{20}x^2}$
$n = 3$	$\dfrac{1}{1-x+\frac{1}{2}x^2-\frac{1}{6}x^3}$	$\dfrac{1+\frac{1}{4}x}{1-\frac{3}{4}x+\frac{1}{4}x^2-\frac{1}{24}x^3}$	$\dfrac{1+\frac{2}{5}x+\frac{1}{20}x^2}{1-\frac{3}{5}x+\frac{3}{20}x^2-\frac{1}{60}x^3}$	$\dfrac{1+\frac{1}{2}x+\frac{1}{10}x^2+\frac{1}{120}x^3}{1-\frac{1}{2}x+\frac{1}{10}x^2-\frac{1}{120}x^3}$
$n = 4$	$\dfrac{1}{1-x+\frac{1}{2}x^2-\frac{1}{6}x^3+\frac{1}{24}x^4}$	$\dfrac{1+\frac{1}{5}x}{1-\frac{4}{5}x+\frac{3}{10}x^2-\frac{1}{15}x^3+\frac{1}{120}x^4}$	$\dfrac{1+\frac{1}{3}x+\frac{1}{30}x^2}{1-\frac{2}{3}x+\frac{1}{5}x^2-\frac{1}{30}x^3+\frac{1}{360}x^4}$	$\dfrac{1+\frac{3}{7}x+\frac{1}{14}x^2+\frac{1}{210}x^3}{1-\frac{4}{7}x+\frac{1}{7}x^2-\frac{4}{210}x^3+\frac{1}{840}x^4}$

As a result, the following relations are obtained for the coefficients of $P_m(x)$ and $Q_n(x)$,

$$\sum_0^n c_{m+n-s-j} b_j = 0, \qquad s = 0, 1, \ldots, n-1$$
$$c_j = 0, \quad \text{if} \quad j < 0; \qquad b_0 = 1 \tag{3.6-14}$$

$$a_r = \sum_0^r c_{r-j} b_j, \qquad r = 0, 1, \ldots, m$$
$$b_j = 0, \qquad j > n \tag{3.6-15}$$

One of the most frequent uses of the Padé approximation is in the approximation of e^x and e^{-x}. Table 3.2 presents the Padé approximations of e^x through $m = 3$ and $n = 4$. A difficulty associated with the Padé approximation is that closed form error expressions are not available. Greater accuracy is obtained for higher values of m and n, and the $m = n$ or $m = n + 1$ approximations in general give the smallest min–max error for a given $m + n$ [19].

The reason for discussing the Padé approximation is that the characteristic roots of several of the implicit Runge–Kutta methods are Padé approximants to $\exp(h\lambda)$. It can be shown that all the diagonal Padé approximants possess the important property of A-stability [2]. This point will become clear shortly.

Let us begin with the three Gauss-type implicit methods (2.5-3)–(2.5-5). The first method, (2.5-3), is

$$y_{n+1} = y_n + k_1$$
$$k_1 = hf(x_n + \tfrac{1}{2}h, y_n + \tfrac{1}{2}k_1) \tag{3.6-16}$$

which when applied to $y' = \lambda y$ yields

$$y_{n+1} = \frac{1 + \tfrac{1}{2}h\lambda}{1 - \tfrac{1}{2}h\lambda} y_n \tag{3.6-17}$$

Equation (3.6-17) is the same relation as for the trapezoidal rule. However, when applied to a nonlinear ODE (2.5-3) will not be the same as the trapezoidal rule because of the derivative evaluation at $x_n + h/2$ and the implicit relation for k_1. Nevertheless, the method is A-stable.

When applied to $y' = \lambda y$, the second method (2.5-4) yields

$$y_{n+1} = y_n + \tfrac{1}{2}k_1 + \tfrac{1}{2}k_2$$
$$k_1 = h\lambda(y_n + \tfrac{1}{4}k_1 + \tfrac{1}{4}k_2 - (\sqrt{3}/6)k_2) \tag{3.6-18}$$
$$k_2 = h\lambda(y_n + \tfrac{1}{4}k_1 + (\sqrt{3}/6)k_2 + \tfrac{1}{4}k_2)$$

Solving the simultaneous equations for k_1 and k_2

$$y_{n+1} = \frac{1 + \tfrac{1}{2}h\lambda + \tfrac{1}{12}h^2\lambda^2}{1 - \tfrac{1}{2}h\lambda + \tfrac{1}{12}h^2\lambda^2} y_n \tag{3.6-19}$$

The characteristic root is

$$\mu_1 = \frac{1 + \frac{1}{2}h\lambda + \frac{1}{12}h^2\lambda^2}{1 - \frac{1}{2}h\lambda + \frac{1}{12}h^2\lambda^2} \tag{3.6-20}$$

for which it can be shown that $|\mu_1| < 1$ for $\mathrm{Re}(\lambda) < 0$. Thus, (2.5-4) is A-stable. In the same way (2.5-5) has the characteristic root

$$\mu_1 = \frac{1 + \frac{1}{2}h\lambda + \frac{1}{10}h^2\lambda^2 + \frac{1}{120}h^3\lambda^3}{1 - \frac{1}{2}h\lambda + \frac{1}{10}h^2\lambda^2 - \frac{1}{120}h^3\lambda^3} \tag{3.6-21}$$

which is also <1 in magnitude for $\mathrm{Re}(\lambda) < 0$.

If we look at (3.6-17), (3.6-20), and (3.6-21), we observe an interesting feature: Instead of the characteristic roots approximating $\exp(h\lambda)$ through higher powers of a Taylor series, e.g., (3.6-7), the characteristic roots of the implicit methods approximate $\exp(h\lambda)$ by rational functions. In fact, looking at Table 3.2 we see that (3.6-17), (3.6-20), and (3.6-21) represent the main diagonal Padé approximants to $\exp(h\lambda)$. Each of these approximations is A-stable for $\mathrm{Re}(\lambda) < 0$ and increase in accuracy with m and n. Table 3.2 contains other interesting information. The $n = 0$ row is composed of the characteristic roots of the explicit Runge–Kutta methods. The $n = 1$, $m = 0$ entry is the characteristic root for the backward Euler method.

Let us continue with the Radau and Lobatto forms in Section 2.5. The Radau implicit methods (2.5-7) and method (2.5-9) have the characteristic root

$$\mu_1 = \frac{1 + \frac{2}{3}h\lambda + \frac{1}{6}h^2\lambda^2}{1 - \frac{1}{3}h\lambda} \tag{3.6-22}$$

which is the $n = 1$, $m = 2$ Padé approximant.

Similarly, the Lobatto implicit method (2.5-12) has the characteristic root

$$\mu_1 = \frac{1 + \frac{3}{4}h\lambda + \frac{1}{4}h^2\lambda^2 + \frac{1}{24}h^3\lambda^3}{1 - \frac{1}{4}h\lambda} \tag{3.6-23}$$

which is the $n = 1$, $m = 3$ Padé approximant.

The Padé approximation can be extended to the case of $\exp(h\mathbf{A})$, as involved in the integration of $\mathbf{y}' = \mathbf{A}\mathbf{y}$ by an implicit Runge–Kutta method. We have

$$\mathbf{F}_{m,n}(h\mathbf{A}) \cong \exp(h\mathbf{A}) \tag{3.6-24}$$

where

$$\mathbf{F}_{m,n}(h\mathbf{A}) = \left(\sum_{i=0}^{n} b_j(h\mathbf{A})^i \right)^{-1} \left(\sum_{i=0}^{m} a_j(h\mathbf{A})^i \right) \tag{3.6-25}$$

If we make the association of $\mathbf{F}_{m,n}(h\mathbf{A}) = \mathbf{M}(h\mathbf{A})$ in the general single-step recursion relation (3.4-14), the stability limit of this class of implicit Runge–Kutta methods is governed by

$$\left| \sum_0^m a_i(h\lambda_1)^i \middle/ \sum_0^n b_i(h\lambda_1)^i \right| \leq 1 \tag{3.6-26}$$

as follows from (3.4-42). λ_1 is taken to be the largest eigenvalue in absolute value.

The implicit Runge–Kutta methods possess very desirable stability characteristics but, being implicit, require iteration at each step. Thus, the computing time saved by use of a large h may be spent in solving nonlinear algebraic equations at each step. In fact, we will see in Chapter 4 that conditions on $h\lambda$ required for the convergence of these iterations may be as strong as the stability conditions on the explicit methods. Thus, at this point it is difficult to predict whether an explicit or implicit Runge–Kutta method will be better in terms of accuracy and computing time for a specific problem. This point will be examined in Section 3.10 through computational examples.

3.6.3. Semi-Implicit Runge–Kutta Methods

In Section 2.5.4 a special class of Runge–Kutta methods were introduced. These methods, which may be classed as semi-implicit, have the desirable feature that each k_i equation in (2.5-16) can be solved explicitly. The characteristic roots of (2.5-18) and (2.5-19) are

$$\mu_1 = \frac{1 - h\lambda - \frac{2}{3}h^2\lambda^2}{1 - 2h\lambda + \frac{5}{6}h^2\lambda^2} \tag{3.6-27}$$

and

$$\mu_1 = \frac{1 + (\sqrt{2} - 1)h\lambda}{1 + (\sqrt{2} - 2)h\lambda + (\frac{3}{2} - \sqrt{2})h^2\lambda^2} \tag{3.6-28}$$

The regions of stability in the complex plane for these two methods are shown in Figure 3.7. For both methods for $\text{Re}(\lambda) > 0$, there is an initial stability limitation.

The characteristic root for the method of Haines, (2.5-20), is

$$\mu_1 = \frac{1 - \frac{8}{3}h\lambda + \frac{2}{9}h^2\lambda^2 + \frac{1}{3}h^3\lambda^3}{1 - \frac{11}{3}h\lambda + 5h^2\lambda^2 - 3h^3\lambda^3 + \frac{2}{3}h^4\lambda^4} \tag{3.6-29}$$

for which there is only a very small region of instability on the imaginary axis.

Also in the semi-implicit format is the method of Calahan, (2.5-21), the characteristic root for which is

$$\mu_1 = \frac{1 - 0.578h\lambda - 0.456h^2\lambda^2}{1 - 1.578h\lambda + 0.622h^2\lambda^2} \tag{3.6-30}$$

and the method is A-stable.

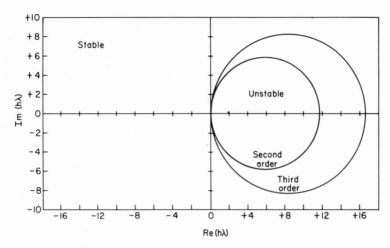

Figure 3.7. Stability regions for second-order and third-order Rosenbrock methods.

Several authors have recently used the A-stability condition of the diagonal Padé approximants to propose semi-implicit Runge–Kutta methods for $y' = Ay$. Davison [7] has suggested using the formula

$$y_{n+1} = [I - \tfrac{1}{2}hA + \tfrac{1}{4}h^2A^2 - \tfrac{1}{12}h^3A^3]^{-1}[I + \tfrac{1}{2}hA + \tfrac{1}{4}h^2A^2 + \tfrac{1}{12}h^3A^3]y_n$$

(3.6-31)

which is accurate to $O(h^5)$. The method is A-stable, and favorable comparisons with the explicit fourth-order Runge–Kutta method and trapezoidal rule were obtained for A of dimension 70×70. Thompson [30] proposed using

$$y_{n+1} = [I - \tfrac{1}{2}hA + \tfrac{1}{12}h^2A^2]^{-1}[I + \tfrac{1}{2}hA + \tfrac{1}{12}h^2A^2]y_n \quad (3.6-32)$$

which is simply the matrix equivalent of (3.6-19) of method (2.5-4). Actually (3.6-32) is $O(h^5)$ and involves less computation than (3.6-31) for the same order of accuracy and A-stability.

The semi-implicit Runge–Kutta methods share the advantages of high stability with the implicit Runge–Kutta methods but do not require iteration. Actually if A is interpreted as the Jacobian matrix of the ODE, then these methods correspond to a first-order linearization of the ODE combined with an implicit Runge–Kutta method.

3.7. STABILITY OF SINGLE-STEP METHODS EMPLOYING SECOND DERIVATIVES

Let us now consider the general class of single-step methods employing second derivatives,

$$y_{n+1} = h\beta_0 y'_{n+1} + h^2\gamma_0 y''_{n+1} + \alpha_1 y_n + h\beta_1 y_n' + h^2\gamma_1 y_n'' \quad (3.7-1)$$

An important point is that the order of accuracy of any A-stable method of the form (3.7-1) is not limited by Theorem 3.3.3 because (3.7-1) contains second derivatives. The parameters in (3.7-1) can be adjusted to achieve desired levels of accuracy and stability. In the cases to be considered $\alpha_1 = 1$, the remaining parameters β_0, γ_0, β_1, and γ_1 being available for adjustment.

Actually (3.7-1) is a special case of a more general class of single-step methods employing higher derivatives,

$$y_{n+1} = y_n - \sum_{i=1}^{r} h^i (\gamma_{0i} y_{n+1}^{[i]} + \gamma_{1i} y_n^{[i]}) \qquad (3.7-2)$$

In this section we will concentrate on the case of $r = 2$. Makinson [24] has considered methods of the class (3.7-2) up to $r = 3$. Obviously for $r = 1$, (3.7-2) reduces to the class of single-step methods discussed previously. We will first outline the general aspects of stability of such methods before citing the individual algorithms proposed.

The characteristic equation of (3.7-1) has the single root,

$$\mu_1 = \frac{1 + \beta_1 h\lambda + \gamma_1 h^2 \lambda^2}{1 - \beta_0 h\lambda - \gamma_0 h^2 \lambda^2} \qquad (3.7-3)$$

Most methods are designed such that the characteristic root matches the Taylor series expansion of $\exp(h\lambda_i)$, $i = 1, 2, \ldots, m$, as $h\lambda_i \to 0$ to an order p. For example, if we want μ_1 given by (3.7-3) to approximate $\exp(h\lambda)$ to order $p = 3$, we let

$$\frac{1 + \beta_1 h\lambda + \gamma_1 h^2 \lambda^2}{1 - \beta_0 h\lambda - \gamma_0 h^2 \lambda^2} = 1 + h\lambda + \tfrac{1}{2}h^2\lambda^2 + \tfrac{1}{6}h^3\lambda^3 \qquad (3.7-4)$$

from which we obtain the conditions

$$\beta_0 + \beta_1 = 1 \qquad (3.7-5)$$

$$\gamma_1 + \beta_0 + \gamma_0 = \tfrac{1}{2} \qquad (3.7-6)$$

$$\gamma_0 = \tfrac{1}{2}(\tfrac{1}{3} - \beta_0) \qquad (3.7-7)$$

wherein (3.7-3) now becomes

$$\mu = \frac{1 + (1 - \beta_0)h\lambda + (\tfrac{1}{3} - \tfrac{1}{2}\beta_0)h^2\lambda^2}{1 - \beta_0 h\lambda - (\tfrac{1}{6} - \tfrac{1}{2}\beta_0)h^2\lambda^2} \qquad (3.7-8)$$

Our next concern is the absolute stability of the methods (3.7-1). Let us now consider some specific algorithms which have been proposed. Lomax [23] has suggested setting $\gamma_1 = 0$, and making the method of order $p = 2$. In terms of β_0, this yields

$$y_{n+1} = y_n + h[\beta_0 y'_{n+1} + (1 - \beta_0)y_n'] + h^2(\tfrac{1}{2} - \beta_0)y''_{n+1} \qquad (3.7-9)$$

The characteristic root of (3.7-9) is a special form of (3.7-3),

$$\mu_1 = \frac{1 + (1 - \beta_0)h\lambda}{1 - \beta_0 h\lambda - (\frac{1}{2} - \beta_0)h^2\lambda^2} \tag{3.7-10}$$

β_0 can now be adjusted to affect both stability and accuracy. For example, if $\beta_0 = \frac{2}{3}$, we can readily see that (3.7-5)–(3.7-7) are satisfied making the method accurate to order $p = 3$. In this case μ_1 is shown in Figure 3.8 for

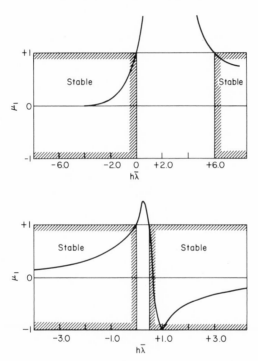

Figure 3.8. Characteristic roots for implicit method employing second derivatives [23].

real values of $h\lambda$. The method is seen to be A-stable and also stable for all positive real $h\lambda \geq 6$. We can observe another very interesting point. For $\beta_0 = \frac{2}{3}$, (3.7-10) becomes identically the $n = 2$, $m = 1$ Padé approximation for $p = 3$ with $n = 2$ and $m = 1$.

The next question to ask is can we adjust β_0 to increase the stability of (3.7-9) for positive real λ. Lomax shows that μ_1 intersects the $+1$ line at

$$h\lambda = 0, \qquad h\lambda = 2/(2\beta_0 - 1) \tag{3.7-11}$$

and the -1 line at

$$h\lambda = 1 \pm \sqrt{[1 - 4/(2\beta_0 - 1)]} \tag{3.7-12}$$

For $\frac{5}{2} \geq \beta_0 \geq \frac{1}{2}$, $|\mu_1| > 1$ only in the region $0 < h\bar{\lambda} < -2/(1 - 2\beta_0)$. The maximum real stability is achieved when $\beta_0 = \frac{5}{2}$. μ_1 for this case is also shown in the lower part of Figure 3.8.

Makinson [24] has considered (3.7-1) of order $p = 3$, yielding (3.7-8). He found that the best stability properties were obtained with $\beta_0 = 1 + \sqrt{3}/3$. However, in this case μ_1 becomes equal to (3.6-31), which Calahan obtained by considering the particular Rosenbrock implicit Runge–Kutta form. Thus, we have used a different basic approach, including second derivatives, and we have obtained the same characteristic root as in an implicit Runge–Kutta method. However, because Calahan's method is in an explicit form, it will be somewhat easier for computation than (3.7-1).

Ehle [9] has unified the ideas relating the methods employing second and higher derivatives and the implicit Runge–Kutta methods the roots of which are diagonal Padé approximants. In particular, the general class of single-step methods given by a special form of (3.7-2).

$$y_{n+1} = y_n + \sum_{i=1}^{r} \gamma_{ir} h^i [(-1)^{i+1} y_{n+1}^{[i]}] \tag{3.7-13}$$

are of order $2r$ when γ_{ir} is the ith coefficient in the numerator of the rth diagonal Padé approximation of $\exp(h\lambda)$. The characteristic root of (3.7-13) is the rth diagonal Padé approximation and is A-stable for all r.

3.8. STABILITY VERSUS ACCURACY

A necessary condition for numerical integration is stability, i.e., a stable value of h must be employed when a finite stability boundary exists. For most explicit methods, the maximum stable value of $|h\lambda|$ is the order of one to six. Nevertheless, there is no guarantee that all values of h from zero to the limiting value will yield accurate integration. For A-stable methods, the choice of h is governed solely by considerations of accuracy. We must face the question: Of what value is a method that is stable in a region where it is inaccurate? The relationship between stability and accuracy is thus fundamental to the choice for h for a particular method.

One indication of the effect of h on the accuracy of a method is by means of an estimate of the local truncation error, the difference between the principal root μ_{1i} and the exact solution component corresponding to that root, $\exp(h\lambda_i)$. This comparison can be visualized by letting $w_i = (\ln \mu_{1i})/h$ and producing the stability chart (see Figure 3.5) for the method. As noted previously, a method can be assumed to be accurate in the region where the hw_R-hw_I (U-V in Figure 3.5) grid matches the $h\lambda_R$-$h\lambda_I$ grid closely.

In the absence of a stability chart, Lomax [21] has suggested that the error in any method, as it is given by the lowest order, nonvanishing truncation terms, loses its significance when these terms exceed about one tenth. To rely on such error estimates, h should be chosen so that $|h\lambda| < 0.1$. Now, we have been continually stressing the desirability of maximizing the stability bounds of methods. However, we can now return to our original question of the value of a method that is stable in a region where it is inaccurate. For the integration of a scalar ODE there is no value, but for the integration of coupled ODE's there is a value.

Let us consider a hypothetical numerical method which has a real stability boundary of one, $|h\lambda_i| \leq 1$, $i = 1, 2, \ldots, m$. With this stability boundary, following the rule of one tenth, it is certainly *sufficient* that $|h\lambda_i| < 0.1$, $i = 1$, $2, \ldots, m$, will yield both a stable and accurate solution. However, this is overly restrictive in that *necessary* conditions are all that are required. The necessary conditions for both stability and accuracy are:

   1. $|h\lambda_i| < 0.1$ for only those i which are to be calculated to the specified accuracy.
   2. $|h\lambda_i| \leq 1$, $i = 1, 2, \ldots, m$.

Thus in some cases it is acceptable to calculate certain undesired solution components inaccurately without destroying the accuracy of the overall computation.

An easy way to visualize the situation is through an example provided by Lomax [21]. Consider the system

$$y_1' = -1.38y_1 - 0.81y_2, \qquad y_2' = -2.16y_1 - 1.92y_2 \qquad (3.8\text{-}1)$$
$$y_1(0) = -2.9995, \qquad\qquad y_2(0) = 4.0010$$

The solution of (3.8-1) is

$$y_1 = 0.0005e^{-3x} - 3e^{-0.3x}$$
$$y_2 = 0.001e^{-3x} + 4e^{-0.3x} \qquad (3.8\text{-}2)$$

Suppose we are not interested in solution components when they are ≤ 0.0005. The stability condition for this problem is $h \leq 0.333$, corresponding to condition 2. As far as the required accuracy is concerned, stability is all that is required for the solution component corresponding to $\lambda_1 = -3$, since it is smaller than the allowable error for $x > 0$ and, being stable, cannot grow. The accuracy of the solution component corresponding to $\lambda_2 = -0.3$ will be acceptable for $0.3h < 0.1$, or $h < 0.333$ also.

If the initial conditions for (3.8-1) are changed to

$$y_1(0) = 2, \qquad y_2(0) = 14$$

the solution is now

$$y_1 = 5e^{-3x} - 3e^{-0.3x}$$
$$y_2 = 10e^{-3x} + 4e^{-0.3x} \qquad (3.8\text{-}3)$$

To achieve the same accuracy as before, $h < 0.0333$, since the solution component corresponding to $\lambda_1 = -3$ is no longer negligible. The point is that considerable savings in computing time for the same level of accuracy can be achieved in cases in which certain solutions decay faster than others by choosing h on the basis of accuracy for the smaller eigenvalues. This subject will be discussed further in Chapter 6.

An alternative approach to accuracy has been taken by Benyon [1], who has introduced accuracy charts based on the steady state sinusoidal response of a linear ODE system. Benyon has pointed out that matching of grids on the stability charts is neither a necessary nor a sufficient condition for accuracy. Close agreement between the eigenvalues of the ODE and the dominant roots of the numerical method is not necessary for those eigenvalues representing resonant modes which, as a result of the particular initial conditions or forcing functions, are not strongly excited. However for those eigenvalues that are strongly excited an accurate dominant root is only one requirement. It is also necessary to consider the effect of both the extraneous roots that are not included in the stability chart and different frequencies present in forcing functions in the ODE.

In order to take into account the above factors, accuracy charts have been used. The accuracy charts are based on the following procedure. A linear ODE with constant coefficients and a linear forcing function $u(t)$ has a solution in the Laplace transform domain of the form $\bar{y}(s)/\bar{u}(s) = G(s)$. The difference equation for the numerical solution of this ODE has a solution in the z-transform domain of the form $\bar{y}(z)/\bar{u}(z) = G'(z)$ [25]. Substituting $s = i\Omega$ in the Laplace transfer function $G(s)$ gives the complex gain $G(i\Omega)$ of the given system. Substituting $z = e^{ih\Omega}$ in the z-transfer function $G'(z)$ gives the complex gain $G'(i\Omega)$ of the numerical method at angular-frequency Ω. The peak error, including both amplitude and phase errors in $G'(i\Omega)$, expressed as a fraction of the peak output is given by $|G'(i\Omega) - G(i\Omega)|/|G(i\Omega)|$. Accuracy charts show this error as a function of $h\Omega$ for various values of the parameters of the given transfer function. The interested reader may consult the paper of Benyon for further analysis. In the absence of forcing functions containing different frequencies, it is probably satisfactory to match $h\lambda_R$ and $h\lambda_I$ to hw_R and hw_I by choice of h to produce an accurate method.

3.9. PUBLISHED NUMERICAL RESULTS

Of direct interest to the material of this chapter are the results of Rosenbrock and Storey [26] on the system (3.4-9) with

$$\mathbf{A} = \begin{bmatrix} -1000 & 0 \\ 0.999 & -1 \end{bmatrix} \tag{3.9-1}$$

and $y(0) = [1, 0.999]$. The solution is

$$y_1 = e^{-1000x}, \qquad y_2 = -.001e^{-1000x} + e^{-x} \qquad (3.9\text{-}2)$$

These authors analyzed the use of Euler's method, the trapezoidal rule, certain Runge–Kutta and Rosenbrock forms on this system. The detail presented is sufficient for an excellent understanding of how the values of h must be constrained to prevent instability.

Distefano [8] has presented results of the use of a variety of numerical integration techniques on a multicomponent distillation simulation. He showed that by increasing h in small increments, a stability bound could be determined when the integration became unstable. The values of the stability limits obtained by numerical experiment agreed closely with those predicted from theory.

3.10. NUMERICAL EXPERIMENTS

In Section 2.15, we have developed many numerical results relating to the integration of the various test systems. However, no attempt was made to obtain an upper bound on the range of valid step size h. This was done purposely since the phenomenon of instability of single-step methods was not detailed until the present chapter.

Table 3.1 presents many of the real negative and imaginary bounds on $h\lambda$ for the explicit Runge–Kutta methods. Such bounds are important when, for example, a single-step method is used with the estimated truncation error monitored. As this error exceeds or drops below a fixed number of decimal places, the corresponding value of h is adjusted, decreased or increased by say $\frac{1}{2}$ or 2. However, care must now be taken that, as one doubles h, the valve of $h\lambda$ does not move into an unstable region rendering the computation worthless. On this basis it is important to have some a priori idea of the upper bound on h.

3.10.1. Preliminaries

In terms of the systems of Chapter 2 we first note that:

1. *System I.* Here the stability bounds for the various integration methods should be as given in Table 3.1 with $\lambda = -1$.

2. *System IV.* Here we may see an effect of the coupling of a system with $\lambda_1 = -1$ and $\lambda_2 = +1$. Even though the first eigenvalue is simple the presence of the second positive value may cause changes in the bounds of Table 3.1.

3. *System V.* Here the eigenvalues are $\lambda_{1,2} = \pm i$ and we must consider the imaginary bounds of Table 3.1.

4. *System* VI. This is the stiff system and the largest negative eigenvalue is $\lambda_3 = -120$. Thus the bounds on h of Table 3.1 will need to be decreased by a factor of $1/120$ (see also Chapter 6).

5. *System* VII. From Section 3.5 we can ascertain that we need to look at $f_{\bar{y}}$ to determine approximate stability bounds. In this case $f_{\bar{y}} = -2$ and the bounds of Table 3.1 need to be decreased by a factor of $\frac{1}{2}$.

3.10.2. Explicit Methods

We now turn to computational verification of the theoretical stability results already presented. In general, the procedure will be to increase h for a particular method and problem until numerical instability ensues. Results for solution of System I by the third-order Heun method and System V by the classic fourth-order Runge–Kutta method are shown in Table 3.3. The third-order Heun method is stable for $h = 1.0$ and becomes unstable at $h = 2.5$, whereas on System V the classic fourth-order Runge–Kutta method is stable

TABLE 3.3

STABILITY BOUNDS FOR SYSTEMS I AND V

System I (third-order Heun)			
$h = 1.0$		$h = 2.5$	
x	R	x	R
5.0	1.63	2.5	$-8.38_{10}{}^{-2}$
10.0	2.68	5.0	$7.02_{10}{}^{-3}$
25.0	12.9	7.5	$-5.89_{10}{}^{-4}$
(stable)		10.0	$4.93_{10}{}^{-5}$
		15.0	$3.47_{10}{}^{-7}$
		20.0	$-2.90_{10}{}^{-8}$
		50.0	$2.93_{10}{}^{-22}$
		(unstable)	

System V (fourth-order classic)		
$(\sin x - y_{1DP})/\sin x$		
x	$h = 1.0$	$h = 3.0$
6.0	$-0.74_{10}{}^{-1}$	2.34
12.0	$-0.24_{10}{}^{-1}$	-2.12
21.0	$-0.59_{10}{}^{-1}$	-16.48
30.0	0.15	45.64
50.0	-0.45	245.4
	(stable)	(unstable)

for $h = 1.0$ and unstable for $h = 3.0$. The stability limits could have been determined more precisely by increasing h in small increments until instability occurs. Often, however, when the stability bound is approached through increasing h the method begins to lose accuracy before the actual stability limit is reached. An inaccurate, though stable, solution may often result in oscillations which appear at first to be actual numerical instability. Actual instability, associated with too large a value of h, usually results in an increasing oscillation in the error or simply the growth of the error in one direction. The particular error behavior depends on the sign of the parasitic root responsible for the instability; a negative root causing oscillations and a positive root causing a uniform error growth. The results in Table 3.3 indicate oscillations when h exceeds the stability limit.

As another indication of the predictive utility of Table 3.1 in dealing with a simple system, we quote some results on System VII using Merson's embedding fourth-order formula (2.8-1) and Butcher's fifth-order formula (2.3-20). In both cases the theoretical negative real stability bound for $y' = -y$ is approximately 3.2 to 3.4. Thus for System VII, the analogous bounds should be about 1.6 to 1.7. Both Runge–Kutta methods were observed to be stable for $h = 1.0$ but unstable for $h = 2.0$. This is an excellent confirmation of the theory.

Turning now to System VI, the stiff system, we quote results using Butcher's fifth-order form (2.3-20) and Lawson's fifth-order form (2.3-30). Based upon $y' = -y$ these should have negative real stability bounds of 3.4 and 5.7 (see Table 3.1) respectively. However, when applied to System VII with its large eigenvalue of $\lambda_3 = -120$, we would then predict bounds of $h = 3.4/120 = 0.028$ and $h = 5.7/120 = 0.0475$. Table 3.4 shows computational data for both those cases. Lawson's method is excellent for $h = 0.025$, but Butcher's is beginning to show increasing errors. When $h = 0.04$ Butcher's method has degenerated completely whereas Lawson's is only beginning to show increasing errors. At $h = 0.1$ Lawson's method has become unstable. These results are completely within the framework of the theoretical bounds.

3.10.3. Implicit Methods

We next consider some of the implicit methods; in particular, we start with Rosenbrock's third-order method which we expect to be A-stable. Systems I, II, IV, VI, and VII have been used to test out this feature.

Table 3.5 presents certain results for Systems I, II, IV, and VII (compare to Tables 2.18 and 2.20) and Table 3.6 presents results for System VI (compare to Tables 2.19 and 3.4). Both Systems I and II yield errors at $h = 2.5$ such that the ratio is oscillating. In the case of System IV no such oscillation occurs, but the errors shown by the values of R_2 make the solution worthless.

TABLE 3.4

SYSTEM VI. DOUBLE PRECISION

	Fifth-order Lawson			Fifth-order Butcher		
x	$h = 0.025$ ε_3	$h = 0.04$ ε_3	$h = 0.1$ ε_3	$h = 0.02$ ε_3	$h = 0.025$ ε_3	$h = 0.04$ ε_3
0.1	$1.04_{10^{-4}}$	—	$-8.96_{10^{+2}}$	$-2.97_{10^{-4}}$	$-5.73_{10^{-2}}$	—
0.2	$1.87_{10^{-6}}$	$1.52_{10^{-3}}$	$-9.03_{10^{+5}}$	$-5.07_{10^{-7}}$	$-3.27_{10^{-3}}$	$-6.89_{10^{+4}}$
0.3	$1.87_{10^{-8}}$	—	$-7.20_{10^{+8}}$	$-4.41_{10^{-9}}$	$-1.87_{10^{-4}}$	—
0.4	$1.66_{10^{-10}}$	$-2.22_{10^{-6}}$	$-6.45_{10^{+11}}$	$-3.96_{10^{-11}}$	$-1.07_{10^{-5}}$	$-4.75_{10^{+9}}$
0.5	$1.39_{10^{-12}}$	—	$-5.78_{10^{+14}}$	$-3.34_{10^{-13}}$	$-6.13_{10^{-7}}$	—
0.6	$1.11_{10^{-14}}$	$3.32_{10^{-9}}$	$-5.18_{10^{+17}}$	$-2.71_{10^{-15}}$	$-3.51_{10^{-8}}$	$-3.28_{10^{+14}}$
0.7	$8.63_{10^{-17}}$	—	$-4.64_{10^{+20}}$	$-2.13_{10^{-17}}$	$-2.01_{10^{-9}}$	—
0.8	$6.57_{10^{-19}}$	$-4.96_{10^{-12}}$	$-4.16_{10^{+23}}$	$-1.65_{10^{-19}}$	$-1.15_{10^{-10}}$	$-2.26_{10^{+19}}$

TABLE 3.5

DOUBLE PRECISION. ROSENBROCK THIRD-ORDER METHOD

x	System I, $h=2.5$ ε	R	System II, $h=2.5$ ε	R	System IV, $h=2.5$ ε	R
2.5	$-1.415_{10}{}^{-1}$	-1.380	$-1.687_{10}{}^{+1}$	-2.597	$9.686_{10}{}^{-1}$	$7.812_{10}{}^{-2}$
5.0	$-3.200_{10}{}^{-3}$	1.904	$-1.264_{10}{}^{+2}$	6.748	$5.484_{10}{}^{-1}$	$1.213_{10}{}^{-2}$
7.5	$-7.635_{10}{}^{-4}$	-2.628	$-1.911_{10}{}^{+3}$	$-1.753_{10}{}^{+1}$	$2.889_{10}{}^{-1}$	$1.910_{10}{}^{-3}$
10.0	$-3.288_{10}{}^{-5}$	3.627	$-2.154_{10}{}^{+4}$	$4.553_{10}{}^{+1}$	$1.490_{10}{}^{-1}$	$3.044_{10}{}^{-4}$
12.5	$-4.471_{10}{}^{-6}$	-5.005	$-2.706_{10}{}^{+5}$	$-1.182_{10}{}^{+2}$	$7.586_{10}{}^{-2}$	$4.912_{10}{}^{-5}$
15.0	$-2.616_{10}{}^{-7}$	6.908	$-3.258_{10}{}^{+6}$	$3.073_{10}{}^{+2}$	$3.816_{10}{}^{-2}$	$8.015_{10}{}^{-6}$
17.5	$-2.774_{10}{}^{-8}$	-9.533	$-3.987_{10}{}^{+7}$	$-7.983_{10}{}^{+2}$	$1.900_{10}{}^{-2}$	$1.321_{10}{}^{-6}$
20.0	$-1.904_{10}{}^{-9}$	$1.315_{10}{}^{+1}$	$-4.849_{10}{}^{+8}$	$2.073_{10}{}^{+3}$	$9.373_{10}{}^{-3}$	$2.198_{10}{}^{-7}$
22.5	$-1.785_{10}{}^{-10}$	$-1.815_{10}{}^{+1}$	$-5.911_{10}{}^{+9}$	$-5.387_{10}{}^{+3}$	$4.586_{10}{}^{-3}$	$3.688_{10}{}^{-8}$
25.0	$-1.333_{10}{}^{-11}$	$2.505_{10}{}^{+1}$	$-7.199_{10}{}^{+10}$	$1.399_{10}{}^{+4}$	$2.228_{10}{}^{-3}$	$6.231_{10}{}^{-9}$

x	System VII, $h=2.5$ ε	R	System VII, $h=1.0$ ε	R
2.5	$8.465_{10}{}^{-1}$	$5.382_{10}{}^{-1}$	$1.957_{10}{}^{-1}$	$7.955_{10}{}^{-1}$
5.0	$-1.439_{10}{}^{-1}$	1.168	$3.430_{10}{}^{-2}$	$9.656_{10}{}^{-1}$
7.5	$5.543_{10}{}^{-2}$	$9.474_{10}{}^{-1}$	$4.878_{10}{}^{-3}$	$9.951_{10}{}^{-1}$
10.0	$-1.765_{10}{}^{-2}$	1.017	$6.680_{10}{}^{-4}$	$9.993_{10}{}^{-1}$
12.5	$6.014_{10}{}^{-3}$	$9.940_{10}{}^{-1}$	$9.068_{10}{}^{-5}$	$9.999_{10}{}^{-1}$
15.0	$-2.004_{10}{}^{-3}$	1.002	$1.228_{10}{}^{-5}$	$9.999_{10}{}^{-1}$
17.5	$6.727_{10}{}^{-4}$	$9.993_{10}{}^{-1}$	$1.662_{10}{}^{-6}$	$9.999_{10}{}^{-1}$

For System VII, we show both the case $h = 1.0$ and $h = 2.5$; the dramatic difference in the deviation is self-evident for these two cases. Note, however, that at $h = 2.5$ the error oscillates, but decreases instead of increasing. In other words, the calculated value oscillates around the true value, but still follows crudely the true value at $h = 2.5$. Since we know that Rosenbrock's method is A-stable, these phenomena which occur at a finite value of h must be due to inaccuracy in the calculation. This is extremely interesting since it confirms our earlier observation that even though a method is A-stable, accuracy considerations prevent the use of an arbitrarily large value of h.

TABLE 3.6

SYSTEM I. DOUBLE PRECISION BUTCHER IMPLICIT METHOD

x	$\varepsilon, h = 2.0$	$\varepsilon, h = 4.0$	$\varepsilon, h = 6.0$
2.0	$0.2001_{10}{}^{-2}$	—	—
4.0	$0.5378_{10}{}^{-3}$	$0.3792_{10}{}^{-1}$	—
6.0	$0.1683_{10}{}^{-3}$	—	0.1563
8.0	$0.1941_{10}{}^{-4}$	$-0.4900_{10}{}^{-4}$	—
10.0	$0.3259_{10}{}^{-5}$	—	—
12.0	$0.5255_{10}{}^{-6}$	$0.1368_{10}{}^{-4}$	$-0.2366_{10}{}^{-1}$
14.0	$0.8237_{10}{}^{-7}$	—	—
16.0	$0.1264_{10}{}^{-7}$	$-0.3527_{10}{}^{-7}$	—
18.0	$0.1911_{10}{}^{-8}$	—	$0.3641_{10}{}^{-2}$
20.0	$0.2853_{10}{}^{-9}$	$0.4959_{10}{}^{-8}$	—
22.0	$0.4217_{10}{}^{-10}$	—	—
24.0	$0.6182_{10}{}^{-11}$	$-0.1907_{10}{}^{-10}$	$-0.5602_{10}{}^{-3}$

Thus we see that in every case tested there does exist a finite h beyond which Rosenbrock's third-order method becomes so inaccurate that it cannot be used. In fact, the bounds noted are not much greater than for the explicit methods discussed previously.

Next we consider the Butcher fully implicit method using Systems I and V. Data in each case for $h = 2.0$, 4.0, and 6.0 are shown in Tables 3.6 and 3.7. For System I, the case $h = 2.0$ yields fine results; for $h = 4.0$, an oscillation occurs, but it is damped, and on an absolute basis the results are not excessively in error. For $h = 6.0$, the oscillations still occur, but the error is quite significant. Table 3.7 using System V behaves essentially the same. The errors tend to grow due to inaccuracies rather than instability.

Finally in this section we turn to a series of implicit methods which use higher-order derivatives. We use

a. Second-order formula, (3.7-9) with $\beta_0 = \frac{5}{2}$.

b. Third-order formula of Makinson, (3.7-8).

TABLE 3.7

SYSTEM V. SINGLE PRECISION BUTCHER IMPLICIT METHOD

x	$\varepsilon_2, h = 2.0$	$\varepsilon_2, h = 4.0$	$\varepsilon_2, h = 6.0$
2.0	$0.5183_{10^{-3}}$	—	—
4.0	$0.1302_{10^{-1}}$	0.2745	—
6.0	$-0.1783_{10^{-1}}$	—	0.4101
8.0	$-0.6429_{10^{-2}}$	-0.4310	—
10.0	$0.3662_{10^{-1}}$	—	—
12.0	$-0.2688_{10^{-1}}$	$0.3952_{10^{-1}}$	$0.3263_{10^{+1}}$
14.0	$-0.2550_{10^{-1}}$	—	—
16.0	$0.6037_{10^{-1}}$	0.9459	—
18.0	$-0.2354_{10^{-1}}$	—	$0.4986_{10^{+1}}$
20.0	$-0.5421_{10^{-1}}$	$-0.1969_{10^{+1}}$	—
30.0	$0.2569_{10^{-1}}$	—	$-0.1574_{10^{+2}}$
40.0	$0.5350_{10^{-1}}$	$-0.5832_{10^{+1}}$	—
50.0	-0.1590	—	—
60.0	0.2443	$-0.1091_{10^{+2}}$	$-0.2535_{10^{+3}}$
66.0	0.2537	—	$-0.1715_{10^{+3}}$

c. Fifth-order formula of Numerov, (1.9-10). This formula evaluates y_{n+1} in terms of y_n, y_{n-1}, and y''_{n+1}, y''_n, y''_{n-1}. As such it can be used directly for second-order differential equations which do not involve y' in the right-hand side.

In order to use (c) we have chosen System V to test these different integration forms. Table 3.8 presents some of the results obtained, where we can see the increased accuracy (as expected) as the order of the integration formulas go up. The case $h = 0.1$ is used since no inaccuracy or instability problems would be predicted at this value of the step size. The third-order

TABLE 3.8

SYSTEM V. DOUBLE PRECISION $h = 0.1$, ONLY ε_2 SHOWN

x	Second-order	Third-order	Fifth-order
0.5	$0.2985_{10^{-2}}$	$0.3658_{10^{-4}}$	$0.3942_{10^{-7}}$
1.0	$0.8507_{10^{-2}}$	$0.3975_{10^{-4}}$	$0.1443_{10^{-6}}$
1.5	$0.1341_{10^{-1}}$	$0.5103_{10^{-5}}$	$0.2602_{10^{-6}}$
2.0	$0.1433_{10^{-1}}$	$0.9143_{10^{-4}}$	$0.3184_{10^{-6}}$
3.0	$0.2414_{10^{-2}}$	$0.2674_{10^{-3}}$	$0.7464_{10^{-7}}$
4.0	$0.3207_{10^{-1}}$	$0.2024_{10^{-3}}$	$0.5355_{10^{-6}}$
5.0	$0.3896_{10^{-1}}$	$0.1722_{10^{-3}}$	$0.8499_{10^{-6}}$
6.0	$0.2116_{10^{-2}}$	$0.5269_{10^{-3}}$	$0.2975_{10^{-6}}$
7.0	$0.5179_{10^{-1}}$	$0.4232_{10^{-3}}$	$0.8170_{10^{-6}}$

method was then used on System V with $h = 0.01, 0.1, 0.5, 1.0, 2.0$, and 5.0. No instabilities were detected, however, the accuracy decreased as h increased. The following conclusions can be drawn:

1. The finite real stability bounds for explicit single step methods, as determined by numerical experiments, are in reasonable agreement with theoretical predictions.

2. Even though implicit integration methods may be A-stable, it is not possible to use an arbitrarily large h and obtain a valid solution, because, as h increases, the accuracy of the numerical solution decreases. The limiting range of h values necessary for accuracy in the A-stable methods are not significantly larger than the stability bounds for the explicit methods.

REFERENCES

1. Benyon, P. R., A Review of Numerical Methods for Digital Simulation, *Simulation*, **11**, 219, 1968.
2. Birkoff, G., and Varga, R. S., Discretization errors for well-set Cauchy problems I, *J. Math. and Phys.* **44**, 1 (1965).
3. Calahan, D. A., Numerical solution of linear systems with widely separated time constants, *Proc. IEEE (Letters)* **55**, 2016 (1967).
4. Dahlquist, G., Convergence and stability in the numerical integration of ordinary differential equations, *Math. Scand.* **4**, 33 (1956).
5. Dahlquist, G., Stability questions for some numerical methods for ordinary differential equations, *Proc. Symposia in Applied Mathematics* **15**, 147 (1963).
6. Dahlquist, G., A special stability problem for linear multistep methods, *BIT* **3**, 27 (1963).
7. Davison, E. J., A high order Crank–Nicholson technique for solving differential equations, *Comput. J.* **10**, 195 (1967).
8. Distefano, G. P., Stability of numerical integration techniques, *AIChE J.* **14**, 946 (1968).
9. Ehle, B. L., High order A-stable methods for the numerical solution of systems of D. E.'s, *BIT* **8**, 276 (1968).
10. Gray, H. J., Numerical methods in digital real-time simulation, *Quart. Appl. Math.* **12**, 133 (1954).
11. Gray, H. J., Propagation of truncation errors in the numerical solution of ordinary differential equations by repeated closures, *J. Assoc. Comput. Mach.* **1**, 5 (1955).
12. Gray, J. J., Digital computer solution of differential equations, Proc. Western Joint Computer Conf., AFIPS, **13**, 87 (1958).
13. Gurk, H. M. and M. Rubinoff, Numerical solution of differential equations, Proc. Eastern Joint Computer Conf., AFIPS, **6**, 58 (1954).
14. Gurk, H. M., The use of stability charts in the synthesis of numerical quadrature formulae, *Quart. Appl. Math.* **13**, 73 (1955).
15. Henrici, P., "Discrete Variable Methods in Ordinary Differential Equations." Wiley, New York, 1962.
16. Henrici, P., "Error Propagation for Difference Methods." Wiley, New York, 1963.

17. Henrici, P., "Elements of Numerical Analysis." Wiley, New York, 1964.
18. Hildebrand, F. B., "Introduction to Numerical Analysis." McGraw-Hill, New York, 1956.
19. Kelly, L. G., "Handbook of Numerical Methods and Applications," Chapter 22. Addison-Wesley, Reading, Massachusetts, 1967.
20. Lawson, J. D., An order five Runge–Kutta process with extended region of stability, *SIAM J. Numer. Anal.* **3**, 593 (1966).
21. Lomax, H., An operational unification of finite difference methods for the numerical integration of ordinary differential equations, *NASA Technical Report*, NASA TR R-262, (May, 1967).
22. Lomax, H., On the construction of highly stable, explicit, numerical methods for integrating coupled ordinary differential equations with parasitic eigenvalues, *NASA Technical Note*, NASA TN D-4547 (April, 1968).
23. Lomax, H., Stable implicit and explicit numerical methods for integrating quasi-linear differential equations with parasitic, stiff and parasitic-saddle eigenvalues, *NASA Technical Note*, NASA TN D-4703 (July, 1968).
24. Makinson, G. J., Stable high order implicit methods for the numerical solution of systems of differential equations, *Comput. J.* **11**, 305 (1968).
25. Ragazzini, J. R. and Franklin, G. F. "Sampled-Data Control Systems." McGraw-Hill, New York, 1958.
26. Rosenbrock, H. H. and Storey, C. "Computational Techniques for Chemical Engineers." Pergamon Press, New York, 1966.
27. Rubinoff, M., Digital computers for real time simulation, *J. Assoc. Comput. Mach.* **1**, 186 (1955).
28. Rubinoff, M., In "Computer Handbook," (Huskey, H. D. and Korn, G. P., eds.) pp. 49–54. McGraw-Hill, New York, 1962.
29. Stiefel, E., "Introduction to Numerical Mathematics." Academic Press, New York, 1963.
30. Thompson. W. E., Solution of linear differential equations, *Comput. J.*, **10**, 417 (1967).
31. Widlund, O. B., A note on unconditionally stable linear multistep methods, *BIT* **7**, 65 (1967).

4

Predictor-Corrector Methods

In the present chapter predictor–corrector (P–C) multistep methods for integrating ODEs will be examined. The basis of many of these methods lies in the linear k-step difference equation with constant coefficients:

$$y_{n+1} = \alpha_1 y_n + \alpha_2 y_{n-1} + \cdots + \alpha_k y_{n+1-k}$$
$$+ h[\beta_0 y'_{n+1} + \beta_1 y'_n + \cdots + \beta_k y'_{n+1-k}] \qquad (1.7\text{-}1)$$

We have already discussed the distinction between the predictor (P) equation with $\beta_0 = 0$ and corrector (C) equation with $\beta_0 \neq 0$. Here we merely introduce the combined P–C pair by stating that the predictor is used to obtain an approximate first value for y_{n+1}; this value of y_{n+1} is then used in the corrector, where an improved or more accurate value of y_{n+1} is obtained.

We shall consider such questions as the advantages and disadvantages of the multistep P–C methods when compared to the single-step methods, the step size to be used in a numerical calculation, the importance of the number of function evaluations, the truncation error, and the stability of various P–C modes of computation.

In general it will be implied that in any application of the P–C equations, the value of h is fixed for any particular step. Without this assumption of a fixed value of h in (1.7-1) the coefficients $\alpha_1, \ldots, \alpha_k, \beta_0, \beta_1, \ldots, \beta_k$ would be functions of h and much of the simplicity and ease of the constant coefficient P–C equations is lost. However, we have detailed in Section 1.7.5 some of the extensions required to generate the variable coefficient cases. Further, in Section 4.11 we shall illustrate the extension to including some (one or two) nonconstant step values of y or y'. In this latter case the methods are referred to as hybrid because they combine many of the features of the P–C and single-step algorithms.

Also we mention that (1.7-1) presupposes that only values of y and y' are used in the P–C formulas. In Section 4.10 we shall indicate the extensions required to include y'', y''', ...

4.1. A SIMPLE PREDICTOR–CORRECTOR SET

To illustrate many of the features of P–C methods, we choose the simple equations (1.6-31) and (1.6-20):

Nystrom midpoint formula

$$y_{n+1} = y_{n-1} + 2hy_n', \qquad T(x, h) = (h^3/3)y^{[3]}(\zeta) \qquad (4.1\text{-}1)$$

Modified Euler formula or trapezoidal rule

$$y_{n+1} = y_n + (h/2)[y_{n+1}' + y_n'], \qquad T(x, h) = -(h^3/12)y^{[3]}(\zeta) \qquad (4.1\text{-}2)$$

The Nystrom formula, an explicit open-end formula, is the predictor, and the implicit or closed-end trapezoidal rule is the corrector. Also observe that from the truncation errors, order P = order C and

$$|T(x, h)|_P > |T(x, h)|_C \qquad (4.1\text{-}3)$$

Finally, we note that the predictor is a two-step ($k = 2$) formula whereas the corrector is a one-step ($k = 1$) formula.

To illustrate the P–C procedure, we assume that by some means we have already calculated $y_1, y_2, \ldots, y_n$ and our next step is to calculate y_{n+1}. We first calculate the predicted value by means of (4.1-1), calling the result $\bar{y}_{n+1}$; then we evaluate $\bar{y}_{n+1}' = f(x_{n+1}, \bar{y}_{n+1})$. Next we use (4.1-2) to compute a corrected value according to

$$y_{n+1} = y_n + (h/2)[\bar{y}_{n+1}' + y_n'] \qquad (4.1\text{-}4)$$

At this point we have two alternatives. First, we may use y_{n+1} from (4.1-4) as the final value for y_{n+1}, reevaluate y_{n+1}', and proceed to the calculation of y_{n+2}. P–C methods used in this way are conveniently referred to as PECE methods, indicating that a *predicted* value of y_{n+1} is followed by a derivative *evaluation*, $\bar{y}_{n+1}'$, and then y_{n+1} is *corrected* and y_{n+1}' *evaluated*. However, the corrected value of y_{n+1} may not be acceptable in the sense that the equality required on both sides of (4.1-4) is not satisfied. In other words we really want that value of y_{n+1}, designated by y^*_{n+1}, such that

$$y^*_{n+1} = y_n + (h/2)[y^{*\prime}_{n+1} + y_n']$$

is satisfied to as many digits as desired. The second alternative is to improve y_{n+1} in the direction of y^*_{n+1} by iterating (4.1-4), by substituting the y_{n+1} just calculated into the right-hand side of (4.1-4), and calculating an improved value of y_{n+1}. This is continued until convergence occurs to as many digits as desired.

We may summarize this procedure in the following manner:

$$
\begin{aligned}
y^{(0)}_{n+1} &= y_{n-1} + 2hf_n \\
f^{(0)}_{n+1} &= f(x_{n+1}, y^{(0)}_{n+1}) \\
y^{(s+1)}_{n+1} &= y_n + (h/2)[f^{(s)}_{n+1} + f_n] \\
f^{(s+1)}_{n+1} &= f(x_{n+1}, y^{(s+1)}_{n+1}), \qquad s = 0, 1, 2, \dots
\end{aligned}
\tag{4.1-5}
$$

The use of a P–C method in this way can be conveniently described as a $PE(CE)^s$ method, where the predicted $\bar{y}_{n+1}$ and evaluated f_{n+1} is followed by s corrections and derivative evaluations.

We might add that there is a third class of methods, called PEC methods, in which the final derivative evaluation after the correction is omitted, and in which the f_{n+1} value using the predicted y_{n+1} is used in the next cycle of computation. Iterative use of such a method is denoted $P(EC)^s$. Such methods save one derivative evaluation per step and will be discussed in Section 4.8.

Let us now discuss a few further points of interest in this P–C scheme.

1. The predictor equation serves as a means of obtaining a good value of y_{n+1} to be used in the corrector equation. This estimate is obtained at very little computing expense since y_{n-1} and y_n' have been calculated previously. Since the local truncation error is lower in the corrector equation, we do not stop at the predictor, but go on to the use of the corrector.

2. The convergence of the iterations in the corrector is an important feature of the computation. In Section 4.3 we shall analyze the necessary conditions for this convergence; as expected it will be a function of h since this factor indicates how far away from y_n we are stepping to obtain y_{n+1}.

3. The number of function evaluations (the number of times we calculate f_{n+1}) is determined by the number of iterations. If we do not iterate, but rather keep the first calculated y_{n+1}, then two function evaluations are needed per calculation. These are $f(x_{n+1}, \bar{y}_{n+1})$ corresponding to the predicted value and the final value $f(x_{n+1}, y_{n+1})$ corresponding to actually completing the calculation. Thus as a minimum the P–C method of (4.1-1) and (4.1-2) requires at least two function evaluations per step forward in x.

4. In order to start the calculation, we need y_0 and y_1. We will have the

initial condition y_0 at x_0, but we need to use some auxiliary method to obtain y_1.

4.2. A MODIFIED PREDICTOR–CORRECTOR SET

The P–C set given by (4.1-1) and (4.1-2) is such that each equation has a local truncation error of the same order. Hamming [19] has used this feature to expand the method of calculation. The local truncation errors for the predictor and corrector are

$$T(x, h)_P = T_{n+1, P} = (h^3/3)y^{[3]}(\zeta_P), \qquad x_{n-1} < \zeta_P < x_{n+1} \qquad (4.2-1)$$

$$T(x, h)_C = T_{n+1, C} = (h^3/12)y^{[3]}(\zeta_C), \qquad x_n < \zeta_C < x_{n+1} \qquad (4.2-2)$$

It follows that the exact value of y at x_{n+1}, $y(x_{n+1})$, is then given by

$$
\begin{aligned}
y(x_{n+1}) &= \bar{y}_{n+1} + (h^3/3)y^{[3]}(\zeta_P) \\
y(x_{n+1}) &= y_{n+1} - (h^3/12)y^{[3]}(\zeta_C)
\end{aligned}
\qquad (4.2-3)
$$

From the equalities in (4.2-3) we obtain

$$\bar{y}_{n+1} + (h^3/3)y^{[3]}(\zeta_P) = y_{n+1} - (h^3/12)y^{[3]}(\zeta_C)$$

or

$$y_{n+1} - \bar{y}_{n+1} = (h^3/3)y^{[3]}(\zeta_P) + (h^3/12)y^{[3]}(\zeta_C) \qquad (4.2-4)$$

Now assuming that

$$y^{[3]}(\zeta_P) \simeq y^{[3]}(\zeta_C) = y^{[3]}(\zeta)$$

over the interval of interest (i.e., the third derivative does not vary greatly over the interval), (4.2-4) becomes

$$y_{n+1} - \bar{y}_{n+1} = (5h^3/12)y^{[3]}(\zeta), \qquad x_{n-1} < \zeta < x_{n+1} \qquad (4.2-5)$$

Even at this point in the development we can identify some interesting features. First we note that (4.2-5) can be obtained *only* in the case where the order of the predictor equation is equal to the order of the corrector equation. Second we now have a measure of the local truncation error in terms of y_{n+1} and $\bar{y}_{n+1}$ which are automatically calculated in the P–C algorithm. This may be seen more explicitly by comparing (4.2-1) and (4.2-2) with (4.2-5) to yield

$$
\begin{aligned}
y_{n+1} - \bar{y}_{n+1} &= \tfrac{5}{12}h^3 y^{[3]}(\zeta) \\
&= \tfrac{5}{4}T_{n+1, P} \\
&= -5T_{n+1, C}
\end{aligned}
\qquad (4.2-6)
$$

Obviously we may use $y_{n+1} - \bar{y}_{n+1}$ to monitor the truncation error in the calculation. (In Section 4.13 we indicate how to adjust h on the basis of this information.) Finally, we see that in P–C methods using equal-order formulas, the local truncation error can be estimated in an almost trivial fashion as compared to the methods required in the single-step methods (Section 2.8–2.10). This point alone makes the present approach most attractive.

Since we may now estimate the local truncation error, the next question is how can we use this information to improve the predicted and corrected values. Considering the predicted value first, we see that from (4.2-6)

$$T_{n+1, P} = \tfrac{4}{5}(y_{n+1} - \bar{y}_{n+1}) \tag{4.2-7}$$

If we assume that the difference between the corrected and predicted values does not change violently from step to step or that the local truncation error remains approximately constant over two steps), then we may write

$$T_{n+1, P} = \tfrac{4}{5}(y_{n+1} - \bar{y}_{n+1}) = \tfrac{4}{5}(y_n - \bar{y}_n) \tag{4.2-8}$$

The predicted value can then be improved or modified by adding the term $\tfrac{4}{5}(y_n - \bar{y}_n)$ to $\bar{y}_{n+1}$ using only information calculated previously.

The corrector can be modified in essentially the same manner since

$$T_{n+1, C} = -\tfrac{1}{5}(y_{n+1} - \bar{y}_{n+1}) \tag{4.2-9}$$

can be added to y_{n+1} to improve this value.

On this basis we may now write the overall P–C calculation in the following steps:

$$
\begin{aligned}
p_{n+1} &= y_{n-1} + 2hf_n & &\text{Predict} \\
m_{n+1} &= p_{n+1} - \tfrac{4}{5}(p_n - c_n) & &\text{Modify} \\
m'_{n+1} &= f(x_{n+1}, m_{n+1}) & &\text{Reevaluate} \\
& & &\text{derivative} \\
c_{n+1} &= y_n + (h/2)[m'_{n+1} + f_n] & &\text{Correct} \\
y_{n+1} &= c_{n+1} + \tfrac{1}{5}(p_{n+1} - c_{n+1}) & &\text{Modify}
\end{aligned}
\tag{4.2-10}
$$

To simplify the presentation the symbols m_{n+1}, p_{n+1}, and c_{n+1} have been used.

The sequence given by (4.2-10) represents one of the most successful means to apply the P–C formulas. It rests on the fact that the original P–C equations had equal-order truncation errors and neglects round-off error. It also differs from the P–C analysis previously given in that there is no iteration in the corrector formula; rather a single application of the corrector formula is used and this value is then modified. As a result, there are two function evaluations per integration step, assuming a final reevaluation of f_{n+1} is required.

Actually there does exist some question as to whether (4.2-10) is the best manner to proceed. Some authors have suggested that the final step in which the corrector value is modified be left out and instead the corrector be iterated or an alternative path be used. The reasoning here is that while we started with formulas of order 2 [or errors of $O(h^3)$] the inclusion of the final step actually leads to an overall procedure of order 3 [or error of $O(h^4)$]. This can be seen by noting that the final step is actually correct for $y = x^3$ [write the formula in $c_{n+1} = y(x_{n+1}) - y_{n+1}$ terms and substitute the truncation error explicitly obtained with $y = x^3$]. Further, the final step may affect the stability of the corrector itself. Thus an alternative path might be to start with a higher-order formula at the beginning and perhaps iterate the corrector in this new sequence. Both the procedures given by (4.2-10) and the higher-order approach seem to have many advocates.

In summary, (4.2-10) represents one of the most feasible approaches (subject to the thoughts above) for handling the P–C algorithm. As such it represents a considerable improvement over that given in Section 4.1 and it can, of course, be extended directly to other equal order P–C formulas. The extension to m first-order equations follows in a direct manner and will not be detailed here.

4.3. CONVERGENCE OF ITERATIONS IN THE CORRECTOR

Since it may be necessary or desirable to iterate for y_{n+1} with the corrector equation, it is important to consider the conditions for convergence of different iterative schemes. Actually the problem of determining y_{n+1} from the corrector equation is the same as that in solving the general implicit form of (1.7-1), namely the solution of a set of m nonlinear algebraic equations. Let us consider the convergence properties of four techniques for the solution of the nonlinear algebraic equations arising in implicit methods for ODEs.

In particular, we want to solve the implicit equation

$$\mathbf{y}_{n+1} - h\beta_0 \mathbf{f}(x_{n+1}, \mathbf{y}_{n+1}) - \sum_{i=1}^{k} (\alpha_i \mathbf{y}_{n+1-i} + h\beta_i \mathbf{f}_{n+1-i}) = 0 \qquad (4.3\text{-}1)$$

Since the last term on the LHS is known, let us denote it by $\mathbf{w}_n$. Then (4.3-1) can be written

$$\mathbf{y}_{n+1} - h\beta_0 \mathbf{f}(x_{n+1}, \mathbf{y}_{n+1}) - \mathbf{w}_n = 0 \qquad (4.3\text{-}2)$$

from which we desire to determine $\mathbf{y}_{n+1}$. Let us consider four common ways to solve (4.3-2): Jacobi iteration, accelerated iteration, Newton–Raphson iteration, and backward iteration.

4.3.1. Jacobi Iteration

A solution of (4.3-2) by repeated substitutions

$$\mathbf{y}_{n+1}^{(s+1)} - h\beta_0 \mathbf{f}(x_{n+1}, \mathbf{y}_{n+1}^{(s)}) - \mathbf{w}_n = 0 \qquad (4.3\text{-}3)$$

is termed a Jacobi iteration [31]. Let us call y_{n+1}^* the theoeretical exact solution of (4.3-2) such that

$$y_{n+1}^* - h\beta_0 \mathbf{f}(x_{n+1}, y_{n+1}^*) - \mathbf{w}_n = 0 \qquad (4.3\text{-}4)$$

is satisfied exactly except possibly for round-off error. Subtracting (4.3-4) from (4.3-3) gives

$$y_{n+1}^{(s+1)} - y_{n+1}^* = h\beta_0(\mathbf{f}_{n+1}^{(s)} - \mathbf{f}_{n+1}^*) \qquad (4.3\text{-}5)$$

Using the mean-value theorem,

$$y_{n+1}^{(s+1)} - y_{n+1}^* = h\beta_0[\mathbf{f_y}]_{\bar{\mathbf{y}}}(y_{n+1}^{(s)} - y_{n+1}^*) \qquad (4.3\text{-}6)$$

where $y_{n+1}^* \leq \bar{\mathbf{y}} \leq y_{n+1}^{(s)}$. If we now assume a Lipschitz bound on $\mathbf{f_y}$, $\|\mathbf{f_y}\| < L$, then (4.3-6) becomes

$$\|y_{n+1}^{(s+1)} - y_{n+1}^*\| \leq h\beta_0 L \|y_{n+1}^{(s)} - y_{n+1}^*\|$$

By induction it follows that

$$\|y_{n+1}^{(s+1)} - y_{n+1}^*\| \leq (h\beta_0 L)^{s+1} \|y_{n+1}^{(0)} - y_{n+1}^*\| \qquad (4.3\text{-}7)$$

A necessary and sufficient condition for convergence of the iterations is then

$$|h\beta_0 L| < 1 \qquad (4.3\text{-}8)$$

For an L_2 norm on $\mathbf{f_y}$, L becomes the largest eigenvalue of $\mathbf{f_y}$, in which case the condition for convergence is

$$h\beta_0 |\lambda_{\max}| < 1 \qquad (4.3\text{-}9)$$

If (4.3-9) is satisfied, the Jacobi iteration will converge to a unique solution in a region encompassing a large enough neighborhood of the solution. Since this bound depends on h, for a small enough h (4.3-9) is satisfied. However, if $|\lambda_{\max}|$ is large, h may have to be so small that an excessive amount of time would be required to integrate over a reasonable range of x. This is precisely the problem faced in stiff equations (Chapter 6).

We would like $h\beta_0 |\lambda_{\max}|$ to be much smaller than one for rapid convergence, e.g. $h\beta_0 |\lambda_{\max}| \cong 0.1$. The condition (4.3-8) can be applied to the methods we have already studied. For example, for the Adams–Moulton formulas of Section 1.6.1, Table 1.2, we may write down immediately:

Adams–Moulton forms

$$q = 1, \qquad hL < 2$$
$$q = 2, \qquad hL < \tfrac{12}{5}$$
$$q = 3, \qquad hL < \tfrac{8}{3}$$
$$q = 4, \qquad hL < \tfrac{720}{251}$$

If we use (1.8-7)

$$y_{n+1} = y_{n-1} + (h/3)[f_{n+1} + 4f_n + f_{n-1}]$$

as a corrector, the condition for convergence would be

$$hL < 3 \qquad (4.3\text{-}10)$$

In order to ascertain when to terminate the iterations note that for $m = 1$,

$$|y^*_{n+1} - y^{(s)}_{n+1}| \le |y^*_{n+1} - y^{(s+1)}_{n+1}| + |y^{(s+1)}_{n+1} - y^{(s)}_{n+1}| \qquad (4.3\text{-}11)$$

from which we obtain

$$(1 - h\beta_0 L)|y^*_{n+1} - y^{(s)}_{n+1}| \le h\beta_0 |f^{(s)}_{n+1} - f^{(s+1)}_{n+1}| \qquad (4.3\text{-}12)$$

If $h\beta_0 L \ll 1$, the right-hand side of (4.3-12) is then an estimate of the difference between the $(s + 1)$ and the exact value of y_{n+1}.

4.3.2. Accelerated Iteration

A modification of Jacobi iteration (4.3-3) is

$$(1 + \alpha)y^{(s+1)}_{n+1} - h\beta_0 f(x_{n+1}, y^{(s)}_{n+1}) - w_n - \alpha y^{(s)}_{n+1} = 0 \qquad (4.3\text{-}13)$$

in which α is an acceleration parameter. If $\alpha = 0$, then (4.3-13) reduces to Jacobi iteration, (4.3-3). Conditions for convergence of (4.3-13) can be developed as for Jacobi iteration. The exact solution of (4.3-13) is given by

$$(1 + \alpha)y^*_{n+1} - h\beta_0 f(x_{n+1}, y^*_{n+1}) - w_n - \alpha y^*_{n+1} = 0 \qquad (4.3\text{-}14)$$

Subtracting (4.3-14) from (4.3-13) and using the mean-value theorem as in (4.3-6), we obtain

$$y^{(s+1)}_{n+1} - y^*_{n+1} = \left(\frac{h\beta_0[f_y]_{\bar{y}} + \alpha I}{1 + \alpha}\right)(y^{(s)}_{n+1} - y^*_{n+1}) \qquad (4.3\text{-}15)$$

The condition for convergence of (4.3-15) is clearly

$$\left\| \frac{h\beta_0 f_y + \alpha I}{1 + \alpha} \right\| < 1 \qquad (4.3\text{-}16)$$

or

$$\left| \frac{h\beta_0 \lambda_{max} + \alpha}{1 + \alpha} \right| < 1 \qquad (4.3\text{-}17)$$

which reduces to (4.3-9) if $\alpha = 0$.

Since the speed of convergence will be governed by the magnitude of the left-hand side of (4.3-17), as discussed in Jacobi iteration, we see that the speed of convergence can be increased over Jacobi iteration by increasing α from zero.

4.3.3. Newton–Raphson Iteration

A well-known method for determining the roots of coupled nonlinear algebraic equations is Newton–Raphson iteration. This method is based on a linearization of f_{n+1} about the previous value of y_{n+1} in the iteration, and is given by

$$y_{n+1}^{(s+1)} = y_{n+1}^{(s)} + [I - h\beta_0 A_{n+1}^{(s)}]^{-1}[h\beta_0 f_{n+1}^{(s)} - y_{n+1}^{(s)} + w_n] \qquad (4.3\text{-}18)$$

where $A_{n+1}^{(s)}$ is the Jacobian matrix f_y evaluated at $y_{n+1}^{(s)}$. Obviously, one application of (4.3-18) corresponds to solution of the linearized form of (4.3-2). The same procedure used in Section 4.3.1 and 4.3.2 to determine conditions for convergence for (4.3-3) and (4.3-13) can also be used with (4.3-18). Although we will not carry out the steps here, the necessary condition for convergence of (4.3-18) is

$$\|[I - h\beta_0 A_{n+1}^{(s)}]^{-1}\| \, \|(\partial/\partial y)[I - h\beta_0 A_{n+1}^{(s)}]\| \, \|y_{n+1}^{(s+1)} - y_{n+1}^{(s)}\| \leq 1 \qquad (4.3\text{-}19)$$

As an example of the application of (4.3-9) to an actual problem consider the ODE system of (6.1-2) for which

$$\|[I - h\beta_0 A_{n+1}^{(s)}]^{-1}\| = 1$$
$$\|(\partial/\partial y)[I - h\beta_0 A_{n+1}^{(s)}]\| \cong 6_{10^7} h\beta_0$$

The condition (4.3-19) becomes

$$h < 10^{-8}\beta_0^{-1} \|y_{n+1}^{(s+1)} - y_{n+1}^{(s)}\| \qquad (4.3\text{-}20)$$

which, unfortunately, is highly conservative. Actual calculations on (4.1-2) presented in Chapter 6 confirm the convergence of Newton–Raphson iteration with h as large as 0.1. More improved convergence conditions for Newton–Raphson iterations are a topic of current study.

In actual use of (4.3-18) it is impractical to recompute A_{n+1} for each iteration. It is usually acceptable to approximate $A_{n+1}^{(s)}$ by $A_{n+1}^{(0)}$, evaluated at the previous step value y_n. y_{n+1} and y_n are normally close enough so that this procedure is valid; however, if too many iterations are required in a given step, either of two alternatives can be followed. First, h can be reduced and the iteration restarted or, second, the matrix A_{n+1} can be reevaluated during the iteration.

4.3.4. Backward Iteration

Up to this point our discussion of iterative methods has centered on forward-type iteration in which we calculate from the right to the left side of the corrector. Alternatively, we can formulate a backward iteration of the form

$$y_{n+1}^{(s)} = h\beta_0 f(x_{n+1}, y_{n+1}^{(s+1)}) + w_n \qquad (4.3\text{-}21)$$

which still requires the solution of nonlinear implicit equations. Proceeding in the usual fashion,

$$\begin{aligned} y_{n+1}^{(s)} - y_{n+1}^* &= h\beta_0[f_{n+1}^{(s+1)} - f_{n+1}^*] \\ &= h\beta_0[f_y]_{\bar{y}}[y_{n+1}^{(s+1)} - y_{n+1}^*] \end{aligned} \qquad (4.3\text{-}22)$$

The condition for convergence of (4.3-21) is then

$$(1/h\beta_0)\|[f_y]^{-1}\| < 1 \qquad (4.3\text{-}23)$$

or, assuming a Lipschitz bound on f_y,

$$(1/h\beta_0)|\lambda_{\min}|^{-1} < 1 \qquad (4.3\text{-}24)$$

Thus, there is a *lower* bound on h for backward iteration rather than an upper bound as in the other methods. A problem will arise, however, if the Jacobian matrix f_y is singular.

4.3.5. Summary

The question of interest is if we decide to iterate a corrector equation or if we are using an implicit method, which technique should be used. Jacobi iteration and accelerated iteration are computationally easy to implement, but have convergence requirements depending on the largest eigenvalue of the Jacobian matrix of the ODE. Thus, if $|\lambda_{\max}|$ is large, an extremely small h is necessary for convergence in these methods. While necessary convergence conditions for Newton–Raphson iteration can be derived, they are extremely conservative. In general, Newton–Raphson iteration has a larger region of convergence than the previous two methods. Backward iteration has a very large region of convergence because of a lower bound on h rather than an upper bound. However, implicit equations still have to be solved in (4.3-21).

Thus, we make the following recommendations for the solution of an implicit multistep equation:

1. If the ratio of the largest to the smallest eigenvalue of A is small, say the order of 10, then Jacobi iteration or accelerated iteration should be used.

2. If the ratio of the largest to the smallest eigenvalue of A is large, say greater than the order of 10, then Newton–Raphson iteration or backward iteration should be used with h selected on the basis of the number of iterations desired per step.

4.4. ACCURACY AND STABILITY FOR SOME SIMPLE
PREDICTOR–CORRECTOR METHODS

Let us analyze the accuracy and stability of two simple P–C sets to illustrate the elements of P–C methods. Consider first the P–C set consisting of an Euler predictor followed by a trapezoidal rule (modified Euler) corrector

$$\bar{y}_{n+1} = y_n + hy_n' \tag{4.4-1}$$

$$y_{n+1} = y_n + (h/2)[\bar{y}_{n+1} + y_n'] \tag{4.4-2}$$

which, by the way, is also the second-order Runge–Kutta formula *if the corrector is used only once.*

The second set to be considered is the P–C pair of (4.1-1) and (4.1-2), the Nystrom predictor (the midpoint rule) and the trapezoidal rule (modified Euler) corrector.

4.4.1. Truncation Error Considerations

First, consider the application of (4.4-1) and (4.4-2) to $y' = \lambda y$ with the corrector used only once. Substituting the predictor into the corrector we obtain

$$y_{n+1} = \left(1 + h\lambda + \frac{(h\lambda)^2}{2}\right)y_n \tag{4.4-3}$$

The single characteristic root is

$$\mu_1 = 1 + h\lambda + \frac{(h\lambda)^2}{2} \tag{4.4-4}$$

Now let us apply the corrector (4.4-2) twice. We obtain

$$
\begin{aligned}
y_{n+1} &= y_n + (h/2)\left[\lambda\left(1 + h\lambda + \frac{h^2\lambda^2}{2}\right)y_n + \lambda y_n\right] \\
&= \left[1 + h\lambda + \frac{h^2\lambda^2}{2} + \frac{h^3\lambda^3}{4}\right]y_n
\end{aligned}
\tag{4.4-5}
$$

By continuing this process we obtain,

$$\mu_1 = 1 + h\lambda \qquad\qquad\qquad\qquad\qquad \text{0 corrector}$$

$$\mu_1 = 1 + h\lambda + \frac{h^2\lambda^2}{2} \qquad\qquad\qquad \text{1 corrector}$$

$$\mu_1 = 1 + h\lambda + \frac{h^2\lambda^2}{2} + \frac{h^3\lambda^3}{4} \qquad\qquad \text{2 correctors} \tag{4.4-6}$$

$$\mu_1 = 1 + h\lambda + \frac{h^2\lambda^2}{2} + \frac{h^3\lambda^3}{4} + \frac{h^4\lambda^4}{8} \qquad \text{3 correctors}$$

as evolving for the single root in this P–C process. However, the true solution of $y' = \lambda y$ is given by

$$e^{h\lambda} = 1 + h\lambda + \frac{h^2\lambda^2}{2} + \frac{h^3\lambda^3}{6} + \frac{h^4\lambda^4}{24} + \cdots \qquad (4.4\text{-}7)$$

and thus the error, given by $\mu_1 - e^{h\lambda}$, is

$-h^2\lambda^2/2$	0 corrector
$-h^3\lambda^3/6$	1 corrector
$h^3\lambda^3/12$	2 correctors
$h^3\lambda^3/12$	3 correctors

Higher-order terms have been omitted. We thus see that the use of the corrector more than twice is wasted since the minimum truncation error has achieved at this point. Experience in many other cases suggests that if iteration of the corrector is used, two applications of the corrector is probably optimum in terms of computer time and results obtained.

We can now analyze the behavior of the predictor and one application of the corrector in the (4.1-1) and (4.1-2) P–C set. This is done exactly as in the Euler predictor–trapezoidal rule corrector combination, i.e.,

$$\bar{y}_{n+1} = y_{n-1} + 2h\lambda y_n$$
$$y_{n+1} = y_n + (h/2)[\lambda\bar{y}_{n+1} + \lambda y_n]$$

or

$$y_{n+1} = (1 + (h\lambda/2) + h^2\lambda^2)y_n + (h\lambda/2)y_{n-1}$$

By solving for the characteristic roots μ_1 and μ_2 and expanding the square roots which result we find (ignoring higher-order terms)

Error	Parasitic root	
$-h^3\lambda^3/6$	$-1 + h\lambda$	0 corrector
$h^3\lambda^3/12$	$-h\lambda/2$	1 corrector

A further application of the corrector yields, in a direct fashion,

$h^3\lambda^3/12$	$-h\lambda/2$	2 correctors

Now we observe that in the Nystrom–trapezoidal rule P–C method only one application (iteration) of the corrector is needed to achieve the minimum truncation error. No further iterations would be useful in a computation.

4.4.2. Stability Considerations

At this point we can state the following very important results:

i. If the corrector in a P–C method is not iterated to convergence, then the stability of the P–C method depends on both the predictor and the corrector equations.

ii.   Even if the corrector in a P–C method is iterated toward convergence, it is not necessarily true that the stability approaches that of the corrector alone.

An important result which we will not prove here is that when used to integrate coupled ODEs, a P–C set based on the general linear multistep formula (1.7-1) yields a characteristic difference equation, the roots of which depend only on the eigenvalues of the system, a result which, of course, has already been shown to hold for single multistep methods.

Let us now analyze the stability characteristics of the two simple P–C sets we are considering. First, for the set (4.4-1) and (4.4-2) we already know that the predictor by itself has a finite range of real stability, that the corrector by itself is A-stable, and that when the PECE mode is used, (4.4-4), the real stability bound becomes that of the predictor, $-2 < h\lambda < 0$.

We can also analyze the P–C stability for any number of iterations of the corrector using the roots given in (4.4-6). Letting s be the number of times the corrector is used, (4.4-6) leads to

$$-2 < h\lambda < 0, \qquad \text{for} \quad s = 0, 1, 2, 3 \qquad (4.4\text{-}8)$$

In other words, the real stability region remains that of the predictor as the corrector is iterated.

We may also proceed further by denoting the sequence of P–C steps as

$$y_{n+1}^{(0)} = y_n + hf_n$$
$$y_{n+1}^{(1)} = y_n + (h/2)[f_{n+1}^{(0)} + f_n]$$
$$y_{n+1}^{(2)} = y_n + (h/2)[f_{n+1}^{(1)} + f_n] \qquad (4.4\text{-}9)$$
$$\vdots$$
$$y_{n+1}^{(s+1)} = y_n + (h/2)[f_{n+1}^{(s)} + f_n]$$

The root of the characteristic equation for s iterations is

$$\mu_1 = \frac{1 + h\lambda/2 - 2(h\lambda/2)^{s+2}}{1 - h\lambda/2} \qquad (4.4\text{-}10)$$

However, as we have seen, if the corrector is iterated to convergence by itself

$$\mu_1 = \frac{1 + h\lambda/2}{1 - h\lambda/2} \qquad (4.4\text{-}11)$$

In order that (4.4-10) converge to (4.4-11) as $s \to \infty$, it is necessary that $|h\lambda| < 2$. This is exactly the condition in (4.3-7) for convergence of the corrector. Thus, for convergence to be assumed it is necessary to iterate until $y_{n+1}^{(s+1)}$ and $y_{n+1}^{(s)}$ differ at most by the truncation error.

Let us now analyze the stability of the other P–C set. The predictor equation alone has been thoroughly studied in Section 3.4, where we found the characteristic roots given by (3.2-28) and (3.2-29) which were expanded to give,

$$\mu_1 = 1 + h\lambda + h^2\lambda^2/2 + O(h^4)$$

$$\mu_2 = -(1 - h\lambda + h^2\lambda^2/2) + O(h^4)$$

where μ_1 is the principal root and μ_2 the parasitic root. If $Re(\lambda) > 0$, μ_1 grows like the exact solution while μ_2 dies out. If $Re(\lambda) < 0$, μ_1 decreases like the exact solution, but μ_2 grows and swamps out the solution. Thus, by itself the predictor is relatively stable for $Re(\lambda) > 0$, but unstable for $Re(\lambda) < 0$.

However, as seen in Section 4.4.1 one application of the corrector changes the parasitic root to $-h\lambda/2$, so that the P–C method in this case is stable for $-2 < h\lambda < 0$. Thus, one application of the corrector converts the method from an unstable to a stable one, although the real stability bounds do not approach those of the corrector alone. Two applications of the corrector do not change the real stability bounds from this value.

4.4.3. Increasing the Stability Bounds

Using the P–C pair of Euler--trapezoidal rule as an example, it is obvious that the real stability bound is seriously constrained, at least as compared to the corrector itself. The extension of this stability bound with a minimum of extra computation would seem to have significant merit.

Stetter [46] has detailed such an extension for this P–C pair which involves a suitable weighted combination of the $PE(CE)^1$ and $PE(CE)^2$ values. Thus he suggests that y_{n+1} be calculated from

$$y_{n+1} = \alpha_1 y_{n+1}(s=1) + (1 - \alpha_1)y_{n+1}(s=2)$$

with α_1 an adjustable parameter in the range $0 \le \alpha_1 \le 1$. By searching for a suitable value of α_1, he shows how the stability bounds may be significantly increased. Figure 4.1 shows the complex stability bounds for

$\alpha_1 = 1.0$: $PE(CE)^1$ algorithm equivalent to the second-order Runge–Kutta formula of Figure 3.4

$\alpha_1 = 0$: $PE(CE)^2$ algorithm

$\alpha_1 = 0.7$: weighted algorithm

As can be seen, the real stability bound is extended to $h\lambda > -5.1$ with $\alpha_1 = 0.7$; the truncation error is also affected (increased) but we shall not detail this here.

Obviously this weighting of $s = 1$ and $s = 2$ values of y_{n+1} has increased the real stability bound by a significant factor. Further results quoted by Stetter indicate that this technique can also be used for many other P–C combinations

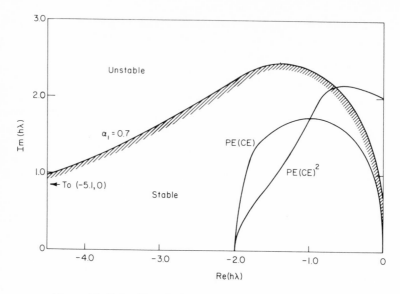

Figure 4.1. Euler (P) and trapezoidal rule (C) stability bounds.

with the same type of stability extension. Further work in this area would seem to be called for.

Finally, we note from Figure 4.1 that, while the real stability bound for the $PE(CE)^1$ and $PE(CE)^2$ algorithms are the same as that for the Euler predictor itself, the complex bounds are quite different.

4.5. MILNE PREDICTOR–CORRECTOR FORMS

Perhaps the earliest P–C form which achieved a significant accuracy was due to Milne [35]. The standard Milne P–C algorithm starts with the explicit 4-step $(k = 4)$ Newton–Cotes formula (1.6-14) as a predictor and the implicit 2-step Newton–Cotes or Milne formula (1.8-7) as a corrector. These are, with local truncation errors,

$$\bar{y}_{n+1} = y_{n-3} + (4h/3) \left[2y'_n - y'_{n-1} + 2y'_{n-2} \right]$$
$$T(x, h) = (28h^5/90)y^{[5]}(\zeta) \tag{4.5-1}$$

and

$$y_{n+1} = y_{n-1} + (h/3)[\bar{y}'_{n+1} + 4y_n' + y'_{n-1}]$$
$$T(x, h) = -(h^5/90)y^{[5]}(\zeta) \tag{4.5-2}$$

Both equations are fourth order and neither has good stability properties. In fact, both are unstable for all negative real $h\lambda$. As such the two formulas when combined would be suspect a priori in terms of stability properties.

We may consider (4.5-1) and (4.5-2) combined as a P–C pair, with and without iteration of the corrector or, by following the material in Section 4.2 which uses the truncation error, evolve the following steps:

$$p_{n+1} = y_{n-3} + (4h/3)[2y_n' - y_{n-1}' + 2y_{n-2}'] \quad \text{Predict}$$
$$m_{n+1} = p_{n+1} - (28/29)(p_n - c_n) \quad \text{Modify}$$
$$c_{n+1} = y_{n-1} + (h/3)[m_{n+1}' + 4y_n' + y_{n-1}'] \quad \text{Correct}$$
$$y_{n+1} = c_{n+1} + (1/29)(p_{n+1} - c_{n+1}) \quad \text{Modify}$$

$$(4.5\text{-}3)$$

4.5.1. Milne Corrector

Considering the corrector (4.5-2) alone, substitution of $y' = \lambda y$ yields

$$\left(1 - \frac{h\lambda}{3}\right)y_{n+1} - \frac{4h\lambda}{3}y_n - \left(1 + \frac{h\lambda}{3}\right)y_{n-1} = 0 \qquad (4.5\text{-}4)$$

and the characteristic equation is second degree with two roots μ_1 and μ_2. In the asymptotic case of $h \to 0$, the characteristic equation becomes

$$\mu^2 - 1 = 0$$

and the roots are $\mu_1 = +1$, $\mu_2 = -1$. Thus we have a weakly stable process.

The characteristic roots of (4.5-4) are given by

$$\mu_1 = \left(1 - \frac{h\lambda}{3}\right)^{-1}\left[\frac{2h\lambda}{3} + \left(1 + \frac{h^2\lambda^2}{3}\right)^{1/2}\right]$$
$$\mu_2 = \left(1 - \frac{h\lambda}{3}\right)^{-1}\left[\frac{2h\lambda}{3} - \left(1 + \frac{h^2\lambda^2}{3}\right)^{1/2}\right]$$

$$(4.5\text{-}5)$$

and expanding the square roots shows that

$$\mu_1 \cong e^{+h\lambda}, \qquad \mu_2 \cong e^{-h\lambda/3} \qquad (4.5\text{-}6)$$

It is, of course, obvious that μ_2 is the parasitic root.

From (4.5-6) we can see that if $\lambda > 0$, μ_1 behaves as the exact solution and μ_2 dies out since $|\mu_2| < 1$. When however $\lambda < 0$, μ_1 decreases as does the exact solution, but μ_2 oscillates with increasing amplitude and eventually overwhelms the correct solution. Thus Milne's corrector is weakly stable and has no real negative stability bound.

Certain authors have actually evaluated the roots μ_1 and μ_2 as a function of $h\lambda$. Milne and Reynolds [36] and Hamming [19] present results such as those given in Figures 4.2 and 4.3, with the latter of interest for relative stability.

Alternatively, we may plot the results as Chase [7] has done in Figure 4.4 with positive roots designated by a circle $\bigcirc$ and negative roots by a square $\square$. It is evident from these plots that there is no region of stability for negative $h\lambda$ for the iterated Milne corrector. By constrast, Figure 4.2 shows that there is a region of relative stability for positive $h\lambda$.

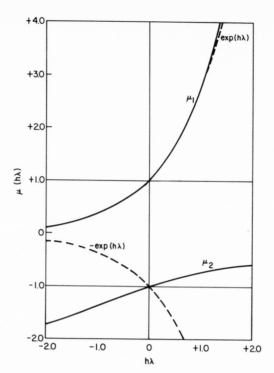

Figure 4.2. Behavior of roots of (4.5-5) as a function of $h\lambda$. [Adapted and reprinted from R. W. Hamming, *Journal of the ACM*, **Volume 6**, pp. 37–47; copyright © 1959, Association for Computing Machinery, Inc.]

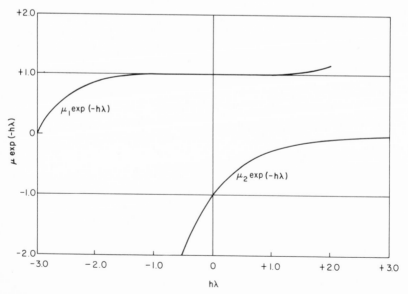

Figure 4.3. Relative stability for (4.5-5). [Adapted and reprinted from R. W. Hamming, *Journal of the ACM*, **Volume 6**, pp. 37–47; copyright © 1959, Association for Computing Machinery, Inc.]

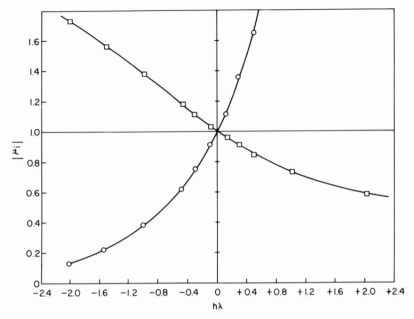

Figure 4.4. Roots of iterated Milne corrector. [Adapted and reprinted from P. E. Chase, *Journal of the ACM*, **Volume 9**, pp. 457–468; copyright © 1962, Association for Computing Machinery, Inc.]

4.5.2. Milne Predictor–Corrector Combination

Next we consider the combined P–C combination of (4.5-1) and (4.5-2). Using $y' = \lambda y$ and substituting the predictor into the corrector, a fourth-degree characteristic equation results

$$\mu^4 - \mu^3 \left[\frac{8(h\lambda)^2}{9} + \frac{4h\lambda}{3} \right] - \mu^2 \left[1 + \frac{h\lambda}{3} - \frac{4(h\lambda)^2}{9} \right] - \mu \left[\frac{8(h\lambda)^2}{9} \right] - \frac{h\lambda}{3} = 0$$

$$(4.5-7)$$

This equation is fourth degree because the combined P–C involves the points from y_{n+1} to y_{n-3} ($k = 3$). There are obviously three parasitic roots. Chase has evaluated the roots of the characteristic equation with the results shown in Figure 4.5 using the cross × to indicate complex roots. It is evident that for negative $h\lambda$, there now does exist a region, $-0.8 < h\lambda < -0.3$ for which the P–C mode is stable (all roots are <1). Thus the P–C Milne mode does extend the interval of stability as compared to the iterated Milne corrector. Also when $h \to 0$ the roots are $\mu_1 = +1$, $\mu_2 = -1$, and $\mu_3 = \mu_4 = 0$, as evident from Figure 4.5.

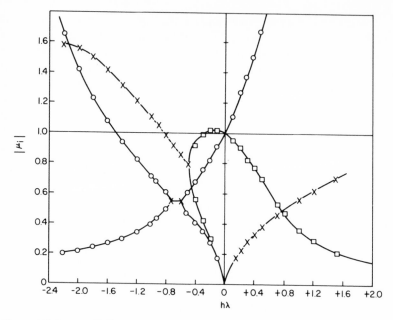

Figure 4.5. Roots of Milne P–C combination. [Adapted and reprinted from P. E. Chase, *Journal of the ACM*, **Volume 9**, pp. 457–468; copyright 1962, Association for Computing Machinery, Inc.]

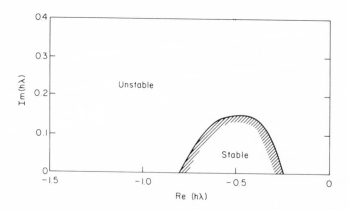

Figure 4.6. Region of stability in complex $h\lambda$ plane for Milne P–C combination. [Adapted and reprinted from R. L. Crane and R. W. Klopfenstein, *Journal of the ACM*, **Volume 12**, pp. 227–241; copyright © 1965, Association for Computing Machinery, Inc.]

Continuing further we may examine the regions of absolute and relative complex stability for the P–C mode. In the figures to be presented the solid curves result from $|\mu| = 1$ and the shaded portion within the solid curves represent the region of absolute stability; the dashed lines result from $|\mu| = |e^{h\lambda}|$ and they bound from the left the region of relative stability. Figure 4.6 shows such a plot from Crane and Klopfenstein [8] for the P–C Milne method. On the real axis of Figure 4.6 we see that the $-0.8 < h\lambda < -0.3$ bound exists. Further, we note that for almost all complex roots with real negative parts the algorithm is unstable; no purely imaginary roots yield stable solutions.

4.5.3. Modified Milne P–C Combinations

Finally, we show in Figure 4.7 a plot of the dominant root for the iterated Milne corrector, the P–C Milne, and the modified P–C Milne (4.5-3) as given by Chase. Here we see again that for negative $h\lambda$, the dominant root for the iterated Milne corrector is always greater than one (unstable) and that for the

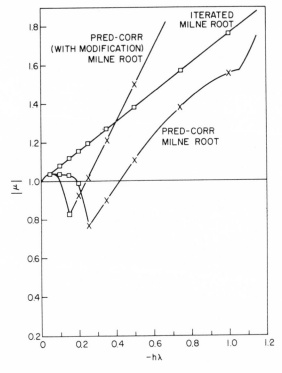

Figure 4.7. Dominant root in different Milne combinations. [Adapted and reprinted from P. E. Chase, *Journal of the ACM*, **Volume 9**, pp. 457–468; copyright © 1962, Association for Computing Machinery, Inc.]

P–C Milne, there is the region $-0.8 < h\lambda < -0.3$ in which the dominant root is less than one (stable). By contrast the modified P–C Milne shows a decreased range of stability as compared to the P–C Milne, namely, $-0.42 < h\lambda < -0.2$. This is due, as was indicated in Section 4.2, to the fact that the modified P–C characteristic equation is fifth degree and not fourth degree. In other words, the modification increases the number of roots by one and this may decrease or increase the stability regions. In the present case the stability is reduced.

As a point of emphasis, in all of the above discussion it is assumed that in a P–C (or modified P–C) mode the corrector is only used one time. If the corrector is used two times in the P–C mode, the real stability bound decreases to zero, i.e., there is no region of negative real $h\lambda$ for which stability exists.

4.5.4. Increased Stability by Averaging

To overcome these stability defects of the Milne P–C forms, Milne and Reynolds [36,37] proposed that after a finite number of steps (say fifty) of the Milne P–C form, a new value of y_{n+1}, call this Y_{n+1}, taken from the implicit Newton–Cotes formula of (1.5-12) be combined with the regular y_{n+1}. In other words, at periodic intervals the value Y_{n+1} is calculated from

$$Y_{n+1} = y_{n-2} + (3h/8)[\bar{y}'_{n+1} + 3y_n' + 3y'_{n-1} + y_{n-2}]$$

and then a new y_{n+1}(new) obtained from

$$y_{n+1}(\text{new}) = \frac{y_{n+1} + Y_{n+1}}{2}$$

As indicated by these authors, this procedure reduces the oscillation due to the parasitic root. Timlake [47] has extended this idea to general weakly stable methods and shown that the unstable component of the error can be reduced by a factor of h.

4.5.5. Hermite Predictor and Milne Corrector

Finally in this section we consider a special P–C pair made up from a Hermite predictor, (1.3-6), and the Milne corrector. As we shall see, the predictor is strongly unstable and it is combined with the weakly stable corrector. As such, they seem like an unlikely pair to use in a computation Nevertheless, as originally developed by Stetter [45], the two together have a stabilizing influence and yield a stable algorithm *if the corrector is not iterated*.

The Hermite predictor is

$$y_{n+1} = -4y_n + 5y_{n-1} + h(4y_n' + 2y'_{n-1}) \tag{4.5-8}$$

Substituting $y' = \lambda y$ yields

$$y_{n+1} + 4(1 - h\lambda)y_n - (5 + 2h\lambda)y_{n-1} = 0 \tag{4.5-9}$$

The roots of the characteristic equation are, as $h \to 0$, $\mu_1 = +1$ and $\mu_2 = -5$. Thus the principal root is on the unit circle but the parasitic root is outside; the predictor is unstable.

In the P–C pair we have

$$\bar{y}_{n+1} = -4y_n + 5y_{n-1} + h[4y_n' + 2y_{n-1}']$$
$$y_{n+1} = y_{n-1} + (h/3)[\bar{y}_{n+1}' + 4y_n' + y_{n-1}']$$

(4.5-10)

and applying $y' = \lambda y$ there results

$$\bar{y}_{n+1} = 4(h\lambda - 1)y_n + (5 + 2h\lambda)y_{n-1}$$
$$y_{n+1} = y_{n-1} + (h\lambda/3)[\bar{y}_{n+1} + 4y_n + y_{n-1}]$$

Substituting the first into the second yields

$$y_{n+1} - \tfrac{4}{3}h^2\lambda^2 y_n - (1 + 2h\lambda + \tfrac{2}{3}h^2\lambda^2)y_{n-1} = 0$$

with the roots of the characteristic equation being

$$\mu_1 = \tfrac{1}{2}\{\tfrac{4}{3}h^2\lambda^2 + [\tfrac{16}{9}h^4\lambda^4 + 4(1 + 2h\lambda + \tfrac{2}{3}h^2\lambda^2)]^{1/2}\}$$
$$\mu_2 = \tfrac{1}{2}\{\tfrac{4}{3}h^2\lambda^2 - [\tfrac{16}{9}h^4\lambda^4 + 4(1 + 2h\lambda + \tfrac{2}{3}h^2\lambda^2)]^{1/2}\}$$

(4.5-11)

As $h \to 0$, $\mu_1 \to 1$ and $\mu_2 \to -1$, and thus asymptotically the P–C form is weakly stable. However, if one analyzes μ_1 and μ_2 for real negative $h\lambda$, one finds that the P–C is stable for $-1.0 < h\lambda < 0$ (see Figure 4.8). Thus this P–C has a finite region of stability for real negative $h\lambda$.

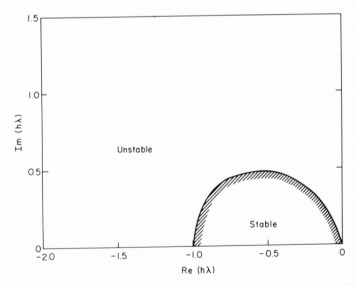

Figure 4.8. Region of stability in complex $h\lambda$ plane for Hermite–Milne combination. [Adapted and reprinted from F. T. Krough, *Journal of the ACM*, **Volume 14**, pp. 351–362; copyright 1967, Association for Computing Machinery, Inc.]

Krogh [28] has analyzed the complex stability features of this P–C form with the results shown in Figure 4.8. Note the real stability bound of $-1.0 < h\lambda < 0$ and that the form is unstable for all pure imaginary values. The relative stability region is also shown with the unusual feature that there is no point on the real axis.

4.6. HAMMING PREDICTOR–CORRECTOR SET

Since the Milne P–C forms have severe stability problems it seems important to revise the equations in some way to remove these restrictions. Hamming [19] was the first to do this. He revised the corrector such that the P–C equations are (with local truncation errors)

$$\bar{y}_{n+1} = y_{n-3} + (4h/3)[2y_n' - y_{n-1}' + 2y_{n-2}']$$
$$T(x, h) = (28h^5/90)y^{[5]}(\zeta) = (112h^5/360)y^{[5]}(\zeta) \qquad (4.6\text{-}1)$$

and

$$y_{n+1} = (1/8)[9y_n - y_{n-2}] + (3h/8)[\bar{y}_{n+1}' + 2y_n' - y_{n-1}']$$
$$T(x, h) = -(9h^5/360)y^{[5]}(\zeta) \qquad (4.6\text{-}2)$$

Alternatively, we may use the truncation error to develop the modified equations

$$p_{n+1} = y_{n-3} + (4h/3)[2y_n' - y_{n-1}' + 2y_{n-2}']$$
$$m_{n+1} = p_{n+1} - (112/121)(p_n - c_n)$$
$$c_{n+1} = (1/8)[9y_n - y_{n-2}] + (3h/8)[m_{n+1}' + 2y_n' - y_{n-1}'] \qquad (4.6\text{-}3)$$
$$y_{n+1} = c_{n+1} + (9/121)(p_{n+1} - c_{n+1})$$

4.6.1. Hamming Corrector

Since the predictor (4.6-1) is the same as the Milne predictor (4.5-1), we have all the necessary information on its behavior. The corrector (4.6-2) is new and we should analyze its stability behavior.

To develop the fourth-order corrector of (4.6-2), Hamming started with the generalized corrector

$$y_{n+1} = \alpha_1 y_n + \alpha_2 y_{n-1} + \alpha_3 y_{n-2} + h[\beta_0 y_{n+1}' + \beta_1 y_n' + \beta_2 y_{n-1}'] \qquad (4.6\text{-}4)$$

Since there are six free parameters and the result is to be fourth order ($p = 4$), one free parameter can be used. As shown previously (Table 1.11) this leads to

$$\alpha_1 = \tfrac{1}{8}(9 - 9\alpha_2), \qquad \beta_0 = \tfrac{1}{24}(9 - \alpha_2)$$
$$\alpha_2 = \alpha_2, \qquad \beta_1 = \tfrac{1}{12}(9 + 7\alpha_2) \qquad (4.6\text{-}5)$$
$$\alpha_3 = -\tfrac{1}{8}(1 - \alpha_2), \qquad \beta_3 = \tfrac{1}{24}(-9 + 17\alpha_2)$$
$$T(x, h) = (-9 + 5\alpha_2)h^5 y^{[5]}(\zeta)/360$$

with α_2 as the free parameter. The case $\alpha_2 = 1$ yields Milne's corrector. The characteristic equation at $h = 0$ is

$$8\mu^3 - 9(1 - \alpha_2)\mu^2 - 8\alpha_2\mu + (1 - \alpha_2) = 0 \qquad (4.6\text{-}6)$$

This equation has three roots with $\mu_1 = 1$ and μ_2 and μ_3 functions of α_2. Figure 4.9 shows the behavior of the roots as a function of α_2 for $h = 0$. It can be seen that for $-0.6 < \alpha_2 < 1.0$ absolute stability results.

Hamming then considered the values of $\alpha_1, \alpha_3, \ldots, \beta_2$ as a function of α_2 for $\alpha_2 = 1, \tfrac{9}{17}, \tfrac{1}{9}, 0, -\tfrac{1}{7}, -\tfrac{9}{31}$, and $-\tfrac{6}{10}$. After looking at the resulting coefficients in terms of equal magnitude, number of zeros, etc., the case $\alpha_2 = 0$ was chosen as the best value. Ralston [40] also showed that this value was

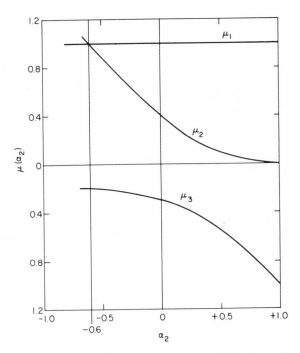

Figure 4.9. Behavior of roots of (4.6-6) as function of α_2. [Adapted and reprinted from R. W. Hamming, *Journal of the ACM*, **Volume 6**, pp. 37–47; copyright © 1959, Association for Computing Machinery, Inc.]

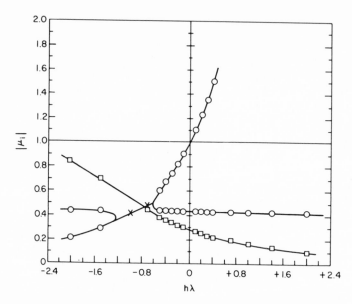

Figure 4.10. Roots of iterated Hamming corrector. [Adapted and reprinted from P. E. Chase, *Journal of the ACM*, **Volume 9**, pp. 457–468; copyright © 1962, Association for Computing Machinery, Inc.]

almost perfect for the widest range of relative stability. The selection of $\alpha_2 = 0$ thus yields (4.6-2) with the local truncation error shown. This error is slightly larger than that in Milne's corrector, but a slight decrease in h (about 15 %) will equalize this error.

Chase analyzed the root behavior of the Hamming corrector ($\alpha_2 = 0$) with $h \neq 0$. Using the nomenclature that positive roots are designated by a circle ○, all negative roots by a square □ and complex roots by ×, Figure 4.10 results from analyzing the characteristic equation

$$\mu^3\left(\frac{3h\lambda}{8} - 1\right) + \mu^2\left(\frac{9}{8} + \frac{3h\lambda}{4}\right) - \mu\left(\frac{3h\lambda}{8}\right) - \frac{1}{8} = 0 \qquad (4.6\text{-}7)$$

When compared to Figure 4.4 for the Milne corrector it becomes obvious that the Hamming corrector has a much greater region of real negative stability. In fact the Hamming corrector is stable for $-2.6 < h\lambda < 0$.

4.6.2. Hamming Predictor–Corrector Combination

When (4.6-1) and (4.6-2) are used in a P–C mode, a fourth-degree characteristic equation results. Three of the roots are parasitic with two of the three not falling on the origin for $h = 0$. While we do not show the detailed root distributions, this characteristic equation shows that the Hamming P–C is stable for $-0.5 < h\lambda < 0$ (see Figure 4.12) and that when two applications of the corrector are used the result is $-0.9 < h\lambda < 0$. These are considerably better than for the Milne P–C mode.

Chase considered the modified P–C form of (4.6-3) and showed that the characteristic equation was given by

$$\mu^5(121) + \mu^4(-126 - 150h\lambda - 112(h\lambda)^2) + \mu^3(54h\lambda + 168(h\lambda)^2)$$
$$+ \mu^2(14 - 24h\lambda - 168(h\lambda)^2) + \mu(-9 - 42h\lambda + 112(h\lambda)^2) + 42h\lambda = 0$$
$$(4.6-8)$$

This equation is fifth degree instead of fourth degree because the modified form always produces a higher order (by one) formula (see Section 4.2). In fact, Lomax [32] shows explicitly that the Hamming modified P–C form is analogous to an unmodified 5-step ($k = 5$) P–C form. The equations for the unmodified form are given by

$$\bar{y}_{n+1} = y_n + y_{n-3} - y_{n-4} + \tfrac{4}{3}h[2y_n' - 3y_{n-1}' + 3y_{n-2}' - 2y_{n-3}']$$
$$y_{n+1} = \tfrac{1}{121}[126y_n - 14y_{n-2} + 9y_{n-3} \qquad\qquad (4.6-9)$$
$$+ h(42\bar{y}_{n+1}' + 108y_n' - 54y_{n-1}' + 24y_{n-2}')]$$

These are equivalent in every respect to (4.6-3). For the characteristic equation of (4.6-8) [or to that for (4.6-9)], Chase shows the root (five roots) plots of Figure 4.11. The range of stability for negative $h\lambda$ is evident from this plot,

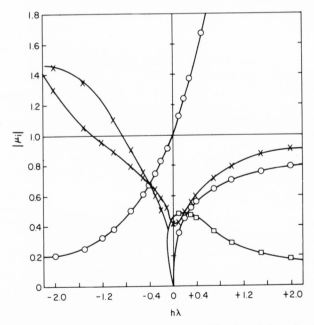

Figure 4.11. Roots of Hamming modified P–C combination. [Adapted and reprinted from P. E. Chase, *Journal of the ACM*, **Volume 9**, pp. 457–468; copyright © 1962, Association for Computing Machinery, Inc.]

i.e., for $-0.85 < h\lambda < 0$. Figure 4.12 shows the dominant root for the iterated Hamming corrector, for the Hamming P–C mode, and for the Hamming modified P–C mode. Here we can see that the iterated Hamming corrector has by far the best stability characteristics and that the modified P–C is better by a factor close to two than the unmodified P–C form. This last point is the exact opposite to that noted in Figure 4.7 for the Milne P–C forms.

Finally, we note in Figure 4.13, from Crane and Klopfenstein, [8], the complex absolute and relatively stable regions for the modified P–C Hamming formulas. As compared to Figure 4.6, the Hamming method has a much greater region of stability and does have finite imaginary bounds of about $0.7i$.

From all these plots, we see that the Hamming corrector is stable for $-2.6 < h\lambda < 0$, the P–C Hamming for $-0.5 < h\lambda < 0$, and the modified Hamming P–C for $-0.85 < h\lambda < 0$. As a result the Hamming formulas are highly recommended for computation as compared to the Milne formulas. Within the Hamming forms, the modified P–C seems best in terms of stability and the number of function evaluations.

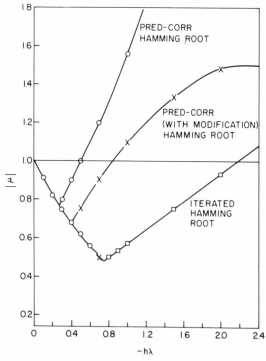

Figure 4.12. Dominant roots in different Hamming combinations. [Adapted and reprinted from P. E. Chase, *Journal of the ACM*, **Volume 9**, pp. 457–468; copyright © 1962, Association for Computing Machinery, Inc.]

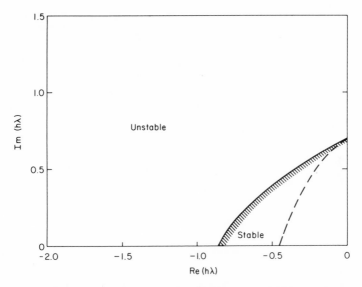

Figure 4.13. Region of stability in complex $h\lambda$ plane for Hamming modified P–C combination. [Adapted and reprinted from R. L. Crane and R. W. Klopfenstein, *Journal of the ACM*, **Volume 12**, pp. 227–241; copyright © 1965, Association for Computing Machinery, Inc.]

4.6.3. Extensions

Before leaving this section we wish to make two further points. First, Ralston [40] suggested that rather than use the modified fourth-order Hamming P–C, one might just as well derive an unmodified fifth-order P–C system. He then proceeded to derive such a pair in a manner very analogous to that used by Hamming, i.e., a fifth-order predictor was selected directly while a generalized corrector was analyzed until the best parameters were obtained. The result is given by the P–C formulas

$$\bar{y}_{n+1} = -18y_n + 9y_{n-1} + 10y_{n-2} + h[9y_n' + 18y_{n-1}' + 3y_{n-2}']$$
$$T(x, h) = (h^6/20)y^{[6]}(\zeta)$$

[See Hermite (1.3-17).] and

$$y_{n+1} = (1/3)[2y_n + 2y_{n-1} - y_{n-2}] + (1/27)[y_{n-3} - y_n]$$
$$+ (h/3)[\bar{y}_{n+1}' + (10/3)y_n' - y_{n-1}'] \qquad (4.6\text{-}10)$$
$$T(x, h) = -(7h^6/540)y^{[6]}(\zeta)$$

Finally, we wish to emphasize that the real stability bounds of the Hamming P–C forms are much smaller than those for the Runge–Kutta algorithms. If it is possible to effectively use these bounds, the often mentioned disadvantage of the Runge–Kutta algorithms, namely that they require more function evaluations than corresponding P–C methods, is bypassed. We may show this by comparing the fourth-order explicit Runge–Kutta with the modified Hamming P–C; the first method has a real stability bound of $h\lambda > -2.785$, while the second method has a bound of $h\lambda > -0.85$. But the Runge–Kutta method requires four stages ($v = 4$) whereas Hamming's method requires two stages ($v = 2$). Thus a stable integration from 0 to 1 requires a minimum of 2.35λ stages for Hamming's method as against 1.44λ for the Runge–Kutta

TABL

AVAILABL

Name	α_1	α_2	α_3	α_4	β_1	β_2	β_3
Euler	1	0	0	0	1	0	
Nystrom	0	1	0	0	2	0	
Hermite	−4	5	0	0	4	2	0
Hermite	−18	9	10	0	9	18	3
Milne	0	0	0	1	8/3	−4/3	8/3
A–B, $k = 2$	1	0	0	0	3/2	−1/	0
A–B, $k = 3$	1	0	0	0	23/12	−16/	5/
A–B, $k = 4$	1	0	0	0	55/24	−59/	37/
A–B, $k = 5$	1	0	0	0	1901/720	−2774/	2616/
A–B, $k = 6$	1	0	0	0	4277/1440	−7923/	9982/
A–B, $k = 7$	1	0	0	0	198,721/60,480	−447,288/	705,549/
A–B, $k = 8$	1	0	0	0	434,241/120,960	−1,162,169/	2,183,877/
Crane–Klopfenstein	1.547	−1.867	2.017	−.6973	2.002	−2.031	1.818
Krogh (A), $k = 3$	1/2	1/2	0	0	119/48	−99/	69/
Krogh (B), $k = 3$	4/7	3/7	0	0	103/42	−88/	61/
Krogh, $k = 4$	−1/31	32/31	0	0	22,321/7440	−21,774	24,216/
Krogh, $k = 5$	−11/12	23/12	0	0	62,249/17280	−62,255/	101,430/
Krogh, $k = 6$	−21/10	31/10	0	0	2,578,907/604,800	−2,454,408/	5,615,199/

AVAILABL

Name	α_1	α_2	α_3	β_0	β_1	β_2
Milne	0	1	0	1/3	4/3	1/3
Hamming	9/8	0	−1/8	3/8	6/8	−3/8
A–M, $k = 1$	1	0	0	1/2	1/	0
A–M, $k = 2$	1	0	0	5/12	8/	−1/
A–M, $k = 3$	1	0	0	9/24	19/	−5/
A–M, $k = 4$	1	0	0	251/720	646/	−264/
A–M, $k = 5$	1	0	0	475/1440	1427/	−798/
A–M, $k = 6$	1	0	0	19,087/60,480	65,112/	−46,461/
A–M, $k = 7$	1	0	0	36,799/120,960	139,849/	−121,797/

[a] When denominators are omitted, they

4.7. Adams Predictor-Corrector Set 181

method. The extra stages per time step in the Runge–Kutta do not seem to be the defect anticipated.

4.7. ADAMS PREDICTOR-CORRECTOR SET

In this section we shall investigate the features of the Adams–Bashforth (A–B) predictors and Adams–Moulton (A–M) correctors. The necessary formulas are derived from (1.6-23) and (1.6-24) and Table 4.1 lists all the

.1

PREDICTORS^a

β_4	β_5	β_6	β_7	β_8	$T(x, h)$
					$(h^2/2)\, y^{[2]}$
					$(h^3/3)\, y^{[3]}$
					$(h^4/6)\, y^{[4]}$
					$(h^6/20)\, y^{[6]}$
					$(28h^5/90)\, y^{[5]}$
					$(5h^3/12)\, y^{[3]}$
					$(9h^4/24)\, y^{[4]}$
-9	0				$(251h^5/720)\, y^{[5]}$
$-1274/$	$251/$	0			$(475h^6/1440)\, y^{[6]}$
$-7298/$	$2887/$	$-475/$	0		$(19{,}087h^7/60{,}480)\, y^{[7]}$
$-688{,}256/$	$407{,}139/$	$-134{,}472/$	$19{,}087/$	0	$(36{,}799h^8/120{,}960)\, y^{[8]}$
$-2{,}664{,}477/$	$2{,}102{,}243/$	$-1{,}041{,}723/$	$295{,}767/$	$-36{,}799/$	$(1{,}070{,}017h^9/3{,}628{,}800)\, y^{[9]}$
$-.7143$	0				$.4016h^5 y^{[5]}$
$-17/$	0				$(161h^5/480)\, y^{[5]}$
$-15/$	0				$(85h^5/252)\, y^{[5]}$
$-12034/$	$2391/$	0			$(13{,}861h^6/44{,}640)\, y^{[6]}$
$-76490/$	$30{,}545/$	$-5079/$	0		$(5977h^7/20{,}736)\, y^{[7]}$
$-5{,}719{,}936/$	$3{,}444{,}849/$	$-1{,}149{,}048/$	$164{,}117/$	0	$(21{,}691h^8/80{,}640)\, y^{[8]}$

CORRECTORS

β_3	β_4	β_5	β_6	β_7	$T(x, h)$
					$-(h^5/90)\, y^{[5]}$
					$-(9h^5/360)\, y^{[5]}$
					$-(h^3/12)\, y^{[3]}$
					$-(h^4/24)\, y^{[4]}$
$/$	0				$-(19h^5/720)\, y^{[5]}$
$06/$	$-19/$	0			$-(27h^6/1440)\, y^{[6]}$
$82/$	$-173/$	$27/$	0		$-(863h^7/60{,}480)\, y^{[7]}$
$7{,}504/$	$-20{,}211/$	$6312/$	$-863/$	0	$-(1375h^8/120{,}960)\, y^{[8]}$
$23{,}133/$	$-88{,}547/$	$41{,}499/$	$-11{,}351/$	$1375/$	$-(339{,}533h^9/3{,}628{,}800)\, y^{[9]}$

.re the same as the first one in that row.

formulas and the local truncation errors. In particular, we shall be interested in the third-order ($p = 3$) P–C formulas and the fourth-order ($p = 4$) P–C formulas. These are, with the local truncation errors,

$$\bar{y}_{n+1} = y_n + (h/12)[23y_n' - 16y_{n-1}' + 5y_{n-2}']$$
$$T(x, h) = (9h^4/24)y^{[4]}(\zeta) \qquad (k = 3) \tag{4.7-1}$$

$$y_{n+1} = y_n + (h/12)[5\bar{y}_{n+1}' + 8y_n' - y_{n-1}']$$
$$T(x, h) = -(h^4/24)y^{[4]}(\zeta) \qquad (k = 2) \tag{4.7-2}$$

and

$$\bar{y}_{n+1} = y_n + (h/24)[55y_n' - 59y_{n-1}' + 37y_{n-2}' - 9y_{n-3}']$$
$$T(x, h) = (251h^5/720)y^{[5]}(\zeta) \qquad (k = 4) \tag{4.7-3}$$

$$y_{n+1} = y_n + (h/24)[9\bar{y}_{n+1}' + 19y_n' - 5y_{n-1}' + y_{n-2}']$$
$$T(x, h) = -(19h^5/720)y^{[5]}(\zeta) \qquad (k = 3) \tag{4.7-4}$$

The modified form of these P–C sets can be directly obtained from the truncation errors to yield

$$p_{n+1} = y_n + (h/12)[23y_n' - 16y_{n-1}' + 5y_{n-2}']$$
$$m_{n+1} = p_{n+1} - (9/10)(p_n - c_n)$$
$$c_{n+1} = y_n + (h/12)[5m_{n+1}' + 8y_n' - y_{n-1}'] \tag{4.7-5}$$
$$y_{n+1} = c_{n+1} + (1/10)(p_{n+1} - c_{n+1})$$

and

$$p_{n+1} = y_n + (h/24)[55y_n' - 59y_{n-1}' + 37y_{n-2}' - 9y_{n-3}']$$
$$m_{n+1} = p_{n+1} - (251/270)(p_n - c_n)$$
$$c_{n+1} = y_n + (h/24)[9m_{n+1}' + 19y_n' - 54y_{n-1}' + y_{n-2}'] \tag{4.7-6}$$
$$y_{n+1} = c_{n+1} + (19/270)(p_{n+1} - c_{n+1})$$

An interesting point regarding the Adams-type formulas can be ascertained by examining (4.7-4), the iterated fourth-order Adams–Moulton (or A–M) corrector. Substituting $y' = \lambda y$ there results

$$\left(1 - \frac{3}{8}h\lambda\right)y_{n+1} - \left(1 + \frac{19}{24}h\lambda\right)y_n + \frac{5h\lambda}{24}y_{n-1} - \frac{h\lambda}{24}y_{n-2} = 0 \qquad (4.7-7)$$

which has a characteristic equation of third degree (note $k = 3$). In other words, the characteristic equation has the form

$$\mu^3 + A(h\lambda)\mu^2 + B(h\lambda)\mu + C(h\lambda) = 0$$

where $A(h\lambda)$, ..., $C(h\lambda)$ are functions of $h\lambda$. In the limiting case $h \rightarrow 0$ this becomes

$$(\mu - 1)\mu^2 = 0$$

and the roots are $\mu_1 = 1$, $\mu_2 = \mu_3 = 0$. Thus we see that the dominant root is 1 while all other roots (the parasitic ones) lie at the origin when $h = 0$. In the general k-step A–M formula the equivalent characteristic equation is

$$(\mu - 1)\mu^{k-1} = 0 \tag{4.7-8}$$

Thus in every case the A–M formulas have one root on the unit circle with all others at the origin when $h = 0$. As such it would seem to have excellent stability characteristics, no matter what k-step formula is used. When h increases from zero, the parasitic roots move toward the unit circle. However, a significant value of h can be reached before instability occurs.

4.7.1. Third–Order Adams P–C Combination

To start the discussion of the Adams P–C forms we turn to the third-order ($p = 3$) formulas of (4.7-1) and (4.7-2). The characteristic equations for the predictor, the corrector, and predictor–corrector are obtained by substituting $y' = \lambda y$. This yields, in order,

$$\mu^3 - \left(1 + \frac{23}{12}h\lambda\right)\mu^2 + \left(\frac{4}{3}h\lambda\right)\mu - \frac{5}{12}h\lambda = 0 \tag{4.7-9}$$

$$\left(1 - \frac{5}{12}h\lambda\right)\mu^2 - \left(1 + \frac{2}{3}h\lambda\right)\mu + \frac{h\lambda}{12} = 0 \tag{4.7-10}$$

$$\mu^3 - \left(1 + \frac{13}{12}h\lambda + \frac{115}{144}h^2\lambda^2\right)\mu^2 + \left(\frac{h\lambda}{12} + \frac{80}{144}h^2\lambda^2\right)\mu - \frac{25}{144}h^2\lambda^2 = 0 \tag{4.7-11}$$

Positive, negative, and complex root plots for these equations have been given by Distefano [12]. We here show Figure 4.14, the complex stability plot, for the PECE set.

From these figures, the real negative stability bounds for the predictor is $-0.55 < h\lambda < 0$, for the corrector is $-6.0 < h\lambda < 0$, and for the PECE is $-1.8 < h\lambda < 0$. Note that these bounds are quite large compared to the previous Milne and Hamming formulas. Another result of interest is that the modified P–C is stable for $-1.6 < h\lambda < 0$. Finally, we note from Figure 4.14 the large region of imaginary stability.

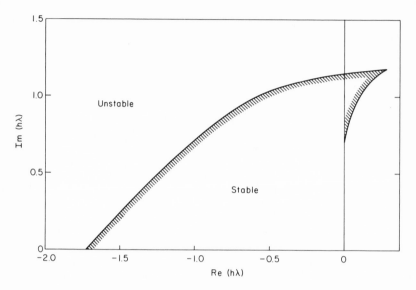

Figure 4.14. Region of stability in complex $h\lambda$ plane for third-order Adams P–C combination.

4.7.2. Fourth-Order Adams P–C Combination

The fourth-order Adams P–C formulas relate to (4.7-3), (4.7-4), and the combination. Typical characteristic equations are, respectively

$$\mu^4 - \left(1 + \frac{55}{24}\,h\lambda\right)\mu^3 + \frac{59}{24}\,h\lambda\mu^2 - \frac{37}{24}\,h\lambda\mu + \frac{9}{24}\,h\lambda = 0 \qquad (4.7\text{-}12)$$

$$\left(1 - \frac{9}{24}\,h\lambda\right)\mu^3 - \left(1 + \frac{19}{24}\,h\lambda\right)\mu^2 + \frac{5}{24}\,h\lambda\mu - \frac{h\lambda}{24} = 0 \qquad (4.7\text{-}13)$$

$$\mu^4 - \left(1 + \frac{7}{6}\,h\lambda + \frac{55h^2\lambda^2}{64}\right)\mu^3 + \left(\frac{5h\lambda}{24} + \frac{59h^2\lambda^2}{64}\right)\mu^2$$

$$- \left(\frac{h\lambda}{24} + \frac{37}{64}\,h^2\lambda^2\right)\mu + \frac{9h^2\lambda^2}{64} = 0 \qquad (4.7\text{-}14)$$

$$\mu^5 - \left(1 + \frac{7h\lambda}{6} + h^2\lambda^2\right)\mu^4 + \left(\frac{5h\lambda}{24} + \frac{95h^2\lambda^2}{64}\right)\mu^3 - \left(\frac{h\lambda}{24} + \frac{91h^2\lambda^2}{64}\right)\mu^2$$

$$+ \frac{45h^2\lambda^2}{64}\,\mu - \frac{9h^2\lambda^2}{64} = 0 \qquad (4.7\text{-}15)$$

corresponding to the P, C, PECE and PE(CE)^2 agorithms respectively.

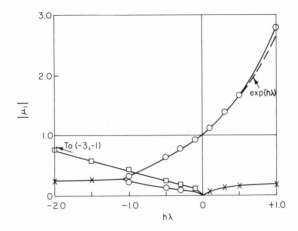

Figure 4.15. Roots of iterated fourth-order Adams corrector. [Adapted and reprinted with permission of the publisher, The American Mathematical Society from *Mathematics of Computation*; copyright © 1965, **Volume 19,** pp. 90–96.]

Brown *et al.* [2] have analyzed the roots of the characteristic equations and we show in Figures 4.15 and 4.16 the results for the corrector (4.7-13) and the PECE (4.7-14). Real positive roots are shown with a circle ○, real negative roots with a square □, and complex roots with a cross ×. Real negative stability bounds are $-0.3 < h\lambda < 0$, $-3.0 < h\lambda < 0$, $-1.3 < h\lambda < 0$, and $-0.9 < h\lambda < 0$ for the order indicated above. Brown *et al.* also considered

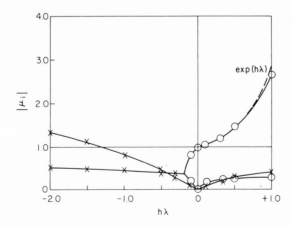

Figure 4.16. Roots of fourth-order Adams P–C combination. [Adapted and reprinted with permission of the publisher, The American Mathematical Society from *Mathematics of Computation*; copyright © 1965, **Volume 19,** pp. 90–96.]

4. **Predictor–Corrector Methods**

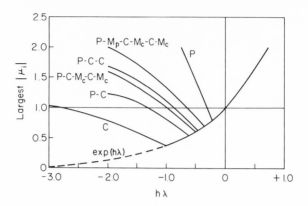

Figure 4.17. Dominant root in different Adams combinations [12].

a number of modified P–C sets. In particular, for the normal modified P–C mode, (4.7-6), but with a second application of the corrector the real stability bound is $-0.7 < h\lambda < 0$. Figure 4.17 shows the dominant root behavior for these cases as given by Distefano [12].

Finally we show the complex plane stability for the iterated corrector, Figure 4.18, and the PECE mode, Figure 4.19, as given by Krogh [27]. The

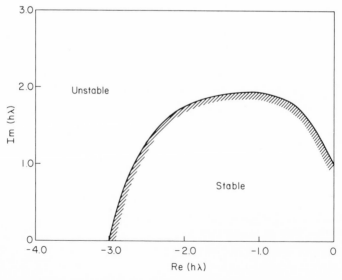

Figure 4.18. Region of stability in complex $h\lambda$ plane for fourth-order Adams corrector. [Adapted and reprinted from F. T. Krogh, *Journal of the ACM*, **Volume 13**, pp. 374–385; copyright © 1966, Association for Computing Machinery, Inc.]

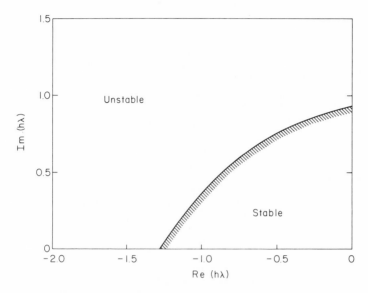

Figure 4.19. Region of stability in complex $h\lambda$ plane for fourth-order Adams P–C combination. [Adapted and Reprinted from F. T. Krogh, *Journal of the ACM*, **Volume 13,** pp. 374–385; copyright © 1966, Association for Computing Machinery, Inc.]

excellent behavior of these algorithms are confirmed from these plots with imaginary roots of approximately i being stable.

It is of importance to note that in all cases the fourth-order Adams forms when compared to the third-order forms are more accurate, i.e., they have lower truncation errors, but at the same time the stability limits become more constrained. Here we have a prime demonstration of the trade-off between the two phenomena.

4.7.3. Higher-Order Adams P–C Combinations

Figures 4.20 and 4.21 show the complex stability bounds for the iterated Adams corrector and the PECE set for orders five through eight. These are taken from Krogh. From these plots it is possible to construct Table 4.2 which lists information (including real relative stability bounds) of interest. We can see that within these possible arrangements an increased accuracy is accompanied by a decrease in all the stability boundaries. Further, the P–C mode of operation would seem to be the best to use even when compared to the iterated corrector. This follows from the two function evaluations in the P–C and the unknown (but probably four or five) evaluations in the iterated corrector.

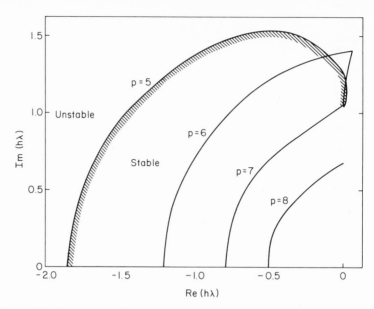

Figure 4.20. Region of stability in complex $h\lambda$ plane for various order Adams correctors. [Adapted and reprinted from F. T. Krogh, *Journal of the ACM*, **Volume 13**, pp. 374–385; copyright © 1966, Association for Computing Machinery, Inc.]

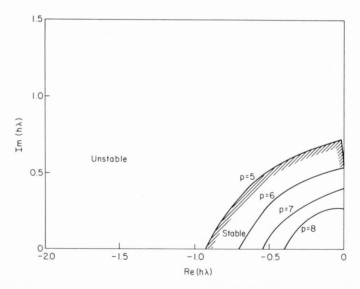

Figure 4.21. Region of stability in complex $h\lambda$ plane for various order Adams P–C combinations. [Adapted and reprinted from F. T. Krogh, *Journal of the ACM*, **Volume 13**, pp. 374–385; copyright © 1966, Association for Computing Machinery, Inc.]

TABLE 4.2
STABILITY BOUNDS OF ADAMS FORMS

Order p	Real stability		Relative stability	Complex stability
	Iterated corrector	PECE pair	Iterated corrector	PECE pair
3	−6.0	−1.8	−1.5	1.3i
4	−3.0	−1.3	−0.92	0.9i
5	−1.8	−0.95	−0.68	0.7i
6	−1.2	−0.7	−0.49	0.55i
7	−0.8	−0.5	−0.35	0.4i
8	−0.5	−0.4	−0.25	0.3i

4.7.4. Crane–Klopfenstein P–C Combination

In this section we wish to detail the approach by Crane and Klopfenstein [8] in which new predictors were proposed to go along with the Adams–Mouton corrector. Krogh [27] has also presented similar work and we shall analyze his results in Section 4.7.5.

Crane and Klopfenstein start with a generalized predictor and a generalized corrector involving a total of fifteen free parameters. By imposing the requirement that the P–C combination be fourth order, five free parameters were left to adjust for maximum stability and minimum truncation error. Included in the free parameters were three in the predictor and two in the corrector. To adjust these parameters the procedure used was to form the gradient of $h\lambda$ with respect to the free parameters and then to minimize this gradient. This would then yield the largest range of absolute stability for the P–C pair. However, it soon became apparent that a fruitful result could be obtained only if the two parameters in the corrector were not adjusted but rather that the corrector be chosen as the standard fourth-order Adams corrector (4.7-4). Thus the end result is a new predictor to go along with the Adams corrector.

In developing the gradient with respect to the three predictor parameters the starting point was the standard A–B, A–M P–C pair (4.7-3) and (4.7-4) with a real absolute stability bound of $h\lambda = -1.3$ and a real relative stability bound of $h\lambda = -0.6$ (see Figure 4.19). After proceeding along the gradient as far as possible the complex-plane plot of Figure 4.22 was obtained. Corresponding to this final converged value is the predictor equation.

$$\bar{y}_{n+1} = 1.547652y_n - 1.867503y_{n-1} + 2.017204y_{n-2} - 0.697353y_{n-3}$$
$$+ h[2.002247y_n' - 2.03169y_{n-1}' + 1.818609y_{n-2}' - 0.714320y_{n-3}']$$

$$(4.7\text{-}16)$$

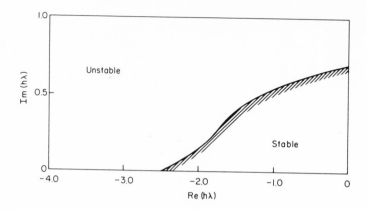

Figure 4.22. Region of stability in complex $h\lambda$ plane for Crane-Klopfenstein predictor with Adams corrector, fourth order. [Adapted and reprinted from R. L. Crane and R. W. Klopfenstein, *Journal of the ACM*, **Volume 12**, pp. 227–241; copyright © 1965, Association for Computing Machinery, Inc.]

which yields real absolute and relative stability bounds of $h\lambda = -2.48$ and $h\lambda = -0.446$ respectively. Thus the real absolute stability bound has been increased considerably as compared to the Adams P–C form; however, the real relative stability has been decreased. Further, on the imaginary axis of Figure 4.22 it can be seen that the bound is $h\lambda = 0.70i$ as compared to $0.92i$ for Figure 4.19.

4.7.5. Krogh P–C Combination

Krogh has developed a set of different order predictors to go along with the corresponding order Adams correctors. These formulas differ from the standard Adams predictors by including the point y_{n-1} as well as y_n. The explicit formulas are listed in Table 4.1 and Figure 4.23 shows the resulting complex-plane plots for the fourth- to eighth-order formulas.

The approximate stability limits are listed in Table 4.3 but it is obvious that improved stability regions have been obtained as compared to the standard Adams P–C and even in some cases to the iterated Adams corrector Thus Krogh obtains a fourth-order real stability boundary of $h\lambda = -1.8$, and an imaginary boundary of $0.95i$; the fourth-order Adams P–C and iterated corrector yield -1.3 and $0.9i$ and -3.0 and $1.0i$ respectively. The eighth-order values in the same sequence are -0.60 and $0.30i$, -0.4 and $0.3i$, and -0.5 and $0.68i$. As such these new P–C formulas would seem very promising.

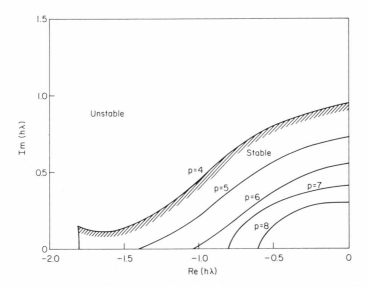

Figure 4.23. Region of stability in complex $h\lambda$ plane for Krogh–Adams P–C combination. [Adapted and reprinted from F. T. Krogh, *Journal of the ACM*, **Volume 13**, pp. 374–385; copyright © 1966, Association for Computing Machinery, Inc.]

4.7.6. Nordsieck Method

We wish to mention the work of Nordsieck [38] which appears as a single-step method but which is equivalent to an iterated Adams corrector. The single-step format allows for convenient step-size changing as compared to the usual corrector or P–C formulas. Basically the method uses values of y_n, y_n', y'', ..., at one value of x_n rather than using a series of back values y_n, y_{n-1}, y_{n-2}, ...

Nordsieck presented third- to seventh-order algorithms and extensions have been made by Osborne [39]. The length of the equations prevents our writing them here, but we refer the reader to the original paper for an interesting analysis of the stability and local truncation error.

4.7.7. Hull Combinations

Hull and his coworkers [21–23] have presented a series of correctors which are related to the fourth-order Adams corrector (4.7-4). The procedure used to develop these formulas is to select the reduced characteristic roots and calculate the α_i and β_i; one parameter c is left free to then adjust for minimum truncation error or to adjust the stability characteristics further.

Three such correctors are given below with (4.7-17) termed a Westward formula with parasitic roots $\mu_2 = -c$ and $\mu_3 = -c$, (4.7-18) is termed an

East–West formula with parasitic roots $\mu_2 = +c$ and $\mu_3 = -c$ and (4.7-19) is termed a radial formula with parasitic roots $\mu_2 = c\exp(2\pi i/3)$ and $\mu_3 = c\exp(4\pi i/3)$. All three reduced to the fourth-order Adams corrector for $c = 0$.

$$y_{n+1} = (1 - 2c)y_n + (2c - c^2)y_{n-1} + c^2 y_{n-2} + (h/24)[(c^2 - 2c + 9)y'_{n+1}$$
$$+ (-5c^2 + 26c + 19)y_n' + (19c^2 + 26c - 5)y'_{n+1}$$
$$+ (9c^2 - 2c + 1)y'_{n-2}] \tag{4.7-17}$$

$$y_{n+1} = y_n + c^2 y_{n-1} - c^2 y_{n-2} + (h/24)[(-c^2 + 9)y'_{n+1} + (5c^2 + 19)y_n'$$
$$- (19c^2 + 5)y'_{n-1} + (-9c^2 + 1)y'_{n-2}] \tag{4.7-18}$$

$$y_{n+1} = (1 - c)y_n + (c - c^2)y_{n-1} + c^2 y_{n-2}$$
$$+ (h/24)[(c^2 - c + 9)y'_{n+1} + (-5c^2 + 13c + 19)y_n'$$
$$+ (19c^2 + 13c - 5)y'_{n-1} + (9c^2 - c + 1)y'_{n-2}] \tag{4.7-19}$$

As shown by Ralston the choice of $c = 11/19$ in (4.7-17) and $c = 11/38$ in (4.7-19) minimizes the truncation error while still maintaining absolute and relative stability. In fact, (4.7-19) then has a local truncation error of

$$T(x, h) = -(147h^5/6080)y^{[5]}(\zeta)$$

which is smaller (slightly) than either the fourth-order Adams corrector or the Hamming corrector.

4.7.8. Extensions

We have already mentioned in Section 4.4.3 that Stetter has shown that significant increases in the stability bounds can be obtained via the use of a suitable weighting procedure. Here we wish to show this feature further using the Adams P–C formulas. In particular, we pick the A–B two-step predictor

$$\bar{y}_{n+1} = y_n + (h/2)[3y_n' - y'_{n-1}] \tag{4.7-20}$$

with a local truncation error of $T(x, h) = (5h^3/12)y^{[3]}(\zeta)$ and the trapezoidal rule corrector (4.4-2). Both formulas are second order.

By substituting $y' = \lambda y$ into (4.7-20) we can show directly that the real negative stability bound is $-1.0 < h\lambda < 0$; in a P–C mode this bound is increased to $-2.0 < h\lambda < 0$. Further, in this case, the complex stability region is much larger for the P–C mode than for the predictor-only mode. Let us now calculate the value of y_{n+1} from

$$y_{n+1} = \gamma_0 \bar{y}_{n+1}(P) + (1 - \gamma_0)y_{n+1}(P\text{–}C) \tag{4.7-21}$$

where $\bar{y}_{n+1}(P)$ is the predictor value, $y_{n+1}(P\text{–}C)$ is the P–C value, and $0 \leq \gamma_0 \leq 1.0$. For $\gamma_0 = 1.0$, we have the predictor alone, while for $\gamma_0 = 0$, we have the P–C combination. Stetter shows that for $\gamma_0 = 0.65$ the real negative stability bound becomes $-5.7 < h\lambda < 0$ representing approximately a three-fold increase over the P–C mode. As indicated previously further work along these lines seems warranted.

As another point of interest we raise the question here of using the predictor itself for calculating y_{n+1} rather than the P–C combination. To use the predictor requires one function evaluation while to use a PECE requires two function evaluations. Thus really to compare the predictor versus the P–1C we should first take two steps of the predictor, each of length h. This covers $2h$ increments, involves two function evaluations, and has twice the local truncation error. For the A–B formula of (4.7-20) this means that

$$T(x, h) = 2[(5h^3/12)y^{[3]}(\zeta)] = (10h^3/12)y^{[3]}(\zeta) \qquad \text{(P only)}$$

For the corrector which already uses two function evaluations we need to take a step of length $2h$. Thus for the trapezoidal rule corrector

$$T(x, h) = (1/12)(2h)^3 y^{[3]}(\zeta) = (8h^3/12)y^{[3]}(\zeta) \qquad \text{(P–C)}$$

On a fair and competitive basis in terms of the number of function evaluations and the distance through the computation, the P–C will yield a lower truncation error and is to be preferred. In many other cases, however, the predictor by itself is better. This, of course, ignores all consideration of stability and we have already seen the significant changes which can occur in a P–C combination. Thus for the third-order Adams combination the predictor has a bound of -0.6 while the P–C has a bound of -1.8; for the fourth-order case the bounds are -0.3 and -1.3. Even when the number of function evaluations are taken into account the P–C arrangement is superior. Nevertheless, it is possible to suggest that if a stable predictor is chosen, the corrector be used only occasionally and then merely to check on the behavior of the local truncation error.

4.8. PE(CE)ˢ VERSUS P(EC)ˢ COMBINATIONS

While we have already mentioned the distinction between the PE(CE)ˢ and P(EC)ˢ algorithms it is of interest to consider this point in more detail. In the P(EC)ˢ form it is not necessary to calculate a new y'_{n+1} after the value of y_{n+1} is calculated in the corrector. If $s = 1$, the computation yields $\{y_{n+1}, \bar{y}'_{n+1}\}$ instead of $\{y_{n+1}, y'_{n+1}\}$ as in the PE(CE) algorithm. As a result only one-function evaluation is needed instead of two. Thus with the P(EC) form y_{n+1} can be calculated over an interval of $2h$ for the equivalent computation time (assuming $s = 1$ and the function evaluations are the major computing factor) as does the PE(CE) for a step size h. Alternately one can advance the same step size in P(EC) in half the computing time as in PE(CE).

Obviously this is a very attractive feature if the function evaluations are time consuming. Intuitively, however, one would guess that a penalty must be paid for this advantage and that this penalty is probably a decreased stability bound. The question to be investigated here is how do these competing factors affect the overall computation?

An interesting and important point can be made regarding the characteristic equations for the PE(CE)^s and P(CE)^s modes. For example, if we use the Euler predictor and the trapezoidal rule corrector, (4.4-1) and (4.4-2), the PE(CE) algorithm yields (with $y' = \lambda y$) (4.4-3). The P(CE) algorithm will require, however, the corrector as

$$y_{n+1} = y_n + (h/2)[\bar{y}'_{n+1} + \bar{y}'_n] \qquad (4.8\text{-}1)$$

and thus also

$$\bar{y}_n = y_{n-1} + h y'_{n-1} \qquad (4.8\text{-}2)$$

When (4.8-2) and the equation for $\bar{y}_{n+1}$ is substituted into (4.8-1) the result will be a characteristic equation of degree two. As a result the P(EC) algorithm will have one parasitic root whereas the PE(CE) algorithm has none. Obviously stability problems can now occur.

Similarly the fourth-order Adams P–C formulas in the PE(CE) form is given by ($s = 1$)

$$
\begin{aligned}
\bar{y}_{n+1} &= y_n + (h/24)[55y_n' - 59y'_{n-1} + 37y'_{n-2} - 9y'_{n-3}] \\
y_{n+1} &= y_n + (h/24)[9\bar{y}'_{n+1} + 19y_n' - 5y'_{n-1} + y'_{n-2}]
\end{aligned}
\qquad \text{PE(CE)} \quad (4.8\text{-}3)
$$

whereas the P(EC) form is

$$
\begin{aligned}
\bar{y}_{n+1} &= y_n + (h/24)[55\bar{y}_n' - 59\bar{y}'_{n-1} + 37\bar{y}'_{n-2} - 9\bar{y}'_{n-3}] \\
y_{n+1} &= y_n + (h/24)[9\bar{y}'_{n+1} + 19\bar{y}_n' - 5\bar{y}'_{n-1} + \bar{y}'_{n-2}]
\end{aligned}
\qquad \text{P(EC)} \quad (4.8\text{-}4)
$$

The corresponding characteristic polynomials are (4.7-14) for (4.8-3), a fourth-degree polynomial, whereas for (4.8-4)

$$\mu^5 - \left(1 + \frac{8h\lambda}{3}\right)\mu^4 + \frac{95h\lambda}{24}\mu^3 - \frac{91}{24}h\lambda\mu^2 + \frac{15}{8}h\lambda\mu - \frac{3h\lambda}{8} = 0 \quad (4.8\text{-}5)$$

results. Here the P(EC) algorithm yields a fifth-degree polynomial as compared to the PE(CE) algorithm with, of course, a resulting extra parasitic root.

We can also consider the local truncation errors of the two alternatives. Assuming that both the P(EC)^s and PE(CE)^s forms have errors of $O(h^5)$ we can compare the two just as we did the explicit and implicit formulas previously. Thus the PE(CE) form, using $s = 1$, takes one step of length $2h$ with two function evaluations while the P(EC) form takes two steps of length h with two function evaluations. The local truncation error in the PE(CE) from is then $(2h)^5 = 32h^5$ while that in the P(EC) from is $2h^5$; the local truncation error on an equal computation time basis is sixteen times greater for the PE(CE) form than the P(EC) form.

Returning to the Adams P–C forms, we have already shown the stability plot for the PE(CE) algorithm. This was Figure 4.19 with a real absolute negative bound of $-1.3 < h\lambda < 0$. Brown [2] and Klopfenstein and Millman [24] have presented the corresponding plots for the P(EC) case and we show this as Figure 4.24. As seen the real negative stability bound is now $-0.16 <$

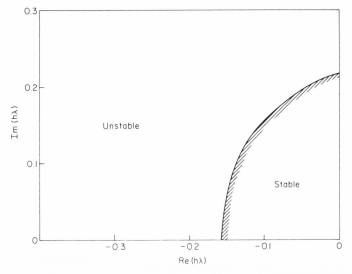

Figure 4.24. Region of stability in complex $h\lambda$ plane for fourth-order Adams P(EC) combination. [Adapted and reprinted with permission of the publisher, The American Mathematical Society from *Mathematics of Computation*; copyright © 1968, **Volume 22**, pp. 557–564.]

$h\lambda < 0$ as compared to $-1.3 < h\lambda < 0$ for the PE(CE) algorithm. While the stability bound has been decreased substantially in the P(EC) algorithm the change in local truncation error compensates for this. The two algorithms seem very competitive. On the complex stability plot, Figure 4.24, we can see a decrease in area to about 10% of that Figure 4.19. This would seem to favor the PE(CE) algorithm.

Klopfenstein and Millman have also presented an alternative predictor to be used with the fourth-order Adams corrector. This predictor is established by seeking the gradient with respect to the free parameters in a manner analogous to the method discussed in Section 4.7.4. The result is the predictor formula

$$\bar{y}_{n+1} = -0.29y_n - 15.39y_{n-1} + 12.13y_{n-2} + 4.55y_{n-3}$$
$$+ h[2.27\bar{y}_n' + 6.65\bar{y}_{n-1}' + 13.91\bar{y}_{n-2}' + 0.69\bar{y}_{n-3}'] \quad (4.8\text{-}6)$$

When combined in a P(EC) mode with the Adams corrector, Figure 4.25 is obtained. As can be seen the real absolute stability boundary has been increased from $h\lambda = -0.16$ to about $h\lambda = -0.78$ by the use of this new predictor. Also the relative stability and the complex stability bounds have been increased significantly. Now this new P–C set operated in a P(EC) mode would seem to be comparable if not better than the PE(CE) mode. The one difficulty with the use of (4.8-6) is the large coefficients and the alternating signs. This will probably mean that round-off error problems will occur and Klopfenstein and Millman confirm this in their numerical experiments.

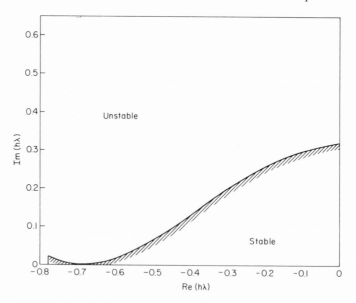

Figure 4.25. Region of stability in complex $h\lambda$ plane for fourth-order Klopfenstein–Millman P(EC) combination. [Adapted and reprinted with permission of the publisher, The American Mathematical Society from *Mathematics of Computation*; copyright © 1968, Volume 22, pp. 557–564.]

Figures 4.26 and 4.27 show the real root plots for the P(EC) form of Milne's and Hamming's P–C method. These are to be compared to Figures 4.5 and 4.12. As can be seen the real negative stability bound for the Milne P(EC) is zero while the Hamming P(EC) bound is $-0.20 < h\lambda < 0$. In both cases the P(EC) algorithm yields a smaller stability bound than does the corresponding PE(CE) algorithm.

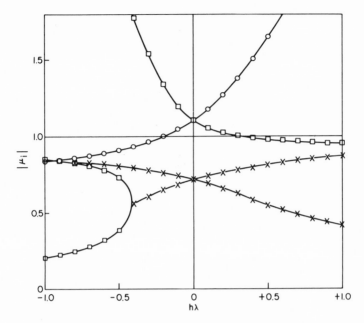

Figure 4.26. Roots of Milne P(EC) combination.

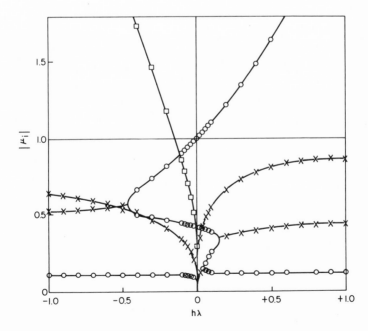

Figure 4.27. Roots of Hamming P(EC) combination.

4.9. SPECIAL SECOND-ORDER DIFFERENTIAL EQUATIONS

In this section we consider the special second-order ODE

$$y'' = f(x, y) \qquad (4.9\text{-}1)$$

in which y' is not included. While many formulas may be developed, only the fifth-order cases derived from (1.9-3) and (1.9-10) will be shown here.

The predictor equation is given by

$$\bar{y}_{n+1} = 2y_{n-1} - y_{n-3} + (4h^2/3)[y_n'' + y_{n-1}'' + y_{n-2}''] \qquad (4.9\text{-}2)$$

with a local truncation error

$$T(x, h) = -(16h^6/240)y^{[6]}(\zeta) \qquad (4.9\text{-}3)$$

and the corrector by

$$y_{n+1} = 2y_n - y_{n-1} + (h^2/12)[\bar{y}_{n+1}'' + 10y_n'' + y_{n-1}''] \qquad (4.9\text{-}4)$$

with error

$$T(x, h) = -(h^6/240)y^{[6]}(\zeta) \qquad (4.9\text{-}5)$$

It follows that we can set up the formulas

$$
\begin{aligned}
&y_n'' = f(x_n, y_n) && \text{Evaluate} \\
&p_{n+1} = 2y_{n-1} - y_{n-3} + (4h^2/3)[y_n'' + y_{n-1}'' + y_{n-2}''] && \text{Predict} \\
&m_{n+1} = p_{n+1} - (16/17)(p_n - c_n) && \text{Modify} \\
&m_{n+1}'' = f(x_{n+1}, y_{n+1}) && \text{Evaluate} \\
&c_{n+1} = 2y_n - y_{n-1} + (h^2/12)[m_{n+1}'' + 10y_n'' + y_{n-1}''] && \text{Correct} \\
&y_{n+1} = c_{n+1} + (1/17)(p_{n+1} - c_{n+1}) && \text{Final}
\end{aligned}
\qquad (4.9\text{-}6)
$$

Obviously the last step (final) can be omitted if desired.

4.10. USE OF HIGHER-ORDER DERIVATIVE P–C COMBINATIONS

An alternative way to solve $y' = f(x, y)$ is to appeal to the use of higher-order derivatives. Many of the formulas of interest have been given in Section 1.10, but it should be realized that the question of the difficulty of calculating the higher derivatives is an important point in the use of the formulas.

While many P–C pairs are possible, only two fourth-order pairs will be given here. In the first case we select the 2-step ($k = 2$) predictor of (1.10-3), or alternatively the 2-step form of (1.10-2)

$$
\begin{aligned}
&\bar{y}_{n+1} = y_{n-1} + 2hy_{n-1}' + (2h^2/3)[2y_n'' + y_{n-1}''] \\
&T(x, h) = (16h^5/360)y^{[5]}(\zeta)
\end{aligned}
\qquad (4.10\text{-}1)
$$

or

$$\bar{y}_{n+1} = y_n + (h/2)[-y_n' + 3y_{n-1}'] + (h^2/12)[17y_n'' + 7y_{n-1}'']$$
$$T(x, h) = (31h^5/720)y^{[5]}(\zeta)$$

(4.10-2)

As a corrector we choose the 1-step (1.10-4)

$$y_{n+1} = y_n + (h/2)[\bar{y}_{n+1}' + y_n'] + (h^2/12)[-\bar{y}_{n+1}'' + y_n'']$$
$$T(x, h) = (h^5/720)y^{[5]}(\zeta)$$

(4.10-3)

If we use (4.10-2) with (4.10-3) the correction to the predictor is $-(31/30)$ $[\bar{y}_n - y_n]$ and this could be used in a modified scheme.

There are two points to note about either (4.10-1) plus (4.10-3) or (4.10-2) plus (4.10-3). First, the local truncation error is very small, in fact smaller than most fourth-order processes. Second, the corrector, being a 1-step process, has no parasitic roots and when iterated has no stability problems. In fact, one can show that the corrector is stable for all real negative $h\lambda$. On this basis this algorithm would appear to be an excellent one with the main defect being the calculation of y''.

The second case we briefly describe is due to Meshaka [34]. He develops the fourth-order P–C pair of

$$\bar{y}_{n+1} = 8y_n - 7y_{n-1} - 2h[2y_n' + y_{n-1}'] + 2h^2 y_n''$$
$$y_{n+1} = (1/7)[8y_n - y_{n-1}] + (2h/7)[\bar{y}_{n+1}' + 2y_n'] + (2h^2/7)y_n''$$

(4.10-4)

Here both the predictor and the corrector are 2-step methods; note that $\bar{y}_{n+1}''$ is not needed in the corrector and this means fewer function evaluations than in the previous algorithm. As $h \to 0$ the corrector leads to

$$7\mu^2 - 8\mu + 1 = 0$$

and the roots $\mu_1 = 1$ and $\mu_2 = 1/7$. Since the second root has a value less than 1 we can say that for a small enough h the corrector will be stable. By contrast the predictor itself is unstable. The numerical examples given by Meshaka confirm the utility of the present approach for those cases where y'' can be calculated without difficulty.

4.11. HYBRID TYPE METHODS

It has already been pointed out in Theorem 3.3.1 that for a linear k-step process of order p, A-stability requires that $p \le k + 2$. This constraint is a severe one, since the generalized corrector (1.7-1) has $2k + 1$ parameters $\alpha_1, \ldots, \alpha_k, \beta_0, \beta_1, \ldots, \beta_k$; in theory at least we should be able to make the corrector of order $p = 2k$ if it were not for the stability consideration.

In an attempt to circumvent this requirement Butcher [4], Gragg and Stetter [17], and Gear [15] developed hybrid P–C methods which can achieve $p = 2k$ or higher. The term hybrid refers to the fact that a "nonstep" point is used in the formulation, i.e., $f(x, y)$ is evaluated at some point *between* x_n and x_{n+1}; as such it bears a resemblance to the Runge–Kutta approach. This connection will become obvious shortly, but it is worth mentioning that the number of function evaluations must also increase in this new formulation over that for the normal P–C mode.

4.11.1. The Butcher Approach

We shall first present the hybrid analysis used by Butcher and then some of the possible variations. The modified k-step corrector is chosen to be

$$
\begin{aligned}
y_{n+1} = \alpha_1 y_n + \alpha_2 y_{n-1} + \cdots + \alpha_k y_{n+1-k} \\
+ h[\beta_0 f_{n+1} + \beta_1 f_n + \cdots + \beta_k f_{n+1-k}] + h\beta f_{n+1-\theta}
\end{aligned}
\tag{4.11-1}
$$

Note that this differs from our previous k-step corrector (1.7-1) by including the term $h\beta f_{n+1-\theta}$. β and θ are new parameters with $\theta \neq 0, 1, 2, \ldots, k$. Now there are $2k + 3$ parameters and the formula can be made of maximum order $p = 2k + 2$. Actually Butcher fixes θ at $\frac{1}{2}$ or $\frac{1}{3}$ and thus the maximum order is $2k + 1$. As variations on this formulation, we mention first that Danchick [11] considered the inclusion of $\alpha y_{n+1-\theta}$ in (4.11-1), but showed that it did not really yield an improvement; second, Gragg and Stetter, as an example, leave θ as a free parameter and a resulting corrector of order $p = 2k + 2$ is achieved.

Ignoring for the moment the question of how to calculate $f_{n+1-\theta}$, we revise the form of (4.11-1) and solve this equation for $f_{n+1-\theta}$. This yields

$$
h f_{n+1-\theta} = \sum_{i=0}^{k} (-\alpha_i/\beta) y_{n+1-i} + h \sum_{i=0}^{k} (-\beta_i/\beta) f_{n+1-i}
\tag{4.11-2}
$$

The coefficients in this equation can be identified with the derivative of the Hermite interpolation polynomial. Writing

$$
M = \{\theta(1-\theta)(2-\theta)\cdots(k-\theta)/k!\}^2
$$

$$
K = -\frac{1}{\theta} + \frac{1}{1-\theta} + \frac{1}{2-\theta} + \cdots + \frac{1}{k-\theta}
$$

$$
H_i = H_{i-1} + (1/i), \qquad i \geq 1, \qquad H_0 = 0
$$

it can be shown [4] that

$$
-\frac{\beta_i}{\beta} = M\binom{k}{i}^2 \frac{1}{i-\theta}\left(2K - \frac{1}{i-\theta}\right)
\tag{4.11-3}
$$

and

$$
-\frac{\alpha_i}{\beta} = M\binom{k}{i}^2 \left[\frac{2}{(i-\theta)^2}\left(K - \frac{1}{i-\theta}\right) - \frac{2}{i-\theta}(H_{k-i} - H_i)\left(2K - \frac{1}{i-\theta}\right)\right]
\tag{4.11-4}
$$

These equations determine all the coefficients in (4.11-1). Butcher also shows that the simplest stable processes in the limit of $h = 0$ are for $k = 1, 2, 3$, with $\theta = \frac{1}{2}$ and for $k = 4, 5, 6$, with $\theta = \frac{1}{3}$. For $k = 8$, no stable process appears to exist.

As a typical example the case $k = 1$, $\theta = \frac{1}{2}$, yields the corrector

$$y_{n+1} = y_n + (h/6)[f_{n+1} + 4f_{n+1/2} + f_n] \qquad (4.11\text{-}5)$$

with $p = 4$ or an error $O(h^5)$. Note this is a compressed form of Milne's corrector. If we visualize the corrector as proceeding from y_n to $y_{n+1/2}$ to y_{n+1} in steps of length $h/2$ then (4.11-5) could be viewed as a $k = 2$ process and the order would be equal to $p = k + 2 = 4$, since $k = 2$ and is even. Thus this formula, while it looks as if it ignores the constraints of Theorem 3.3.3, really does not but rather changes the frame of reference of h.

Next we consider how to obtain the term $f_{n+1-\theta}$. Since, in a sense, we are faced with the same question for f_{n+1} if we do not iterate the corrector it seems natural to write two predictors [in a PE(CE) mode]

$$\bar{y}_{n+1-\theta} = \bar{\alpha}_1 y_n + \bar{\alpha}_2 y_{n-1} + \cdots + \bar{\alpha}_k y_{n+1-k}$$
$$+ h[\bar{\beta}_1 f_n + \cdots + \bar{\beta}_k f_{n+1-k}] \qquad (4.11\text{-}6)$$

$$\bar{y}_{n+1} = \alpha_1 y_n + \alpha_2 y_{n-1} + \cdots + \alpha_k y_{n+1-k}$$
$$+ h[\beta \bar{f}_{n+1-\theta} + \beta_1 f_n + \cdots + \beta_k f_{n+1-k}] \qquad (4.11\text{-}7)$$

The idea is to use (4.11-6) to predict $\bar{y}_{n+1-\theta}$ and to evaluate $\bar{f}_{n+1-\theta}$; then $\bar{f}_{n+1-\theta}$ is used in (4.11-7) to predict $\bar{y}_{n+1}$; this value is used to evaluate $\bar{f}_{n+1}$ and the corrector used for y_{n+1} and f_{n+1}. Thus in a PE(CE) mode two predictors and one corrector are required with three function evaluations as contrasted to two function evaluations in previous PE(CE) schemes. Obviously there are $2k$ parameters in (4.11-6) and $2k + 1$ in (4.11-7) and we could make them of order $p = 2k - 1$ and $p = 2k$ respectively. Rather than do this, Butcher makes both $p = 2k - 1$ with a special added relationship used to simplify the formulas and the entire P–1C hybrid formulas are determined. Typical examples are given below; the local truncation errors are so complicated that we do not include them:

$$\bar{y}_{n+1/2} = y_n + (h/2)f_n$$
$$\bar{y}_{n+1} = y_n + h[2\bar{f}_{n+1/2} - f_n]$$
$$y_{n+1} = y_n + (h/6)[\bar{f}_{n+1} + 4\bar{f}_{n+1/2} + f_n]$$
$$k = 1, \qquad \theta = \tfrac{1}{2}, \qquad p = 3 \qquad (4.11\text{-}8)$$

$$\bar{y}_{n+1/2} = y_{n-1} + (h/8)[9f_n + 3f_{n-1}]$$
$$\bar{y}_{n+1} = (1/5)[28y_n - 23y_{n-1}] + (h/15)[32\bar{f}_{n+1/2} - 60f_n - 26f_{n-1}]$$
$$y_{n+1} = (1/31)[32y_n - y_{n-1}] + (h/93)[64\bar{f}_{n+1/2} + 15\bar{f}_{n+1} + 12f_n - f_{n-1}]$$
$$k = 2, \qquad \theta = \tfrac{1}{2}, \qquad p = 5 \qquad (4.11\text{-}9)$$

$$\bar{y}_{n+1/2} = (1/128)[-225y_n + 200y_{n-1} + 153y_{n-2}]$$
$$+ (h/128)[225f_n + 300f_{n-1} + 45f_{n-2}]$$
$$\bar{y}_{n+1} = (1/31)[540y_n - 297y_{n-1} - 212y_{n-2}]$$
$$+ (h/155)[384\bar{f}_{n+1/2} - 1395f_n - 2130f_{n-1} - 309f_{n-2}]$$
$$y_{n+1} = (1/617)[783y_n - 135y_{n-1} - 31y_{n-2}]$$
$$+ (h/3085)[2304\bar{f}_{n+1/2} + 465\bar{f}_{n+1} - 135f_n - 495f_{n-1} - 39f_{n-2}]$$
$$k = 3, \qquad \theta = \tfrac{1}{2}, \qquad p = 7 \qquad\qquad (4.11\text{-}10)$$

Figure 4.28 shows the real stability plot for the Butcher fifth-order equation (4.11-9). As can be seen this hybrid P–C has an excellent bound of $-1.45 < h\lambda < -0.1$.

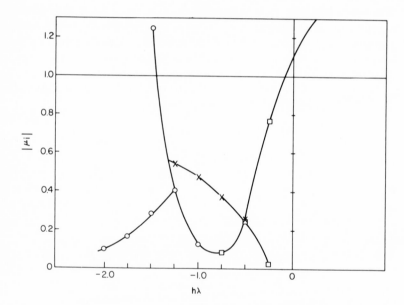

Figure 4.28. Roots of Butcher Hybrid P–C combination.

At this point some further interesting comparisons can be made between (4.11-9), the fifth-order Adams P–C form, and the fifth-order Runge–Kutta form:

	k	Function evaluations
(4.11-9)	2	3
Adams	5	2
Runge–Kutta	1	6

Thus as expected (4.11-9) has properties between the two extremes of the usual P–C and the Runge–Kutta single-step methods. As a result it is called a hybrid method.

4.11.2. Different Predictors

A natural extension of the work of Butcher would be to make the two predictors of order $p = 2k + 2$ and make the corrector of the same order by leaving θ free. Danchick has done just this and shows that it is possible to achieve strongly stable maximal order $(2k + 2)$ methods for $k \le 6$. The corrector formulas are equivalent to (4.11-1) but with θ also free; since there are $2k + 3$ parameters the formulas can be made of order $p = 2k + 2$. The predictors are slightly different than (4.11-6) and (4.11-7). On one hand the term $\bar{f}_{n+1-\theta}$ is not included in the $\bar{y}_{n+1}$ predictor, while on the other hand, enough additional back values are added to make $2k + 3$ parameters, i.e., in (4.11-6) two more $y(x)$ points and one more $f(x, y)$ point is added. The same idea is used for (4.11-7). The end result is a series of 2P–1C hybrid formulas which are all of order $2k + 2$. One predictor is used for $\bar{y}_{n+1-\theta}$, the other for $\bar{y}_{n+1}$ and the corrector then includes both $\bar{f}_{n+1-\theta}$ and $\bar{f}_{n+1}$. Two of these are listed below along with the modulus of the characteristic roots to show that the equations are stable (for $h \to 0$).

$$\bar{y}_{n+1} = -9y_n + 9y_{n-1} + y_{n-2} + h[6f_n + 6f_{n-1}]$$
$$\bar{y}_{n+1/2} = -\tfrac{45}{64}y_n + \tfrac{25}{16}y_{n-1} + \tfrac{9}{64}y_{n-2} + h[\tfrac{90}{64}f_n + \tfrac{15}{16}f_{n-1}]$$
$$y_{n+1} = y_n + (h/6)[\bar{f}_{n+1} + 4\bar{f}_{n+1/2} + f_n]$$

$$k = 1, \qquad \theta = \tfrac{1}{2}, \qquad |\mu_1| = 1 \qquad\qquad (4.11\text{-}11)$$

$$\bar{y}_{n+1} = -28y_n + 28y_{n-2} + y_{n-3} + h[12f_n + 36f_{n-1} + 12f_{n-2}]$$
$$\bar{y}_{n+0.5773} = -4.5550y_n + 0.8369y_{n-1} + 4.565y_{n-2} + 0.1530y_{n-3}$$
$$\qquad\qquad + h[2.844f_n + 6.247f_{n-1} + 1.911f_{n-2}]$$
$$y_{n+1} = 0.9603y_n + 0.03963y_{n-1} + h[0.1293\bar{f}_{n+1} + 0.6237\bar{f}_{n+0.5773}$$
$$\qquad\qquad + 0.2772f_n + 0.009285f_{n-1}]$$

$$k = 2, \qquad \theta = 0.5773502, \qquad |\mu_1| = 1, \qquad |\mu_1| = 0.039630 \quad (4.11\text{-}12)$$

Other formulas are given by Danchick with the coefficients to ten digit accuracy.

4.11.3. Further Predictors

Kohfeld and Thompson [25] pointed out that in the Gragg–Stetter formulation for the predictors (like those of Danchick, but only of order $p = 2k + 1$) the truncation error may be so large that it invalidates the accuracy of the corrector. In the case $k = 2$ they showed that the corrector

was about eight hundred times as accurate as the equivalent sixth-order
Adams formula. However, they also showed that the Gragg–Stetter predictor
for $\bar{y}_{n+1}$ is about seventeen hundred times less accurate than that of the
corrector. Thus the remarkable accuracy of the corrector is largely lost owing
to the relatively large error in the predictor. This contrasts with the Adams
P–C in which the local truncation error of the predictor is only about twenty-
one times that of the corrector. Note that this comparison is made with a predictor of the form of (4.11-7) but without the $b\bar{f}_{n+1-\theta}$ term; thus in Butcher's
approach this disparity between predictor and corrector error does not occur.

As a result, Kohfeld and Thompson suggest that a more accurate predictor be used to calculate $\bar{y}_{n+1}$; this predictor uses another "nonstep"
point θ_1. Thus their method involves:

1. a predictor to generate $\bar{y}_{n+\theta_1}$ using only previous values of y_i and f_i (P_1)
2. a predictor to generate $\bar{y}_{n+\theta}$ using previous values of y_i and f_i and a
    weighted $\bar{f}_{n+\theta_1}$ (P_2)
3. a predictor to generate $\bar{y}_{n+1}$ using previous values of y_i and f_i and a
    weighted $\bar{f}_{n+\theta_1}$ (P_3)
4. a corrector to generate y_{n+1} using previous values of y_i and f_i and
    weighted $\bar{f}_{n+\theta}$ (C).

The corrector is the same as that of Danchick and Gragg–Stetter, of order
$p = 2k + 2$ (equivalent to Butcher's if θ were not fixed). The predictors are all
made of order $p = 2k + 1$ by selecting enough past values of y_i and f_i and
the weighting factors to achieve this accuracy. Thus they use

(P_1) $k + 2$ of the y_i
        $k + 2$ of the f_i $2k + 4$ parameters

(P_2) $k + 2$ of the y_i
        $k + 1$ of the f_i $2k + 4$ parameters
        1 weight factor

(P_3) k of the y_i
        $k + 1$ of the f_i $2k + 3$ parameters
        1   θ_1
        1 weight factor

(C) k of the y_i
        $k + 1$ of the f_i $2k + 3$ parameters
        1   θ
        1 weight factor

By this approach Kohfeld and Thompson decrease the predictor error to
about that in the Adams P–C method. Numerical values to about twelve
decimals are presented for the cases $k \le 6$ with two sets of formulas given

below

$$\bar{\bar{y}}_{n+0.7} = -5.793y_n + 3.572y_{n-1} + 3.221y_{n-2}$$
$$+ h[3.686f_n + 6.072f_{n-1} + 0.9558f_{n-2}]$$
$$\bar{\bar{y}}_{n+1/2} = 0.5336y_n + 0.4494y_{n-1} + 0.01695y_{n-2}$$
$$+ h[0.7024f_n + 0.1942f_{n-1} + 0.08660\bar{\bar{f}}_{n+0.7}]$$
$$\bar{y}_{n+1} = y_n + h[0.3095f_n - 0.009803f_{n-1} + 0.7002\bar{\bar{f}}_{n+0.7}]$$
$$y_{n+1} = y_n + (h/6)[\bar{f}_{n+1} + 4\bar{f}_{n+1/2} + f_n]$$
$$k = 1, \qquad \theta = \tfrac{1}{2}, \qquad \theta_1 = 0.7 \qquad (4.11\text{-}13)$$

$$\bar{\bar{y}}_{n+.7566} = -16.30y_n - 11.61y_{n-1} + 23.41y_{n-2} + 4.509y_{n-3}$$
$$+ h[6.955f_n + 26.96f_{n-1} + 17.18f_{n-2} + 1.400f_{n-3}]$$
$$\bar{\bar{y}}_{n+.5773} = -.5709y_n + .8758y_{n-1} + .6808y_{n-2} + .01430y_{n-3}$$
$$+ h[1.262f_n + 1.266f_{n-1} + .2466f_{n-2} + .08253\bar{\bar{f}}_{n+.7566}]$$
$$\bar{y}_{n+1} = .7195y_n + .2804y_{n-1} + h[.5929f_n + .09455f_{n-1}$$
$$- .002092f_{n-2} + .5941\bar{\bar{f}}_{n+.7566}]$$
$$y_{n+1} = .9603y_n + .03963y_{n-1} + h[.1293\bar{f}_{n+1} + .2772f_n$$
$$+ .009285f_{n-1} + .6237\bar{\bar{f}}_{n+.5773}]$$
$$k = 2, \qquad \theta = 0.5773, \qquad \theta_1 = .7566 \qquad (4.11\text{-}14)$$

Note that now the P–C process requires four function evaluations as compared to two for the Adams P–C and three for the Butcher hybrid. However, because of the vastly increased accuracy, these formulas only require about two-thirds as much computing time as the Adams formulas. Thus the method is to be recommended.

4.11.4. The Special Second-Order ODE

Dyer [13] has recently extended the hybrid approach with a nonstep point to the special second-order equation with no first derivative. He attains an accuracy of order $p = 2k$ and the formulas given by

Predictor, $p = 6$, $\theta = 0.3$

$$\bar{y}_{n+1-\theta} = \sum_{i=1}^{4} \bar{\alpha}_i y_{n+1-i} + h^2 \sum_{i=1}^{4} \bar{\gamma}_i y''_{n+1-i} \qquad (4.11\text{-}15)$$

with

$$\bar{\alpha}_1 = -1.9320993 \qquad \bar{\gamma}_1 = 0.862500525$$
$$\bar{\alpha}_2 = 6.4719549 \qquad \bar{\gamma}_2 = 3.030380475$$
$$\bar{\alpha}_3 = -3.4476119 \qquad \bar{\gamma}_3 = 0.433679775$$
$$\bar{\alpha}_4 = -0.0922437 \qquad \bar{\gamma}_4 = -0.007217775$$

and
  Predictor

$$\bar{y}_{n+1} = -16y_n + 34y_{n-1} - 16y_{n-2} - 1y_{n-3}$$
$$+ h[2.66 \cdots 67y_n'' + 14.66 \cdots 67y_{n-1}'' + 2.66 \cdots 67y_{n-2}''] \quad (4.11\text{-}16)$$

and
  Corrector $p = 7$

$$y_{n+1} = \sum_{i=1}^{4} \alpha_i y_{n+1-i} + h^2 \sum_{i=0}^{4} \gamma_i y_{n+1-k}'' + h^2 \gamma y_{n+1-\theta}'' \quad (4.11\text{-}17)$$

with

$$\alpha_1 = \quad 2.05804967222800 \qquad \gamma_0 = \quad 0$$
$$\alpha_2 = -0.963457924410107 \qquad \gamma_1 = \quad 0.704431783153855$$
$$\alpha_3 = -0.247233242848194 \qquad \gamma_2 = \quad 0.0872587915934802$$
$$\alpha_4 = \quad 0.152641470035498 \qquad \gamma_3 = \quad 0.154458906599228$$
$$\gamma = \quad 0.161595975532325 \qquad \gamma_4 = -0.951881093873159$$

Note that (4.11-17) has the unusual feature that the corrector is explicit because $\beta_0 = 0$. This means that only two function evaluations per step are required.

4.11.5. More Than One Nonstep Point

Finally we consider a logical extension of the Butcher approach. Since the cases (4.11-8)–(4.11-10) require three function evaluations with one nonstep point, they resemble a third-order Runge–Kutta formula. If we were to use two nonstep points *in the corrector* the result would resemble a fourth-order Runge–Kutta (four function evaluations). Note, however, that this is different than the use of two nonstep formulas such as (4.11-13) with one step used for the predictor and the other for the corrector.

The analysis we briefly present on this case follows Butcher [6] although Brush, Kohfeld, and Thompson [3] have also contributed in this area. As indicated previously for zero nonstep points, stability limits the formula to $k < 3$; with one nonstep point, stability has been indicated as limited to $k \leq 7$ with $p = 2k + 1$ realizable; now we shall have two nonstep points and stability allows $k \leq 15$ with $p = 2k + 2$. The use of the nonstep points also increases the available accuracy by a significant margin.

We now define a corrector which has two nonstep terms

$$y_{n+1} = \sum_{i=1}^{k} \alpha_i y_{n+1-i} + h\left(b_1 \bar{f}_{n+1-\theta} + b_2 \bar{\bar{f}}_{n+1-\theta_2} + \sum_{i=0}^{k} \beta_i f_{n+1-i}\right) \quad (4.11\text{-}18)$$

If θ and θ_2 are fixed, there are $2k + 3$ parameters $\alpha_1, \ldots, \alpha_k, b_1, b_2, \beta_0, \beta_1,$ $\ldots, \beta_k$. Thus we can achieve $p = 2k + 2$. Obviously $\theta \neq \theta_2$ with neither an integer. Butcher shows that for $k \leq 15$ the parasitic roots (for $h = 0$) as a function of θ and θ_2 are all less than one and the resulting formulas are A-stable. To go along with (4.11-18) we use three predictors with the following intent:

1. Predict $\bar{y}_{n+1-\theta}$ from past values of $y_i(x)$ and $f_i(x, y)$. This yields $\bar{f}_{n+1-\theta}$.
2. Predict $\bar{y}_{n+1-\theta_2}$ from past values of $y_i(x)$ and $f_i(x, y)$ and $\bar{f}_{n+1-\theta}$. This yields $\bar{\bar{f}}_{n+1-\theta_2}$.
3. Predict $\bar{y}_{n+1}$ from past values of $y_i(x)$ and $f_i(x, y)$ and $\bar{f}_{n+1-\theta}$ and $\bar{\bar{f}}_{n+1-\theta_2}$. This yields $\bar{\bar{\bar{f}}}_{n+1}$.

Typical examples are given below for $\{\theta, \theta_2\} = \{\frac{2}{3}, \frac{1}{3}\}$ and $\{\frac{1}{2}, \frac{1}{4}\}$.

$$\bar{y}_{n+1/3} = (1/27)[16y_n + 11y_{n-1}] + [h/27)(16f_n + 4f_{n-1}]$$

$$\bar{y}_{n+2/3} = (1/27)[47y_n - 20y_{n-1}] + (h/27)[27\bar{f}_{n+1/3} - 22f_n - 7f_{n-1}]$$

$$\bar{y}_{n+1} = (1/10)[-13y_n + 23y_{n-1}] + (h/80)[108\bar{\bar{f}}_{n+2/3} - 189\bar{f}_{n+1/3}$$
$$+ 284f_n + 61f_{n-1}]$$

$$y_{n+1} = (1/49)[48y_n + y_{n-1}] + (h/1470)[160\bar{\bar{\bar{f}}}_{n+1} + 648\bar{\bar{f}}_{n+2/3}$$
$$+ 405\bar{f}_{n+1/3} + 280f_n + 7f_{n-1}]$$

$$k = 2, \qquad \theta = \tfrac{2}{3}, \qquad \theta_2 = \tfrac{1}{3} \qquad\qquad (4.11\text{-}19)$$

$$\bar{y}_{n+1/2} = y_{n-1} + (h/8)[9f_n + 3f_{n-1}]$$

$$\bar{y}_{n+3/4} = (1/256)[1309y_n - 1052y_{n-1}]$$
$$+ (h/512)[756\bar{f}_{n+1/2} - 1659f_n - 819f_{n-1}]$$

$$\bar{y}_{n+1} = (1/53)[-140y_n + 193y_{n-1/2}] + (h/1113)[512\bar{\bar{f}}_{n+3/4} - 560\bar{f}_{n+1/2}$$
$$+ 3640f_n + 1574f_{n-1/2}]$$

$$y_{n+1} = (1/33)[32y_n + y_{n-1}] + (h/10395)[1113f_n + 2048\bar{\bar{f}}_{n+3/4} + 4928\bar{f}_{n+1/2}$$
$$+ 2548f_n + 73f_{n-1}]$$

$$k = 2, \qquad \theta = \tfrac{1}{2}, \qquad \theta_2 = \tfrac{1}{4} \qquad\qquad (4.11\text{-}20)$$

Further formulas, including $k \leq 4$, are explicitly presented by Butcher.

In summary, the formulas of the present section result in a high accuracy (large p) for moderate k but at the expense of the number of function evaluations. However, no work has been carried out relating to the stability bounds ($h \neq 0$) for these formulas; further, the question of the feasibility of P(EC) algorithms to decrease the number of function evaluations has not been investigated. Such results would be of interest.

4.12. STARTING THE P–C COMPUTATION

It is readily apparent that one of the computational defects of the P–C methods is that the methods are not self-starting. This is in contrast to the single-step methods where only the single value $y = y_0$ is needed to start the computation. For the P–C methods, an additional set of points $y_1, y_2, \ldots$ are required in addition to y_0 before the methods can be used. The recommended procedure is to use a single-step formula to step from y_0 to $y_1, \ldots$. Obviously, if the P–C set is pth order, at least a pth-order single-step formula should be used.

4.13. ADJUSTMENT OF THE STEP SIZE DURING THE P–C SOLUTION

Once the method for starting the solution is used, the predictor–corrector algorithm can be evoked and the solution continued as far into the x domain as desired. However, in the second and major portion of the computation, a most important parameter is the stepsize to use. In fact, the maximum step size $h_{\max}$ as constrained by the accuracy and stability is the item which is desired.

4.13.1. Stability Considerations

The upper bound on the stepsize $h_{\max}$ is determined by the stability bounds of the particular P–C algorithm. From the root and root locus plots, the upper bound on $(-\lambda h)_{\text{calc}}$ can be ascertained. Using this number, with a degree of caution since it is developed on the basis of the use of linear differential equations, the next step is to calculate $h_{\max}$.

For a set of simultaneous first-order ODEs the most obvious approach is to linearize the differential equations to yield the Jacobian matrix. From the Jacobian matrix, the largest negative eigenvalue can be calculated, $\lambda_{i,\max}$. Actually this is not a significant computation problem, since eigenvalue determination for the single largest value can be obtained rather easily by the power method [31]. A bound on $h_{\max}$ follows from

$$h_{\max} \le (-\lambda h)_{\text{calc}}/|\lambda_{i,\max}| \tag{4.13-1}$$

and all computations must operate within this bound. Distefano [12] presents many interesting results along this line for the special case where the Jacobian matrix is a tridiagonal matrix.

When only a single ODE is being integrated, the linearization yields $\lambda = f_y$ immediately and the corresponding $h_{\max}$ follows directly. Alternatively f_y can be calculated from the computation by the use of a finite difference approximation

$$\lambda = f_y \simeq \frac{\bar{f}_n - f_n}{\bar{y}_n - y_n} \qquad (4.13\text{-}2)$$

A certain amount of caution must be used in this analysis because of the linear approximation. But as long as one is conservative, this analysis provides a useful upper bound on the allowable stepsize in the P–C computation. Within this framework the computation must be adjusted for accuracy.

4.13.2. Truncation Error Considerations

In terms of the accuracy or the truncation error we turn to the use of $\bar{y}_n - y_n$ or $\bar{y}_{n+1} - y_n$ or the modified P–C equations to ascertain the magnitude of the h to be used in the calculation. As the computation proceeds, we monitor this difference between the predicted and corrected values. If this value is too large, as determined by some external condition, the step size h is too large and we step back and repeat the calculation with a smaller step, say $h/2$ for convenience. Note we do not go back to the beginning of the calculation but only to the point at which the truncation error became too large. If the truncation error is too small, we are wasting computation time and the h should be increased, say to $2h$ for convenience (assuming that $h_{\max}$ is not reached). If we can devise some simple means to carry out these calculations for $h/2$ or $2h$ a full running P–C system is available which starts and then monitors itself as the calculation proceeds to maintain always an established bound on the local truncation error. One word of caution is in order however. This adjustment down and up of h should not be done too often; this follows from the fact that in deriving the measure of the local truncation error we assumed that $\bar{y}_{n+1} - y_{n+1}$ did not vary significantly from step to step. Experience would seem to indicate that once h has been changed about ten steps be taken at the new step size before the error calculation is repeated.

The easiest case of changing the step size is when h is to be doubled. All that is needed is to be sure that sufficient data are retained in the computer memory to allow the change. Thus if information for y_n, y_{n-1}, and y_{n-2} associated with the use of h has been retained, when we change to $2h$ the points in the P–C formula (for $k = 2$) to be used are y_n and y_{n-2}. At the same time, since we are talking about a formula with h^3 in its error term, the value of $\bar{y}_n - y_n$ previously calculated should be multiplied by $2^3 = 8$. Except for the special case where the calculation has just started and the back values are not available, this scheme for doubling h is quite easy to implement.

In order to halve the interval, a more complicated procedure may be necessary. For the $k = 2$ P–C set, we need only replace y_{n-1} with $y_{n-1/2}$

and then use $y_{n-1/2}$ and y_n to calculate $y_{n+1/2}$. This can be done by the formulas

$$y_{n-1/2} = (1/2)[y_n + y_{n-1}] - (h/8)[y_n' - y_{n-1}']$$
$$y_{n-1/2}' = f(x_{n-1/2}, y_{n-1/2})$$

(4.13-3)

where the first equation is a suitable interpolation equation. At the same time, we need to use $\frac{1}{8}(\bar{y}_n - y_n)$ in the error calculation. If, however, we were using a fourth-order P–C set, the interpolation requires a considerable number of points and complex formulas. Further, the programming associated with the change in h may become complicated.

As a result many P–C schemes use a suitable Runge–Kutta formula to generate the necessary additional points as the h is decreased or increased. This is particularly recommended if the Runge–Kutta routine is used to start the calculation. Thus, many operating P–C computer routines use the Runge–Kutta to start the calculation and then to generate any point needed as the local truncation error suggests a shift up or down in the value of h.

4.13.3. Further Stability Details

The work of Lambert [30] removes one defect of the stability analysis we have previously presented and thus bears directly on the specification of h_{max}. Rather than assume that $\lambda = f_y = $ a constant in the analysis, Lambert suggests the use of $\lambda = \lambda_n$ and $\lambda_n' = f_y'$ in the linear stability analysis. When a suitable combination of λ_n and λ_n' is substituted in the linear k-step difference equation, a characteristic equation is obtained which is a function of $h\lambda$ and $\sigma = \lambda_n'/\lambda_n$, $\lambda_n \neq 0$. Now it is possible to determine the usual regions of stability but as a function of $h\lambda$ and σ.

Lambert shows plots of this approach for the Hermite predictor–Milne corrector and the two-step Adams predictor and corrector. σ is plotted against $h\lambda$ and the case $\sigma = 0$ corresponds to the analysis given previously. Now there are substantial regions with $\sigma \neq 0$ where the P–C systems are stable. By means of an example, Lambert also shows that the case $\sigma = 0$ ($\lambda_n = $ constants would lead to a conclusion of stability whereas the present analysis show) instability; this latter point is then confirmed numerically. Obviously work such as this will prove of importance in terms of more realistic determination of stability regions.

4.14. TABULATION OF EQUATIONS AND STABILITY BOUNDS

Table 4.1 contains a tabulation of most of the equations used in this chapter. Table 4.3 contains a summary of much of the stability information of this chapter plus certain other values not discussed explicitly.

TABLE 4.3

THEORETICAL STABILITY LIMITS FOR DIFFERENT P–C COMBINATIONS

Method	Real negative stability limit	Imaginary stability limit
1. Euler P	-2	0
Modified Euler C	$-\infty$	
PE(CE)^s	$-2, \quad s = 1$	0
	$-2, \quad s = 2$	$2i$
	$-2, \quad s = 3$	
2. Nystrom P	unstable	
PE(CE)^s with modified		
Euler C	$-2, \quad s = 1$	
	$-2, \quad s = 2$	
3. Euler P-backward		
Euler C, PE(CE)^s	$-1, \quad s = 1$	
	$-1, \quad s = 2$	
	$-1, \quad s = 3$	
4. Milne P	unstable	
Milne C	unstable	
PE(CE)	$-0.8 < h\lambda < -0.3$	0
P–M_p–C–M_c	$-0.42 < h\lambda < -0.2$	
5. Hermite P	unstable	
Hermite P–Milne C	-1	0
6. Hamming C	-2.6	
Milne P–Hamming C as		
PE(CE)^s	$-0.5, \quad s = 1$	
	$-0.9, \quad s = 2$	
P–M_p–C–M_c	-0.85	$0.7i$
7. Third-order Adams		
P	-0.55	$0.72i$
C	-6.0	
PE(CE)^s	$-1.8, \quad s = 1$	$1.3i$
	$-1.3, \quad s = 2$	$1.6i$
	$-2.0, \quad s = 3$	$0.7i$
P–M_p–C–M_c	-1.6	
8. Fourth-order Adams		
P	-0.3	$0.42i$
C	-3.0	$1i$
PE(CE)^s	$-1.3, \quad s = 1$	$0.9i$
	$-0.9, \quad s = 2$	
P–M_p–(CE)^2	-0.7	
9. Fifth-order Adams		
P	-0.2	$0.25i$
C	-1.8	$1.3i$
PE(CE)	-0.95	$0.7i$

TABLE 4.3 (continued)

Method	Real negative stability limit	Imaginary stability limit
10. Higher-order Adams		
Sixth-order C	-1.2	$1.3i$
PE(CE)	-0.7	$0.55i$
Seventh-order C	-0.8	$1.05i$
PE(CE)	-0.5	$0.4i$
Eighth-order C	-0.5	$0.68i$
PE(CE)	-0.4	$0.3i$
11. Crane–Klopfenstein		
PE(CE)	-2.48	$0.7i$
Fourth-order Krogh PE(CE)	-1.8	$0.95i$
Fifth-order	-1.45	$0.72i$
Sixth-order	-1.05	$0.55i$
Seventh-order	-0.80	$0.42i$
Eighth-order.	-0.60	$0.3i$
12. P(EC) Combination		
Fourth-order Adams	-0.16	
Milne	0	
Hamming	-0.20	
Klopfenstein–Millman	-0.78	
13. Butcher Fifth-order hybrid	$-1.45 < h\lambda < -0.1$	

4.15. PUBLISHED NUMERICAL RESULTS

The amount of numerical results discussed in published papers on P–C methods is extensive. Here we will outline some of those which make points of definite interest and importance.

4.15.1. Stability of P–C Methods

Both Chase [7] and Ralston [41] have shown the computational instabilities in the use of Milne's and Hamming's methods. Chase used the system

$$y' = +100 - 100y, \qquad y(0) = 0 \qquad (4.15\text{-}1)$$

and Ralston used

$$y' = \pm y, \qquad y(0) = 1 \qquad (4.15\text{-}2)$$

and

$$y' = \frac{1}{1 + \tan^2 y}, \qquad y(0) = 0 \qquad (4.15\text{-}3)$$

Both authors showed explicitly that when λ was negative, (4.15-1) and (4.15-2) with the negative sign, Milne's method exhibited instability. By contrast Hamming's method tended to remain stable for x distances which are much greater than for Milne's method. When, however, λ was positive, (4.15-2) with the plus sign, Milne's method performed quite well and because of its lower truncation error may be preferred to the Hamming approach.

In addition Ralston showed that when the higher-order derivatives can be calculated easily [such as for (4.15-2) but not (4.15-3)] P–C methods which employ these higher derivatives are extremely powerful.

Babuska (see Chapter 2[1]) has shown some further interesting computational points. First he illustrated the inherent instability of the Nystrom predictor (4.1-1) when used alone while the Adams–Bashforth predictor, of the equivalent order, remained stable. When the latter was applied to the system

$$y' = 1 - y, \qquad y(0) = 2 \qquad (4.15\text{-}4)$$

he showed that a plot of the error versus $1/h$ exhibited the characteristic minimum point resulting from a decreasing truncation error followed by an increasing round-off error. Second, Babuska considered the system

$$y'' = -y, \qquad y(0) = 0, \quad y'(0) = 1 \qquad (4.15\text{-}5)$$

and solved this with a number of correctors or predictors [such as (4.9-4)] which take advantage of the lack of a first derivative; this was contrasted with the solution via an Adams predictor with (4.15-5) treated as two first-order equations, i.e.,

$$y_1' = y_2, \qquad y_2' = y_1 \qquad (4.15\text{-}6)$$

In each case the solution of (4.15-6) gave more accurate and better results than when (4.15-5) was solved directly.

4.15.2. Adams P–C Combinations

Hull and co-workers have developed results which indicate that high order ($p = 6$ or 7) Adams P–C methods in which the corrector is iterated twice are perhaps the best general integration formulas to use. For test systems of the form

$$y' = ay + b \sin \omega x \qquad (4.15\text{-}7)$$

they first analyzed the correctors of the form of (4.7-17)–(4.7-19). For $c = 0$, these are the Adams correctors and if iterated to convergence, the largest errors as a function of k were calculated to be

k	Error $\times 10^{-6}$	k	Error $\times 10^{-6}$
2	1500	6	17
3	380	7	6
4	130	8	2
5	43		

These comparisons, however, do not consider the number of iterations on a cost [number of times to evaluate $f(x, y)$] basis and thus further calculations were performed with s, the number of corrector iterations, as a variable. These computations were summarized in plots of $\log_{10}$ error versus $\log_2(s/h)$ or $\log_2(\text{cost})$; here s/h is taken as a measure of the relative cost of the calculation. On the basis of many calculations and plots the following results were indicated: (1) better stability was achieved with $s = 2$ than $s = 1$ and the relative cost was frequently lower, (2) in no case was $s = 3$ or $s = 4$ better than $s = 2$, (3) the optimum answers were obtained with $p = 6$ or 7 Adams forms in which the corrector was iterated only twice.

4.15.3. Variable Coefficient and Hybrid Methods

Van Wyck (see Chapter 1[9]) also presented computational results for a P–C mode with variable coefficients (Section 1.7.5). When compared on a basis of equivalent computations the error associated with this procedure was lower for a wide variety of test ODEs than the Adams P–C, the Runge–Kutta, and the Nordsieck approaches. Because of the ease of changing h in this variable coefficient method, further experimental results would seem indicated to establish the superiority over a wide variety of conditions.

Actually the variable coefficient method is merely one approach for developing a nonstandard linear difference equation as a predictor or corrector. A second such approach is to use the hybrid methods of Section 4.11. Extensive results are available by Gragg and Stetter [17], Gear [15], Kohfeld and Thompson [24] and Brush, et al. [3] on variations on this hybrid approach. The extreme difficulties in comparing these hybrid methods with the more standard approaches results in a lack of clarity in any analysis of cost or efficiency. While high accuracy can be obtained with the hybrid formulas, it still is not clear if these have a real advantage over the more standard methods such as the Adams P–C.

4.15.4. Comparison of P–C and Runge–Kutta Methods

Finally, we turn to the work of Gallaher and Perlin [14] and Waters (see Chapter 2[73]) and Benyon [1]. In the first two cases a wide variety of high-order methods were analyzed with almost equivalent results. In the Gallaher–Perlin case, Adams P–C forms, hybrid P–C forms, and a variety of Runge–Kutta forms were tested on some orbit calculations. Using orders up to fifteen the end result of a variety of calculations was that a Runge–Kutta Fehlberg form was, in general the best method in terms of accuracy and the number of function evaluations. By contrast, Waters analyzed a set of sixty-four simultaneous equations which had the feature of an almost trivial form of $f(x, y)$; thus the number of function evaluations was not important in this case. Using many of the fourth- to sixth-order Runge–Kutta methods, the

Adams P–C forms, and some hybrid forms the result was that the Butcher fifth-order Runge–Kutta method, (2.3-20), was always best in terms of computation time and accuracy. Thus in these two series of computations the Runge–Kutta forms are more attractive than the P–C methods.

Benyon also showed that the fourth-order Runge–Kutta method was much faster than the fourth-order Adams predictor or corrector on a problem in which the function evaluations were not complicated. This was due to the larger stability bound for the Runge–Kutta method allowing correspondingly larger values of the integration step size.

From these results one may reach the conclusions:

1. Comparing only the P–C modes, the Adams forms have perhaps the best features associated with such methods, i.e., low cost per computation, moderate stability, and excellent accuracy.

2. In a direct comparison to the Runge–Kutta methods, the optimum mode of operation is not clear, but the P–C methods do not seem superior.

4.16. NUMERICAL EXPERIMENTS

In this section we continue the procedure adopted in Chapters 2 and 3 of analyzing the results of numerical calculations on Systems I–IX of Chapter 2. Here we shall consider the various multiple-step methods discussed in this chapter. By contrast, Section 4.17 will detail direct comparisons between the single- and multiple-step methods.

4.16.1. Milne Methods

Because of the theoretically known deficiencies (instability) of the Milne method, we shall only present a limited amount of calculational results. Typical selected results in single precision for $h = 0.01$ and 1.0 are given in Table 4.4 for System IV. The P–M_p–C–M_c or PMCM mode (the modified P–C mode), the PMC mode (the PECE mode) and the PC$_\infty$ mode (the iterated corrector mode) of Milne's equation have been used, but we only report the first mode and the number of iterations (convergence to 10^{-6}) in the last.

The three modes of operation are essentially identical for $h = 0.01$ except at large values of x. At $h = 0.1$, the differences are not excessive although it is interesting to note that the number of iterations in the PC$_\infty$ mode is beginning to increase. Even $h = 1.0$ does not cause the PMCM mode to oscillate and show instability, although PC$_\infty$ is obviously in an unstable condition requiring an excessive number of iterations. When the PMCM results are compared to the equivalent fifth-order Runge–Kutta results, it can be seen that the present errors are considerably larger.

These Milne modes have also been used on System VII and VI respectively. In general terms one can say that at small values of h the three modes

TABLE 4.4

SYSTEM IV. SINGLE PRECISION MILNE FORMS

x	ε_1	ε_2	PC_∞ iterations
	PMCM ($h = 0.01$)		
0.1	$0.1907_{10}{}^{-5}$	$0.3476_{10}{}^{-6}$	1
0.3	$0.5722_{10}{}^{-5}$	$0.1370_{10}{}^{-5}$	1
0.5	$0.9536_{10}{}^{-5}$	$0.2562_{10}{}^{-5}$	1
0.7	$0.1144_{10}{}^{-4}$	$0.3516_{10}{}^{-5}$	1
1.0	$0.1239_{10}{}^{-4}$	$0.4708_{10}{}^{-5}$	1
2.5	$-0.1662_{10}{}^{-2}$	$0.1406_{10}{}^{-4}$	1
5.0	$-0.6498_{10}{}^{-1}$	$0.3427_{10}{}^{-5}$	2
7.5	$-0.1525_{10}{}^{+1}$	$0.5082_{10}{}^{-6}$	2
	PMCM ($h = 0.1$)		
0.5	$0.2861_{10}{}^{-5}$	$-0.3576_{10}{}^{-6}$	
0.7	$0.4768_{10}{}^{-5}$	$-0.1788_{10}{}^{-6}$	2
1.0	$0.7629_{10}{}^{-5}$	$0.3576_{10}{}^{-6}$	
2.5	$-0.1459_{10}{}^{-3}$	$0.1490_{10}{}^{-5}$	3
5.0	$-0.7049_{10}{}^{-2}$	$0.3956_{10}{}^{-5}$	
7.5	-0.1718	$0.8055_{10}{}^{-7}$	3
10.0	$-0.5265_{10}{}^{+1}$	$-0.4787_{10}{}^{-8}$	4
	PMCM ($h = 1.0$)		
4.0	$0.1336_{10}{}^{+2}$	$-0.2814_{10}{}^{-3}$	15
5.0	$0.6708_{10}{}^{+2}$	$0.4386_{10}{}^{-2}$	13
6.0			23
7.0	$0.1214_{10}{}^{+4}$	$0.4510_{10}{}^{-2}$	7
8.0			24
10.0	$0.2180_{10}{}^{+5}$	$0.1369_{10}{}^{-1}$	6

behave identically, but that as h increases the PC_∞ mode becomes unstable first. At a large enough value of h ($h = 0.1$ for System VII and $h = 0.01$ for System IV) the PMCM mode breaks down and instability occurs. Even in the small h case the present results are not as accurate as one might desire.

Nevertheless it is important to note that while Figure 4.7 predicts unstable conditions for essentially all values of h for the three Milne modes of operation, the actual computation results tend to hold up better than predicted. This is probably due to the fact that a sufficient number of steps for instability to show up have not been taken.

4.16.2. Hamming Methods

Next we turn to the use of the Hamming forms for integrating Systems VI and VII. The PMCM, PMC, and PMC_3 modes have been used.

Table 4.5 presents results for PMCM on System VII using $h = 0.02, 0.1,$

TABLE 4.5

SYSTEM VII. SINGLE PRECISION HAMMING FORMS IN PMCM

x	ε		
	$h = 0.02$	$h = 0.1$	$h = 0.5$
0.1	$-0.1430_{10}-5$	$0.1788_{10}-6$	
0.2	$-0.1609_{10}-4$	$0.2093_{10}-6$	
0.3	$-0.3319_{10}-4$	$-0.2886_{10}-6$	
0.5	$-0.6395_{10}-4$	$-0.1907_{10}-5$	$-0.2965_{10}-3$
0.7	$-0.8481_{10}-4$	$-0.4172_{10}-5$	
0.9	$-0.9787_{10}-4$	$-0.4529_{10}-5$	
1.0		$-0.3755_{10}-5$	$-0.1171_{10}-2$
1.5		$-0.5960_{10}-7$	$-0.1555_{10}-2$
2.0		$-0.1037_{10}-7$	$-0.1507_{10}-1$
3.0		$-0.8344_{10}-6$	$0.1896_{10}-1$
4.0		$-0.6556_{10}-6$	$-0.1217_{10}-2$
5.0		$-0.8940_{10}-6$	$-0.1571_{10}-1$

and 0.5 in single precision. At the small value of h, the PMCM, PMC, and PMC$_3$ modes are essentially equivalent. For a larger h, the PMCM is the best, but at $h = 0.5$ this mode is beginning to break down. The case $h = 1.0$ was completely unstable and in fact could not be handled in single precision. In general the answers seem to be better than the equivalent Milne calculations.

Results on System VI in the PMCM, PMC, and PMC$_3$ mode have also been obtained. Both single and double precision calculations were carried out. In general terms the PMCM mode is once again the best, but the advantage is small. Also double precision is significantly better than single precision. When compared to the equivalent Milne calculations, Hamming's is many orders of accuracy better.

Thus, on the problems tested, Hamming's method yields excellent results, with the PMCM mode in double precision yielding the most accurate answers.

4.16.3. Adams Methods

Next we turn to the use of the Adams forms for integrating Systems III, IV, VI, and VII. In particular, we shall use the third-, fourth-, fifth-, and sixth-order P–C arrangement with PE(CE)^s, $s = 1$, 2, and 3.

To start the analysis, we turn to System III with $b = -1$. Selected results are shown in Table 4.6 for the overall interval of $\{0, 20\}$; only the maximum error in each twenty steps of $h = 0.25$ calculation are shown. The answers are

TABLE 4.6

System III. $b = -1$, Adams Method, $h = 0.25$.
Maximum Error in Intervals

$p = 3$			
$s = 1$		$s = 3$	
ε_{SP}	ε_{DP}	ε_{SP}	ε_{DP}
$0.2372_{10^{-2}}$	$0.2372_{10^{-2}}$	$0.4550_{10^{-2}}$	$0.4550_{10^{-2}}$
$0.2166_{10^{-3}}$	$0.2167_{10^{-3}}$	$0.3976_{10^{-3}}$	$0.2861_{10^{-3}}$
$0.2682_{10^{-5}}$	$0.2829_{10^{-5}}$	$0.4887_{10^{-5}}$	$0.5028_{10^{-5}}$
$0.2384_{10^{-6}}$	$0.2766_{10^{-7}}$	$0.2384_{10^{-6}}$	$0.4772_{10^{-7}}$

$p = 6$	
$s = 1$	$s = 3$
ε_{SP}	ε_{SP}
$0.3392_{10^{-2}}$	$0.4028_{10^{-2}}$
$0.3977_{10^{-3}}$	$0.3345_{10^{-3}}$
$0.3278_{10^{-5}}$	$0.3874_{10^{-5}}$
$0.2384_{10^{-6}}$	$0.2384_{10^{-6}}$

excellent; with $h = 0.015625$ the error at $x = 20.0$ remains at tolerable levels of about 10^{-6} in single precision and 10^{-9} in double precision. Figures 4.29–4.31 show plots of these data in terms of the $\log_{10}$ of the largest total absolute error versus $\log_2 s/h$. The case $s = 1$ is better than $s = 2$ or 3 and on this basis the third-order method ($p = 3$) is best in terms of the different order forms. Finally, we mention that stability is maintained until about $h = 2.0$. We have also analyzed System VII in the interval $\{0, 20\}$ and System IV

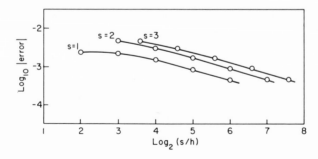

Figure 4.29. Error versus s/h for System III. Third-order Adams P–C combination in form PE(CE)^s.

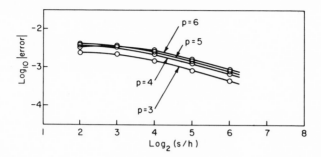

Figure 4.30. Error versus s/h for System III. Various order Adams P–C combinations in form $PE(CE)^s$.

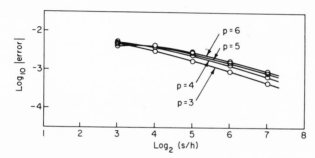

Figure 4.31. Error versus s/h for System III. Various order Adams P–C combinations in form $PE(CE)^2$.

in the interval $\{0, 10\}$. The results are basically equivalent to those for System III with the best data associated with $s = 1$ and $p = 3$. The stability bounds are also of the same order of magnitude as for System III.

Finally, we turn to the single precision analysis of the third- through sixth-order Adams P–C methods on the stiff equations of System VI. Of particular interest was the use of different values of h and the subsequent stability of the different Adams formulas. The values of $h = 1/100$, $1/50$, $1/40$, $1/35$, and $1/16$ were used for each of the different orders.

Table 4.7 present some of the results obtained on System VI for $p = 4$. For the case $h = 1/100 = 0.01$ other data show that an excellent behavior is obtained for y_2 and y_3 but only moderate behavior for y_1. In particular, y_1 begins to decrease at first, but then in single precision begins to increase again. At large values of x, $x \simeq 10$–20, this trend is reversed and the error continues to decrease. When the case $h = 1/50 = 0.02$ is used, the behavior of y_1 and y_2 are approximately as before; y_3, by contrast, lags well behind y_2, but eventually shows a decrease to zero. The case $h = 1/40 = 0.025$, $p = 4$, shows that this behavior of y_3 is due to the integration being close to insta- bility with $h = 1/40$ being unstable (note the error in y_3).

TABLE 4.7

SYSTEM VI. SINGLE PRECISION ADAMS FORMS, $p = 4$

x	ε_1	ε_2	ε_3
		$h = 1/50$	
0.16	$0.4025_{10}{}^{-3}$	$0.4000_{10}{}^{-3}$	$0.4305_{10}{}^{-3}$
0.26	$0.7629_{10}{}^{-3}$	$0.2739_{10}{}^{-5}$	$0.1861_{10}{}^{-4}$
0.36	$0.7450_{10}{}^{-5}$	$0.1878_{10}{}^{-7}$	$0.4104_{10}{}^{-5}$
0.46	$0.9894_{10}{}^{-5}$	$0.1289_{10}{}^{-9}$	$0.1036_{10}{}^{-5}$
0.56	$0.1227_{10}{}^{-4}$	$0.8867_{10}{}^{-12}$	$0.2627_{10}{}^{-6}$
0.76	$0.1704_{10}{}^{-4}$	$0.4213_{10}{}^{-16}$	$0.1689_{10}{}^{-7}$
0.96	$0.2145_{10}{}^{-4}$	$0.2015_{10}{}^{-20}$	$0.1085_{10}{}^{-8}$
1.26	$0.2789_{10}{}^{-4}$	$0.6768_{10}{}^{-27}$	$0.1770_{10}{}^{-10}$
1.56	$0.3379_{10}{}^{-4}$	$0.2316_{10}{}^{-33}$	$0.2885_{10}{}^{-12}$
1.96	$0.4118_{10}{}^{-4}$	$0.5719_{10}{}^{-42}$	$0.1192_{10}{}^{-14}$
5.96	$0.8672_{10}{}^{-4}$	$0.2485_{10}{}^{-78}$	$0.1733_{10}{}^{-38}$
10.96	$0.9953_{10}{}^{-4}$	$0.2485_{10}{}^{-78}$	$0.2768_{10}{}^{-68}$
15.96	$0.9059_{10}{}^{-4}$	$0.2485_{10}{}^{-78}$	$0.1839_{10}{}^{-77}$
25.96	$0.6049_{10}{}^{-4}$	$0.2485_{10}{}^{-78}$	$0.1839_{10}{}^{-77}$
45.96	$0.1424_{10}{}^{-4}$	$0.2485_{10}{}^{-78}$	$0.1839_{10}{}^{-77}$
		$h = 1/40$	
0.20	$0.6080_{10}{}^{-3}$	$0.6038_{10}{}^{-3}$	$0.1680_{10}{}^{+2}$
0.325	$0.1049_{10}{}^{-4}$	$0.2477_{10}{}^{-5}$	$0.8257_{10}{}^{+2}$
0.45	$0.1186_{10}{}^{-4}$	$0.1039_{10}{}^{-7}$	$0.4058_{10}{}^{+3}$
0.70	$0.1913_{10}{}^{-4}$	$0.1877_{10}{}^{-12}$	$0.9803_{10}{}^{+4}$
1.20	$0.3236_{10}{}^{-4}$	$0.6272_{10}{}^{-22}$	$0.5720_{10}{}^{+7}$
1.70	$0.4428_{10}{}^{-4}$	$0.2101_{10}{}^{-31}$	$0.3337_{10}{}^{+10}$
2.20	$0.5513_{10}{}^{-4}$	$0.7042_{10}{}^{-41}$	$0.1947_{10}{}^{+13}$
2.70	$0.6484_{10}{}^{-4}$	$0.2359_{10}{}^{-50}$	$0.1136_{10}{}^{+16}$
3.70	$0.8141_{10}{}^{-4}$	$0.2649_{10}{}^{-69}$	$0.3869_{10}{}^{+21}$
4.70	$0.9405_{10}{}^{-4}$	$0.2975_{10}{}^{-78}$	$0.1317_{10}{}^{+27}$

The behavior of the Adams forms as a function of k, the number of back points, is not significant. As long as $h \leq 1/50$, $k = 4$, 5, and 6 yield essentially identical results. When $h \geq 1/40$, the three forms all become unstable and useless for computation. Actually stability is maintained for a larger h than one would normally predict from a theoretical stability analysis, but the difference is not too large.

4.16.4. Hermite–Milne Methods

Now we turn to the use of the Hermite–Milne P–C set of Section 4.5.5 in a PE(CE)^s mode. Table 4.8 shows some results for Systems I and V. Note first that as theoretically predicted the case $s = 0$ is completely unstable; $y(x)$ is a decreasing function whereas the calculation results increase and oscillate with increasing x. For the case $s = 1$, the algorithm yields excellent results on

TABLE 4.8

HERMITE–MILNE P–C PAIR

System I			System I		
$h = 0.5, s = 0$			$s = 1$, at $x = 1.0$		
x	y_{DP}	$1/h$	$(e^{-x} - y_{DP})/e^{-x}$	$(e^{-x} - y_{SP})/e^{-x}$	
0	1.0	2.0	$-0.2838_{10}{}^{-2}$	$-0.2838_{10}{}^{-2}$	
0.5	0.6067	4.0	$-0.1215_{10}{}^{-3}$	$-0.1216_{10}{}^{-3}$	
1.0	0.3593	8.0	$-0.6376_{10}{}^{-5}$	$-0.6318_{10}{}^{-5}$	
1.5	0.2708	10.0	$-0.2526_{10}{}^{-5}$	$-0.2430_{10}{}^{-5}$	
2.0	-0.1875	20.0	$-0.1479_{10}{}^{-6}$	$-0.1620_{10}{}^{-6}$	
2.5	2.2083	100.0	$-0.2250_{10}{}^{-9}$	$0.2268_{10}{}^{-5}$	
3.0	-14.0000				
3.5	92.8333				
4.0	-613.000				
4.5	4049.33				
5.0	-26748.0				

System V

	$s = 1$ $(\sin x - y_{1DP})/\sin x$			
h	$x = 0.5$	$x = 2.0$	$x = 6.0$	$x = 20.0$
0.50	$0.5400_{10}{}^{-3}$	$0.9118_{10}{}^{-3}$	$-0.2151_{10}{}^{-1}$	14.88
0.25	$0.8731_{10}{}^{-4}$	$0.4050_{10}{}^{-4}$	$-0.9727_{10}{}^{-3}$	0.3899
0.125	$0.5340_{10}{}^{-5}$	$0.2029_{10}{}^{-5}$	$-0.4973_{10}{}^{-4}$	$0.1186_{10}{}^{-1}$
0.10	$0.1879_{10}{}^{-5}$	$0.7903_{10}{}^{-6}$	$-0.1946_{10}{}^{-4}$	$0.3886_{10}{}^{-2}$
0.05	$0.1353_{10}{}^{-6}$	$0.4427_{10}{}^{-7}$	$-0.1103_{10}{}^{-5}$	$0.1222_{10}{}^{-3}$
0.01	$0.2154_{10}{}^{-9}$	$0.6427_{10}{}^{-10}$	$-0.1616_{10}{}^{-8}$	

System I, especially in double precision. In addition, we have also tried the cases of $s = 2$ and $s = 3$; as theoretically predicted these calculations completely confirm the concepts mentioned above since both cases yield unstable solutions.

The case $s = 0$ for System V yields unstable solutions and thus we present only the $s = 1$ data. As can be seen the algorithm is once again an effective one with double precision yielding excellent answers, especially at small values of h. Further calculations not shown here on Systems IV and VII also show the same qualitative behavior. The algorithm seems to be an effective one.

4.16.5. $P(EC)^s$ Modes

All the previous results have been based upon the use of the various forms operated in a $PE(CE)^s$ mode. Since we have pointed out the computational advantages of the corresponding $P(EC)^s$ mode in Section 4.8, it would seem natural to use this latter mode to integrate some of the systems. Thus we shall now detail some results on a number of systems using the Milne P–C pair, the Hamming P–C pair and a fourth-order Adams P–C pair operating as $P(EC)^s$. No modifiers were used in any of the P–C pairs. The influence of variations in h, in s, and in single and double precision will be of major interest. Further, these results will allow a comparison between three different types of multiple-step methods.

In Table 4.9 and Figure 4.32 we present selected results for System I when solved by the $P(EC)^s$ algorithm in $\{0, 15\}$. Using some of data not shown explicitly we can state that for $h = 0.5$ and $s = 1$, all the algorithms are unstable; in fact only for $h = 0.125$ (or smaller) do Hamming's and Adams' algorithms yield excellent, highly accurate results. By contrast, Milne's method is still inaccurate. From these points we can identify that as long as $s = 1$, the stability bound in the P(EC) forms has been decreased considerably over that in the PE(CE) form; in fact, decreased from about $h = 1.0$ to $h = 0.125$. When s is increased to 2 and 3 this difference tends to disappear, since for $h = 0.5$ and $s = 3$, only the Milne form is unstable. However, it should be realized that the case $s = 1$ is really the method of interest since it has the advantage, for P(EC), of fewer function evaluations as compared to PE(CE).

When comparing the behavior of the three algorithms the Milne algorithm

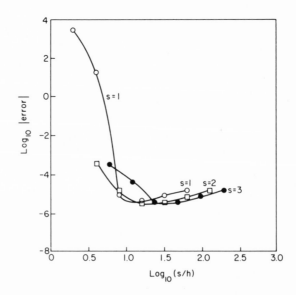

Figure 4.32. Error versus s/h for System I. Hamming P–C combination in form $P(EC)^s$.

TABLE 4.9

System I. P(EC)^s Mode[a]

h	ε_H	ε_M	ε_A
	$x = 5.0$, SP, $s = 1$		
0.5	-0.2041_{10}^{-2}	-0.1547_{10}^{-1}	-0.2972_{10}^{-2}
0.25	-0.2404_{10}^{-3}	-0.1604_{10}^{-2}	-0.1921_{10}^{-3}
0.125	0.1102_{10}^{-5}	-0.4485_{10}^{-5}	0.6742_{10}^{-6}
0.0625	0.5252_{10}^{-6}	-0.8195_{10}^{-7}	0.7078_{10}^{-7}
0.03125	0.8381_{10}^{-6}	-0.7823_{10}^{-7}	0.1043_{10}^{-6}
0.015625	0.1575_{10}^{-5}	0.5215_{10}^{-7}	0.2495_{10}^{-6}
	$x = 5.0$, DP, $s = 1$		
0.5	-0.2041_{10}^{-2}	-0.1547_{10}^{-1}	-0.2972_{10}^{-2}
0.125	0.8304_{10}^{-6}	-0.4369_{10}^{-5}	0.6674_{10}^{-6}
0.03125	0.1551_{10}^{-8}	0.1483_{10}^{-8}	0.1239_{10}^{-8}
	$x = 5.0$, SP, $s = 3$		
0.5	-0.1120_{10}^{-3}	0.2631_{10}^{-3}	0.8105_{10}^{-4}
0.25	0.5070_{10}^{-5}	0.1785_{10}^{-4}	0.3568_{10}^{-5}
0.125	0.4619_{10}^{-6}	0.9424_{10}^{-6}	0.2235_{10}^{-6}
0.0625	0.4060_{10}^{-6}	-0.5960_{10}^{-7}	0.6705_{10}^{-7}
0.03125	0.7860_{10}^{-6}	-0.1341_{10}^{-6}	0.1080_{10}^{-6}
0.015625	0.1523_{10}^{-5}	-0.2235_{10}^{-7}	0.2495_{10}^{-6}
	$x = 10.0$, SP, $s = 1$		
0.5	-0.3058_{10}^{+1}	-0.1771_{10}^{+2}	-0.3200_{10}^{+1}
0.25	-0.7248_{10}^{-1}	-0.8299	-0.2915_{10}^{-1}
0.125	0.1447_{10}^{-7}	-0.2045_{10}^{-3}	0.4575_{10}^{-8}
0.0625	0.6766_{10}^{-8}	-0.1267_{10}^{-5}	0.1004_{10}^{-8}
0.03125	0.1052_{10}^{-7}	-0.9539_{10}^{-6}	0.1586_{10}^{-8}
0.015625	0.1960_{10}^{-7}	-0.4230_{10}^{-6}	0.3536_{10}^{-8}
	$x = 15.0$, SP, $s = 1$		
0.5	-0.3170_{10}^{+4}	-0.1939_{10}^{+5}	-0.2764_{10}^{+4}
0.25	-0.1970_{10}^{+2}	-0.4248_{10}^{+3}	-0.3999_{10}^{+1}
0.125	0.4939_{10}^{-10}	-0.8519_{10}^{-2}	0.9788_{10}^{-10}
0.0625	0.6622_{10}^{-10}	-0.1436_{10}^{-4}	0.1080_{10}^{-10}
0.03125	0.1084_{10}^{-9}	-0.6984_{10}^{-5}	0.1648_{10}^{-10}
0.015625	0.1987_{10}^{-9}	-0.2612_{10}^{-5}	0.3740_{10}^{-10}
	$x = 15.0$, DP, $s = 1$		
0.5	-0.3170_{10}^{+4}	-0.1939_{10}^{+5}	-0.2764_{10}^{+4}
0.125	0.2957_{10}^{-10}	-0.8340_{10}^{-2}	0.9705_{10}^{-10}
0.03125	0.2152_{10}^{-12}	0.5159_{10}^{-7}	0.1715_{10}^{-12}

[a] The subscript on ε indicates the specific algorithm.

is the worst. Double precision is very helpful to all three forms, but the Milne algorithm never approaches the excellent behavior of the other two. As in previous analyses, the double precision calculations reverse the trends for a minimum in the error versus h data. There can be no question of the vast improvement which can occur when double precision is used for large values of x as compared to that in single precision. Note, however, that double precision does not convert an unstable process to a stable one. In terms of a further comparison between the three algorithms, it seems that the Adams form is best, the Hamming is next, and the Milne form is the worst.

We have also solved Systems II, IV, VI, and VII using the three algorithms. In terms of System II, we have a difficulty in the numerical results not following closely the rapidly increasing function. If double precision and a small h are used, all three algorithms, although the Milne form is the best, behave excellently even at large values of x. Thus at $x = 15.0$, at which the exact value of $y(x) = 0.33_{10+7}$, the Milne error is only 0.197 for $h = 0.03125$. By contrast, the single precision results show an error of 0.26_{10+3} at this same point.

In terms of System IV, the case $h = 0.50$ yields extremely inaccurate results (unstable) and even $h = 0.125$, while much better, does show large errors in y_1 at large x. In double precision and small h's the methods all approximate the true solutions rather well. Once again the Adams mode seems the best.

For $s = 1$, all three algorithms are unstable until approximately $h = 0.0626$ using System VII. For $s = 2$, stability does occur at $h = 0.25$. Thus this case also shows a marked decrease in stability bound when operated in a P(CE) mode. Double precision improves the results materially for a small h (stable case), but does not really affect the unstable cases. In double precision and a small h all the algorithms are excellent, but the Adams forms are the best.

We have also tested the three algorithms on System VI, the stiff system. The results were an upper stability bound of about $h = 0.00125$ which is much smaller (by about a factor of ten) than the corresponding PE(CE) algorithm. Further, the Adams and Hamming forms were far superior to the Milne form. As an illustration at $x = 0.5$, the Adams and Hamming errors were on the order of 10^{-9}, whereas the Milne error was approximately 10^{-2}.

As a result of these calculations certain points can be made.

1. For double precision and a small h the P(EC) algorithms work well.

2. The stability bounds are decreased considerably compared to the usual PE(CE) algorithm.

3. The order of decreasing efficiency is Adams, Hamming, and Milne P–C pairs.

4.16.6. Hybrid Methods

Here we present some brief results obtained using the fifth-order Butcher hybrid form of (4.11-9). Table 4.10 shows selected values for Systems I, II, IV, and VI using $h = 0.1$ and single precision. The results are quite good and competitive with other methods. Of special interest is that the case $h = 1.0$ leads, for all systems, to unstable integration. This is probably lower than expected and is certainly not better than the alternative methods.

TABLE 4.10

SINGLE PRECISION BUTCHER HYBRID FORM, $h = 0.1$

	System I	System II
x	ε	ε
0.2	$-0.6735_{10^{-5}}$	$-0.1430_{10^{-4}}$
0.4	$0.1126_{10^{-4}}$	$-0.1041_{10^{-4}}$
1.0	$-0.8165_{10^{-5}}$	$0.1144_{10^{-4}}$
2.0	$-0.2264_{10^{-5}}$	$-0.7629_{10^{-5}}$
5.0	$0.4087_{10^{-7}}$	$0.2380_{10^{-2}}$

	System IV	
x	ε_1	ε_2
0.2	$0.1074_{10^{-5}}$	$0.1178_{10^{-6}}$
0.4	$0.1110_{10^{-5}}$	$0.2965_{10^{-6}}$
0.6	$0.1047_{10^{-5}}$	$0.4246_{10^{-6}}$
0.8	$0.8287_{10^{-6}}$	$0.5110_{10^{-6}}$
1.0	$0.3767_{10^{-6}}$	$0.5641_{10^{-6}}$
2.0	$0.1121_{10^{-4}}$	$0.5412_{10^{-6}}$
4.0	$0.3962_{10^{-3}}$	$0.2080_{10^{-6}}$
5.0	$0.1683_{10^{-2}}$	$0.1094_{10^{-6}}$

	System VI
x	ε
0.2	$< 10^{-6}$
0.4	$0.7152_{10^{-6}}$
0.6	$0.1788_{10^{-5}}$
0.8	$0.4827_{10^{-5}}$
1.0	$0.6556_{10^{-5}}$
2.0	$0.8821_{10^{-5}}$
5.0	$0.1001_{10^{-4}}$

4.16.7. Higher-Derivative Methods

Finally we consider the use of higher derivatives in the P–C mode. In particular we use

(a) P: (4.10-2)
  C: (4.10-3)

(b) P: $\bar{y}_{n+1} = y_n + hf_n$ (Euler method)
  C: $y_{n+1} = y_n + (h/2)[f_n + \bar{f}_{n+1}]$ (Trapezoidal rule)

(c) P: $\bar{y}_{n+1} = y_n + hf_n + (h^2/2)f_n'$
  C: (4.10-3)

(d) P: (1.10-6)
  C: (1.10-8)

(e) P: $\bar{y}_{n+1} = y_n + hf_n + (h^2/2)f_n' + (h^3/6)f_n'' + (h^4/24)f_n'''$
  C: $y_{n+1} = y_n + (h/2)[f_n + \bar{f}_{n+1}] + (3h^2/28)[f_n' - \bar{f}_{n+1}']$
       $+ (h^3/84)[f_n'' + \bar{f}_{n+1}''] + (h^4/1680)[f_n''' - \bar{f}_{n+1}''']$

(f) P: $\bar{y}_{n+1} = y_{n-1} + 2hf_n$ (Midpoint method)
  C: $y_{n+1} = y_{n-1} + (h/3)f_{n-1}' + (4/3)hf_n + (1/3)h\bar{f}_{n+1}$

(g) P: $\bar{y}_{n+1} = 2y_n - y_{n-1} + h[f_{n-1} - f_n] + (h^2/2)[f_{n-1}' + 3f_n']$
  C: $y_{n+1} = 2y_n - y_{n-1} + (3h/8)[\bar{f}_{n+1} - f_{n-1}]$
       $+ (h^2/24)[-f_{n-1}' + 6f_n' - \bar{f}_{n+1}']$

Algorithm (a) has been presented in this chapter, whereas algorithms (b)–(g) all follow from the table given by Lambert and Mitchell as mentioned in Chapter 1. Algorithms (a)–(e) are really single-step modes whereas (f) and (g) are two-step formulas. In particular we have applied these formulas to System V in double precision using h's ranging from 0.001 to 2.0.

Table 4.11 presents results for the fourth-order combination [Method (a)] of (4.10-2) and (4.10-3). The effect of the value of h is apparent with $h = 0.01$ yielding superb results. When compared to the fourth-order single-step (not P–C) method of Table 2.8, it can be seen that at $h = 0.1$ the present P–C is better by orders of magnitude. On the negative side, however, the present P–C shows instability at $h = 2.0$ whereas the single-step method did not.

Table 4.12 shows results at $x = 5.0$ for Methods (b)–(g) using different values of h. Stability information is also presented. In addition, Figures 4.33–4.36 show plots of error ε_1 versus x for some of the methods and for different values of h. Exact solutions are also shown on the coordinates. These figures show quite nicely the effect of increased accuracy associated with decreasing values of h.

TABLE 4.11

SYSTEM V. DOUBLE PRECISION, HIGHER DERIVATIVES
P = (4.10-2); C = (4.10-3)

x	ε_2			x	ε_2	
	$h = 0.01$	$h = 0.1$	$h = 0.5$		$h = 1.0$	$h = 2.0$
0.5	$0.4134_{10}{}^{-11}$	$0.1018_{10}{}^{-6}$		5.0	$0.6331_{10}{}^{-1}$	
1.0	$0.1265_{10}{}^{-10}$	$0.2240_{10}{}^{-6}$	$0.3284_{10}{}^{-3}$	8.0		$0.5132_{10}{}^{+1}$
1.5	$0.2083_{10}{}^{-10}$	$0.2519_{10}{}^{-6}$	$0.5945_{10}{}^{-3}$	10.0	$0.1235_{10}{}^{-1}$	
2.0	$0.2338_{10}{}^{-10}$	$0.1246_{10}{}^{-6}$	$0.5231_{10}{}^{-3}$	12.0		$0.6236_{10}{}^{+2}$
2.5	$0.1653_{10}{}^{-10}$	$0.1575_{10}{}^{-6}$	$0.2706_{10}{}^{-5}$	15.0	0.2027	
3.0	$0.3029_{10}{}^{-12}$	$0.5262_{10}{}^{-6}$	$0.9098_{10}{}^{-3}$	16.0		$0.5138_{10}{}^{+2}$
3.5	$0.2380_{10}{}^{-10}$	$0.8609_{10}{}^{-6}$	$0.1939_{10}{}^{-2}$	20.0	0.1521	$0.3624_{10}{}^{+5}$
4.0	$0.4736_{10}{}^{-10}$	$0.1026_{10}{}^{-5}$	$0.2717_{10}{}^{-2}$	24.0		$0.1636_{10}{}^{+7}$
5.0	$0.6351_{10}{}^{-10}$	$0.5063_{10}{}^{-6}$	$0.2194_{10}{}^{-2}$			
6.0	$0.1127_{10}{}^{-10}$	$0.8955_{10}{}^{-6}$	$0.1376_{10}{}^{-2}$			

TABLE 4.12

SYSTEM V. DOUBLE PRECISION. VARIOUS HIGH DERIVATIVE METHODS

	ε_2 at $x = 5.0$			Stability comments
Method	$h = 0.01$	$h = 0.1$	$h = 1.0$	
(b)	$0.8000_{10}{}^{-4}$	$0.8139_{10}{}^{-2}$	0.8679	Almost unstable at $h = 1.0$. Completely unstable at $h = 2.0$.
(c)	$-0.1188_{10}{}^{-6}$	$-0.8178_{10}{}^{-4}$	0.1731	Stable at $h = 1.0$. Unstable at $h = 2.0$.
(d)	$-0.1017_{10}{}^{-8}$	$-0.1023_{10}{}^{-4}$	$-0.7387_{10}{}^{-1}$	Stable at $h = 1.0$. Unstable at $h = 2.0$.
(e)	$-0.2326_{10}{}^{-10}$	$0.3544_{10}{}^{-7}$	$-0.1438_{10}{}^{-1}$	Stable at $h = 1.0$. Unstable, but just at $h = 2.0$.
(f)	$-0.8921_{10}{}^{-6}$	$-0.7948_{10}{}^{-4}$	0.3644	Stable at $h = 1.0$. Unstable at $h = 2.0$. Much worse at $h = 2.0$, than any of above.
(g)	$-0.5977_{10}{}^{-7}$	$-0.3096_{10}{}^{-4}$	-0.2462	Stable at $h = 1.0$. Unstable at $h = 2.0$.

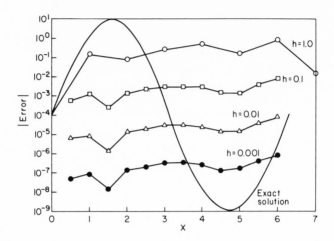

Figure 4.33. Error versus x for System V using different h values. Algorithm b, first derivative and single step.

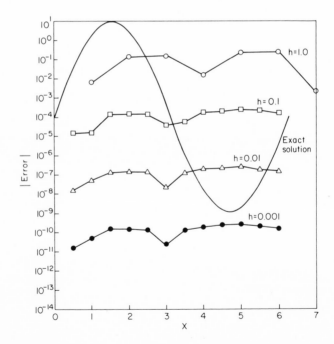

Figure 4.34. Error versus x for System V using different h values. Algorithm a, second derivative and single step.

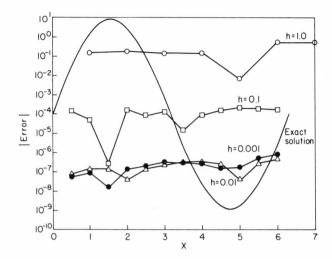

Figure 4.35. Error versus x for System V using different h values. Algorithm f, first derivative and two steps.

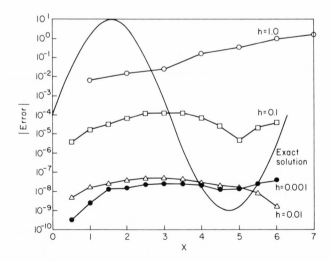

Figure 4.36. Error versus x for System V using different h values. Algorithm g, second derivative and two steps.

In general terms it is apparent from Table 4.12 that as a P–C algorithm uses higher derivatives the accuracy and the stability bounds go up. However, as the algorithm involves more back points ($k = 2$ rather than $k = 1$) the accuracy goes up, but the stability bounds decrease. The major factor is, however, the accuracy rather than the stability bounds. Finally, we note that algorithms (a) and (c) both use the same corrector, but different predictors; (a) is much better than (c) because of the better predictor.

4.16.8. Recommendations

A number of recommendations and observations may be made on the basis of these data:

1. The Adams algorithm when operated in a PE(CE) mode is the best of the different P–C methods.
2. Double precision is advantageous whenever it can be attained without increasing the computation time significantly.
3. The computations remain stable for larger bounds on h than are predicted theoretically.
4. Higher-order forms are not necessarily the best.
5. Higher-order derivatives, when easily included, are quite advantageous.

4.17. NUMERICAL COMPARISONS OF SINGLE- AND MULTIPLE-STEP METHODS

To this point we have treated the single-step and multiple-step methods separately with no numerical comparison between the two approaches. Now we wish to make such comparisons using two different problems. The first problem is the solution of a linear and nonlinear partial differential equation (PDE) via the "method of lines" which converts the PDE into m coupled ODEs; the second problem uses System IX as the test system.

4.17.1. Solution of PDEs by the Method of Lines

It is common practice in analog computation to solve PDEs by converting them into coupled sets of ODEs. Hicks and Wei [20] have recently illustrated the digital features of this approach as applied to parabolic PDEs. Here we shall use this method to generate large sets of linear and nonlinear ODEs which can then serve as test systems for comparing the various integration methods.

To illustrate the procedure consider the linear parabolic PDE system

$$y_t = y_{xx}$$
$$y(1, t) = 0, \qquad t > 0 \qquad\qquad (4.17\text{-}1)$$
$$y_x(0, t) = 0, \qquad t > 0$$
$$y(x, 0) = -1, \qquad 0 < x < 1$$

By carrying out a finite difference approximation to the derivative y_{xx} we obtain a set of m ODEs which represent a number of fixed points in x space and are integrated along these lines of constant x in t space. Thus the term "method of lines." Using the boundary conditions and the central difference approximation

$$\frac{\partial^2 y_{k-1}}{\partial x^2} = \frac{1}{12 \, \Delta x^2} \left[-y_{k-3} + 16y_{k-2} - 30y_{k-1} + 16y_k - y_{k+1} \right] \qquad (4.17\text{-}2)$$

(4.17-1) may be converted to the linear ODEs,

$$\mathbf{y}' = \mathbf{A}\mathbf{y}, \qquad \mathbf{y}(0) = -1 \qquad\qquad (4.17\text{-}3)$$

where $\mathbf{y}$ is a vector of dimension ($m \times 1$) and $\mathbf{A}$ is a quintidiagonal matrix with constant coefficients. Further details on this approach and the properties of $\mathbf{A}$ can be obtained from Hicks and Wei.

As a nonlinear PDE we will take [20]

$$y_t = \left[1 + \frac{2x^2}{(x + y + 1)^2} \right]^{-1} y_{xx}$$
$$y(1, t) = 0, \qquad t > 0 \qquad\qquad (4.17\text{-}4)$$
$$y_x(0, t) = 0, \qquad t > 0$$
$$y(x, 0) = -1, \qquad 0 < x < 1$$

Conversion of this system leads to (4.17-3) with the one change that all coefficients in the $\mathbf{A}$ matrix are multiplied by

$$\left[1 + \frac{2x_i^2}{(x_i + y_i + 1)^2} \right]^{-1}, \qquad i = 1, \dots, m \qquad (4.17\text{-}5)$$

We should point out that two classes of errors exist in solving (4.17-3). The first is due to the finite differencing and can be minimized by increasing m. We shall not consider this problem here. The second occurs in integrating the ODEs in t space for a fixed value of m; these errors will be of concern here.

As methods of integration we shall use:

1. The single-step embedding formula of Sarafyan, (2.8-14) and (2.8-15).
2. The single-step Rosenbrock, third-order form (2.5-18).

3. The Hamming multiple-step method in the P–M_p–C–M_C mode, (4.6-3).

Except for the Sarafyan form, the procedures used are obvious. In the Sarafyan form the following process for implementing the equations was used. Given y_n and h, the method calculates y_{n+1} via the fourth-order Runge–Kutta formula and y_{n+2} via the fifth-order formula; then using y_{n+1} and the fourth-order formula, a second value of y_{n+2} is calculated and compared with that y_{n+2} already calculated. This comparison (made every step in the calculation) is used to adjust h up or down depending on the digits of agree-

TABLE 4.13

EQUATION (4.17-3). DOUBLE PRECISION

Errors at value of x	Sarafyan embedding, 50 points	
	$t = 0.10$	$t = 0.20$
0.8	$3.2_{10}{}^{-4}$	$2.7_{10}{}^{-4}$
0.6	$4.8_{10}{}^{-4}$	$4.9_{10}{}^{-4}$
0.4	$4.5_{10}{}^{-4}$	$6.4_{10}{}^{-4}$
0.2	$3.5_{10}{}^{-4}$	$7.0_{10}{}^{-4}$
0	$2.9_{10}{}^{-4}$	$7.2_{10}{}^{-4}$
h	10^{-5} to 10^{-3}	
Computing time	~ 150 sec/100 steps.	
Comments	Because of manner of operation, computing time is very high. No stability problems.	

Errors at value of x	Rosenbrock, 50 points				
	$t = 0.001$	$t = 0.01$	$t = 0.10$	$t = 0.20$	$t = 0.50$
0.8	$5.2_{10}{}^{-5}$	$3.2_{10}{}^{-2}$	$1.4_{10}{}^{-4}$	$1.2_{10}{}^{-1}$	$1.7_{10}{}^{-2}$
0.6	$1.9_{10}{}^{-5}$	$3.0_{10}{}^{-3}$	$9.1_{10}{}^{-5}$	$3.4_{10}{}^{-2}$	$3.5_{10}{}^{-3}$
0.4	$1.5_{10}{}^{-6}$	$2.4_{10}{}^{-4}$	$1.7_{10}{}^{-5}$	$3.3_{10}{}^{-2}$	$3.2_{10}{}^{-3}$
0.2	$3.0_{10}{}^{-8}$	$4.1_{10}{}^{-6}$	$5.1_{10}{}^{-5}$	$9.9_{10}{}^{-3}$	$9.0_{10}{}^{-4}$
0	$1.0_{10}{}^{-8}$	$2.6_{10}{}^{-6}$	$5.6_{10}{}^{-5}$	$1.3_{10}{}^{-4}$	$3.0_{10}{}^{-4}$
h	10^{-3}	10^{-2}	10^{-2}	10^{-1}	10^{-1}
Computing time	~ 81 sec/100 steps				
Comments	No stability problems although inaccuracy occurs at $h = 10^{-1}$.				

TABLE 4.13 (continued)

EQUATION (4.17-3). DOUBLE PRECISION

Errors at value of x	Rosenbrock		Hamming	
	30 points		30 points	50 points
	$t = 0.10$	$t = 0.20$	$t = 0.01$	
0.8	$1.2_{10^{-5}}$	$2.4_{10^{-5}}$	$4.2_{10^{-4}}$	
0.6	$9.1_{10^{-5}}$	$2.2_{10^{-5}}$	$5.2_{10^{-5}}$	
0.4	$1.7_{10^{-5}}$	$7.0_{10^{-6}}$	$5.3_{10^{-7}}$	
0.2	$6.1_{10^{-5}}$	$1.3_{10^{-5}}$	$1.0_{10^{-8}}$	
0	$5.6_{10^{-5}}$	$2.2_{10^{-5}}$	$1.0_{10^{-9}}$	
h	10^{-2}	10^{-2}	10^{-4}	
Computing time	~ 54 sec/100 steps.		~ 18 sec/100 steps.	
Comments			Unstable at $h = 10^{-3}$	Stable for $h = 10^{-5}$. Unstable for $h = 10^{-4}$.

ment. In addition a maximum value of h was used to bound the value which the program could use. Because of this procedure, the Sarafyan embedding method used ten function evaluations per $2h$. On this basis the computing times will be greater by about 50% than should be quoted on a basis comparative to the other modes of integration. Other possible sequences could have been used which would have minimized the number of function evaluations.

Certain selected double precision results for the linear PDE are given in Table 4.13. The errors listed are obtained by comparing the numerical solution with the analytical solution of (4.17-1). On the simple basis of computing time, the best method is the Hamming P–C; this is followed in order by Rosenbrock's method and finally, although we repeat that this figure is excessive because of the specific computational procedure, the embedding method. Also note that the ratio of computing times $81/54 = 1.5$ in the Rosenbrock method is of the order of the ratio of the number of points used, i.e., $50/30 = 1.7$. Thus most of the calculation time is involved in the function evaluations.

However, this computing time comparison is deceptive. Stability considerations show that the Rosenbrock method can operate at h values which are much larger, by at least ten–one hundred times, than the Hamming method.

Table 4.14 presents analogous results for the nonlinear PDE system.

TABLE 4.14

EQUATION (4.17-4). DOUBLE PRECISION

Deviation from 30 point Sarafyan	Sarafyan embedding 15 points		
x	$t = 0.10$	$t = 0.20$	$t = 0.40$
0.8	$7.9_{10^{-3}}$	$6.8_{10^{-3}}$	$1.0_{10^{-2}}$
0.6	$1.3_{10^{-2}}$	$1.2_{10^{-2}}$	$1.9_{10^{-2}}$
0.4	$7.2_{10^{-3}}$	$1.2_{10^{-2}}$	$2.3_{10^{-2}}$
0.2	$1.4_{10^{-3}}$	$8.7_{10^{-3}}$	$2.3_{10^{-2}}$
0.0	$1.8_{10^{-4}}$	$4.6_{10^{-3}}$	$1.9_{10^{-2}}$
h	10^{-3} to 10^{-2}		
Computing time	~ 77 sec/100 steps		
Comments	No stability problems		

	Rosenbrock			
	30 points	15 points		
x		$t = 0.10$	$t = 0.20$	$t = 0.40$
0.8		$2.6_{10^{-3}}$	$4.7_{10^{-3}}$	$2.7_{10^{-3}}$
0.6		$5.5_{10^{-3}}$	$8.3_{10^{-3}}$	$4.3_{10^{-3}}$
0.4		$2.7_{10^{-3}}$	$8.7_{10^{-3}}$	$3.6_{10^{-3}}$
0.2		$2.1_{10^{-4}}$	$5.7_{10^{-3}}$	$2.4_{10^{-4}}$
0.0		$4.8_{10^{-5}}$	$2.1_{10^{-3}}$	$4.6_{10^{-3}}$
h	$10^{-3}, 10^{-2}$	10^{-2}		
Computing time	65 sec/100 steps	43 sec/100 steps		
Comments	10^{-2} begins to show inaccuracy	10^{-1} begins to show inaccuracy		

	Hamming 15 points	
x	$t = 0.10$	$t = 0.20$
0.8	$4.2_{10^{-3}}$	$5.4_{10^{-3}}$
0.6	$8.1_{10^{-3}}$	$9.4_{10^{-3}}$
0.4	$4.9_{10^{-3}}$	$1.0_{10^{-2}}$
0.2	$8.7_{10^{-4}}$	$6.9_{10^{-3}}$
0.0	$7.6_{10^{-5}}$	$3.2_{10^{-3}}$
h	10^{-3}	
Computing time	20 sec/100 steps	
Comments	Unstable at 10^{-2}	
	30 Points unstable at 10^{-3}	

Here we do not have an anlytical solution. The 30-point Sarafyan has been used as the standard; all deviations shown represent a comparison to this particular solution. These results show approximately the same behavior as for the linear system, although the stability differences are not as great as before.

From the results of computations carried out on the linear and nonlinear PDEs we can draw the following conclusions:

1. The best method to use from the standpoint of both accuracy and computing time is Rosenbrock's, followed by Hamming's and Sarafyan's, in that order.

2. It should be emphasized that the test on the Sarafyan method is probably not a fair one because of the computing routine used.

3. In terms of accuracy versus computing time, although the Rosenbrock method is superior, the distinction between the Rosenbrock and Hamming methods is not major.

4.17.2. Solution of System IX

The second system chosen as a test for comparing the various single- and multiple-step methods is System IX. Starting with the initial conditions of

$$y_1(0) = -0.1111889, \qquad y_2(0) = 0.3233580_{10^{-1}}$$

this system was integrated until $x = 5.0$. This was sufficient to have made one traverse around the limit cycle. Using both single and double precision and step sizes ranging from $h = 0.001$ to $h = 3.0$, the following algorithms were used in the integrations:

1. Classic fourth-order Runge–Kutta, (2.3-13).
2. Butcher fifth-order Runge–Kutta, (2.3-20).
3. Butcher sixth-order Runge–Kutta, (2.3-32).
4. Rosenbrock third-order, (2.5-18).
5. Fourth-order Milne PE(CE), (4.5-1) and (4.5-2).
6. Fourth-order Hamming PE(CE), (4.6-1) and (4.6-2).
7. Fourth-order Adams PE(CE), (4.7-3) and (4.7-4).

Selected double precision results for $h = 0.1$ and 0.5 are presented in Table 4.15. In this table the Butcher fifth-order Runge–Kutta $h = 0.001$ double precision data has been taken as correct and all other results compared to it. The symbol E indicates at least an eight digit agreement with the standard values. This procedure is necessary since analytical solutions are not known. Table 4.16 tabulates computer time, number of function evaluations, and some stability limits.

TABLE 4.15

SYSTEM IX. DOUBLE PRECISION. ABSOLUTE ε_2 FOR DIFFERENT METHODS

x	Fourth-order R–K	Fifth-order R–K	Sixth-order R–K	Rosenbrock	Milne	Hamming	Adams
				$h = 0.1$			
0.5	$2.2_{10^{-7}}$	E	$4.2_{10^{-4}}$	$2.2_{10^{-6}}$	$2.2_{10^{-8}}$	$3.9_{10^{-7}}$	$4.6_{10^{-7}}$
1.0	$8.0_{10^{-8}}$	E	$1.4_{10^{-3}}$	$2.1_{10^{-5}}$	$2.7_{10^{-7}}$	$2.4_{10^{-6}}$	$3.0_{10^{-6}}$
2.0	$2.9_{10^{-7}}$	E	$1.8_{10^{-3}}$	$3.0_{10^{-5}}$	$1.0_{10^{-6}}$	$5.6_{10^{-6}}$	$4.7_{10^{-6}}$
3.0	$5.6_{10^{-7}}$	$2.0_{10^{-9}}$	$1.5_{10^{-4}}$	$1.9_{10^{-7}}$	$7.3_{10^{-7}}$	$3.7_{10^{-6}}$	$1.7_{10^{-6}}$
4.0	$2.8_{10^{-7}}$	E	$6.6_{10^{-3}}$	$1.0_{10^{-4}}$	$6.3_{10^{-7}}$	$4.0_{10^{-6}}$	$9.1_{10^{-7}}$
5.0	$1.3_{10^{-6}}$	E	$3.9_{10^{-3}}$	$8.5_{10^{-5}}$	$6.0_{10^{-8}}$	$2.8_{10^{-6}}$	$2.5_{10^{-6}}$
				$h = 0.5$			
0.5	$1.3_{10^{-3}}$	$1.5_{10^{-5}}$	$1.4_{10^{-4}}$	$7.9_{10^{-4}}$			
1.0	$1.1_{10^{-4}}$	$5.3_{10^{-6}}$	$5.3_{10^{-3}}$	$1.9_{10^{-3}}$			
2.0	$1.0_{10^{-4}}$	$7.6_{10^{-6}}$	$9.8_{10^{-3}}$	$9.8_{10^{-4}}$	$4.8_{10^{-4}}$	$4.5_{10^{-4}}$	$7.0_{10^{-4}}$
3.0	$3.2_{10^{-4}}$	$1.5_{10^{-5}}$	$3.3_{10^{-3}}$	$2.7_{10^{-3}}$	$4.4_{10^{-4}}$	$9.1_{10^{-4}}$	$1.5_{10^{-3}}$
4.0	$1.1_{10^{-4}}$	$1.3_{10^{-5}}$	$2.5_{10^{-2}}$	$3.5_{10^{-3}}$	$4.2_{10^{-3}}$	$1.5_{10^{-3}}$	$2.3_{10^{-3}}$
5.0	$7.7_{10^{-4}}$	$2.1_{10^{-5}}$	$8.5_{10^{-3}}$	$9.2_{10^{-3}}$	$5.5_{10^{-3}}$	$6.7_{10^{-4}}$	$6.2_{10^{-3}}$

TABLE 4.16

SYSTEM IX. DIFFERENT METHODS TO REACH $x = 5.0$

Method	$h = 0.001$			$h = 0.01$			$h = 0.1$			Stability (DP)
	v^a	$t_{SP}{}^b$	$t_{DP}{}^c$	v	t_{SP}	t_{DP}	v	t_{SP}	t_{DP}	
Fourth-order R–K	20,000	16.1	17.3	2000	1.6	1.7	200	0.13	0.13	Stable $h = 1.0$ Unstable $h = 3.0$
Fifth-order R–K	30,000	21.4	22.8	3000	2.1	2.4	300	0.18	0.34	Stable $h = 1.0$ Unstable $h = 3.0$
Sixth-order R–K	35,000	25.3	27.5	3500	2.5	2.7	350	0.21	0.36	Stable $h = 1.0$ Unstable $h = 3.0$
Rosenbrock	10,000		12.9	1000		1.04	100		0.08	Stable $h = 3.0$
Milne	10,007	6.4	7.3	1007	0.71	0.80	113	0.11	0.12	Stable $h = 0.5$ Unstable $h = 1.0$
Hamming	10,007	5.8	7.5	1007	0.56	0.68	113	0.09	0.11	Stable $h = 0.5$ Unstable $h = 1.0$
Adams	10,007	6.3	7.7	1007	0.57	0.77	113	0.10	0.11	Stable $h = 1.0$ Unstable $h = 3.0$

[a] Number of function evaluations.
[b] Time in seconds to reach $x = 5.0$ in single precision (SP).
[c] Time in seconds to reach $x = 5.0$ in double precision (DP).

The following conclusions can be drawn and recommendations made from the results in these tables:

1. In double precision and for $h = 0.01$ all the algorithms yield either exact or almost exact results to the standard fifth-order Butcher form. The Rosenbrock form, since it is only third order, is not quite exact but the error is small. Even in single precision, the various methods are in good agreement with the standard. The only algorithm which shows any significant deviation is the sixth-order Butcher method. Note we have previously encountered difficulties in the use of a sixth-order method. For $h = 0.01$ and single precision, typical errors are $5.1_{10^{-6}}$, for the fourth- and fifth-order Runge–Kutta, $5.0_{10^{-4}}$ for the sixth order Runge–Kutta and $2.4_{10^{-6}}$, $1.1_{10^{-5}}$, $5.3_{10^{-6}}$ for the Milne, Hamming, and Adams methods respectively.

2. For $h = 0.1$ the various methods are beginning to show errors in double precision. The sixth-order Butcher method is still the worst, and the Rosenbrock method is not quite as accurate as the other remaining methods.

3. For $h = 0.5$, Table 4.15 shows a continuation of the general deterioration of the accuracy. However, many of the algorithms are still about equal in terms of accuracy.

4. Table 4.16 shows that the P–C methods are the fastest (use the smallest amount of computing time), Rosenbrock's method is next, and the Runge–Kutta methods are the slowest. The ratio of times is roughly $1 : 2 : 3.5$ in the given order. The double precision times are so close to those for single precision that double precision is the suggested mode of operation. The Runge–Kutta methods are stable at $h = 1.0$, but unstable at $h = 3.0$. Of the three P–C methods, the Adams from has the greatest range of stability —namely to $h = 1.0$—whereas the other two are unstable at $h = 1.0$. Further internal details on these three methods indicate that the Milne method breaks down first, the Hamming method next, and finally the Adams method.

Based on the results of System IX, we make the following recommendations:

1. For accurate results the most efficient methods are the P–C forms followed by Rosenbrock's method and finally the fifth- and fourth-order Runge–Kutta algorithms.

2. The sixth-order Runge–Kutta method is not competitive under any measure.

4.17.3. Recommendations

1. For ODE systems of moderate size ($m < 10$), multiple-step P–C and implicit single-step methods such as Rosenbrock's seem to give the best results from the point of view of accuracy and efficiency, when compared to explicit single-step methods. However, such features as the ease of pro-

gramming and the ability to change step size during a calculation for the explicit single-step methods may indicate that a fifth-order Runge–Kutta algorithm is competitive.

2. For large systems of ODE, stability considerations become more important than for small systems because of the possibility of a wide variation in the eigenvalues. For this reason, an implicit single-step method like Rosenbrock's is the preferred method.

REFERENCES

1. Benyon, P. R., A review of numerical methods for digital simulation, *Simulation* **31**, 219 (1968).
2. Brown, R. R., Riley, J. D. and Bennett, M. M., Stability properties of Adams–Moulton type methods, *Math. Comp.* **19**, 90 (1965).
3. Brush, D. G., Kohfeld, J. J. and Thompson, G. T., Solution of ordinary differential equations using two "off-step" points, *J. Assoc. Comput. Mach.* **14**, 769 (1967).
4. Butcher, J. C., A modified multistep method for the numerical integration of ordinary differential equations, *J. Assoc. Comput. Mach.* **12**, 124 (1965).
5. Butcher, J. C., On the convergence of numerical solutions to ordinary differential equations, *Math. Comp.* **20**, 1 (1966).
6. Butcher, J. C., A multistep generalization of Runge–Kutta methods with 4 or 5 stages, *J. Assoc. Comput. Mach.* **14**, 84 (1967).
7. Chase, P. E., Stability properties of predictor–corrector methods for ordinary differential equations, *J. Assoc. Comput. Mach.* **9**, 457 (1962).
8. Crane, R. L. and Klopfenstein, R. W., A predictor–corrector algorithm with an increased range of absolute stability, *J. Assoc. Comput. Mach.* **12**, 227 (1965).
9. Crane, R. L. and Lambert, R. J., Stability of a generalized corrector formula, *J. Assoc. Comput. Mach.* **9**, 104 (1962).
10. Dahlquist, G. G., A special stability problem for linear multistep methods, *BIT* **3**, 27 (1963).
11. Danchick, R., Further results on generalized predictor–corrector methods, *J. Comp. Syst. Sci.* **2**, 203 (1968).
12. Distefano, G. P., Stability of numerical integration techniques, *AIChE J.* **14**, 946 (1689).
13. Dyer, J., Generalized multistep methods in satellite orbit computation, *J. Assoc. Comput. Mach.* **15**, 712 (1968).
14. Gallaher, L. J. and Perlin, I. E., "A Comparison of Several Methods of Numerical Integration of Nonlinear Differential Equations," Presented at SIAM Meeting, University of Iowa, March, 1966.
15. Gear, C. W., Hybrid methods for initial value problems in ordinary differential equations, *SIAM J. Numer. Anal.* **2**, 69 (1965).
16. Gear, C. W., The numerical integration of ordinary differential equations, *Math. Comp.* **21**, 146 (1967).
17. Gragg, W. B., and Stetter, H. J., Generalized multistep predictor–corrector methods, *J. Assoc. Compact. Mach.* **11**, 188 (1964).
18. Hall, G., The stability of predictor–corrector methods, *Comput. J.* **9**, 410 (1967).

19. Hamming, R. W., Stable predictor–corrector methods for ordinary differential equations, *J. Assoc. Comput. Mach.* **6**, 37 (1959).
20. Hicks, J. S. and Wei, J., Numerical solution of parabolic partial differential equations with two point boundary conditions by use of the method of lines, *J. Assoc. Comput. Mach.* **14**, 549 (1967).
21. Hull, T. E. and Newbery, A. C. R., Integration procedures which minimize propagated errors, *J. SIAM* **9**, 31 (1961).
22. Hull, T. E. and Newbery, A. C. R., Corrector formulas for multistep integration methods, *J. SIAM* **10**, 351 (1962).
23. Hull, T. E. and Creemer, A. L., Efficiency of predictor–corrector procedures, *J. Assoc. Comput. Mach.* **10**, 291 (1963).
24. Klopfenstein, R. W. and Millman, R. S., Numerical stability of a one-evaluation predictor–corrector algorithm for numerical solutions of ordinary differential equations, *Math. Comp.* **22**, 557 (1968).
25. Kohfeld, J. J. and Thomspon, G. T., Multistep methods with modified predictors and correctors, *J. Assoc. Comput. Mach.* **14**, 155 (1967).
26. Kohfeld, J. J. and Thomspon, G. T., A modification of Nordsieck's method using an "off-step" points, *J. Assoc. Comput. Mach.* **15**, 390 (1968).
27. Krogh, F. T., Predictor–corrector methods of high order with improved stability characteristics, *J. Assoc. Comput. Mach.* **13**, 374 (1966).
28. Krogh, F. T., A test for instability in the numerical solution of ordinary differential equations, *J. Assoc. Comput. Mach.* **14**, 351 (1967).
29. Krogh, F. T., A note on the effect of conditionally stable correctors, *Math. Comp.* **21**, 717 (1967).
30. Lambert, R. J., An analysis of the numerical stability of predictor–corrector solutions of nonlinear ordinary differential equations, *SIAM J. Numer. Anal.* **4**, 597 (1967).
31. Lapidus, L., "Digital Computation for Chemical Engineers." McGraw-Hill, New York, 1962.
32. Lomax, H., An operational unification of finite difference methods for the numerical integration of O.D.E., *NASA Technical Report*, NASA TR R-262 (May, 1967).
33. Martens, H. R., A comparative study of digital integration methods, *Simulation* **12**, 87 (1969).
34. Meshaka, P., Arux Methodes D'Integration Numerique Pour Systems Differentiels, *Rev. Franc. de Traitement de L'Information*, 135 (1964).
35. Milne, W. E., "Numerical Solution of Differential Equations." Wiley, New York, 1953.
36. Milne, W. E. and Reynolds, R. R., Stability of a numerical solution of differential equations, *J. Assoc. Comput. Mach.* **6**, 196 (1959); **7**, 46 (1960).
37. Milne, W. E. and Reynolds, R. R., Fifth-order methods for the numerical solution of ordinary differential equations, *J. Assoc. Comput. Mach.* **9**, 64 (1962).
38. Nordsieck, A., Numerical integration of ordinary differential equations, *Math. Comp.* **16**, 22 (1962).
39. Osborne, M. R., On Nordsieck's method for the numerical solution of ordinary differential equations, *BIT* **6**, 51 (1966).
40. Ralston, A., Some theoretical and computational matters relating to predictor–corrector methods of numerical integration, *Comput. J.* **4**, 64 (1962).
41. Ralston, A., Relative stability in the numerical solution of ordinary differential equations, *SIAM Rev.* **7**, 114 (1965).
42. Ralston, A., "A First Course in Numerical Analysis." McGraw-Hill, New York, 1965.
43. Reimer, M., An integration procedure involving error estimation, *BIT* **5**, 164 (1965).

44. Stetter, H. J., A study of strong and weak stability in discretization algorithms, *SIAM J. Numer. Anal.* **2**, 265 (1965).
45. Stetter, H. J., Stabilizing predictors for weakly unstable correctors, *Math. Comp.* **19**, 84 (1965).
46. Stetter, H. J., " Improved Absolute Stability of Predictor–Corrector Schemes," Computing (*Arch. Elektron. Rechmen.*) **3**, 286 (1968).
47. Timlake, W. P., On an algorithm of Milne and Reynolds, *BIT* **5**, 276 (1965).

5

Extrapolation Methods

5.1. EXTRAPOLATION TO THE LIMIT

5.1.1. Accumulated Truncation Error of Single-Step Methods

The reader will recall that for certain mild assumptions on $y(x)$ and $\Phi(x, y; h)$ we were able to show that the accumulated truncation error for a single-step method can be bounded by (1.11-25). As Henrici [9] is careful to point out, (1.11-25) merely serves to prove the convergence of (1.11-2) and to indicate the order of magnitude of the error. Actually, (1.11-25) states that there exists a constant C such that $|\varepsilon_n| \leq Ch^p$, indicating that $\varepsilon_n \to 0$ as $h \to 0$.

It is more useful to have a statement about *how* the error ε_n tends to zero. Henrici has proved the following theorem for computing ε_n.

Theorem 5.1.1. If f is sufficiently differentiable, ε_n will satisfy

$$\varepsilon_n = h^p \varepsilon(x_n) + O(h^{p+1}) \tag{5.1-1}$$

where $\varepsilon(x)$, the *magnified error function*, is the solution of the initial value problem,

$$\varepsilon'(x) = f_y(x, y(x))\varepsilon(x) - \frac{1}{(p+1)!} y^{[p+1]}(x) \tag{5.1-2}$$

$$\varepsilon(x_0) = 0 \tag{5.1-3}$$

If we denote by $y_n(h)$ the approximation to $y(x_n)$ calculated from (1.11-2) with step h, then (5.1-1) can be written

$$y_n(h) = y(x_n) + h^p \varepsilon(x_n) + O(h^{p+1}) \tag{5.1-4}$$

Even though both $y(x)$ and $\varepsilon(x)$ are unknown in general, (5.1-4) will form the basis for the extrapolation procedure.

Say we perform the same integration with (1.11-2) with step length $h/2$. The relation analgous to (5.1-4) is

$$y_n(h/2) = y(x_n) + (h/2)^p \varepsilon(x) + O(h^{p+1}) \tag{5.1-5}$$

We now have two computed values for y_n. The objective is to use these two values to produce a more accurate approximation to $y(x_n)$. Let us form

$$\frac{2^p y_n(h/2) - y_n(h)}{2^p - 1} \tag{5.1-6}$$

This quantity, which is a new approximation to $y(x_n)$, differs from $y(x_n)$ by only $O(h^{p+1})$. Thus, by forming an appropriate linear combination of the numerical results evaluated using h and $h/2$ we have eliminated the leading error term in the asymptotic error expansion. This procedure which can be repeated indefinitely with $h/4$, $h/8$, etc., each time removing the leading error term, is called extrapolation to the limit.

In Section 2.9.1 this idea was used to provide an estimate of the local truncation error. By considering only the calculation from x_{n-1} to x_{n+1} by two steps of length h and one step of length $2h$, and assuming that the local truncation error is given by (2.9-1), the truncation error could be estimated. From (2.9-4) we obtain

$$y_{n+1}^{(1)} \cong \frac{2^p y_{n+1}^{(h)} - y_{n+1}^{(2h)}}{2^p - 1} \tag{5.1-7}$$

which is a specific form of (5.1-6) that was derived by considering only the local truncation error. It is easily seen that $y_{n+1}^{(1)}$ is a more accurate approximation to $y(x_{n+1})$,

$$y_{n+1}^{(1)} = y(x_{n+1}) + O(h^{p+1}) \tag{5.1-8}$$

Let us consider this procedure in more detail.

5.1.2. Extrapolation to the Limit

In the preceding subsection the elements of extrapolation were outlined. Let us now consider the problem from a slightly more general point of view. The solution of a continuous problem by a numerical method, e.g., the integration of a differential equation or the evaluation of an integral, involves the generation of a discrete approximation the accuracy of which depends on the discretization step used. Let us denote such a numerical approximation by $T(h)$. In the terminology of the preceding subsection $y_n(h) = T(h)$. Let us assume that associated with the discrete approximation is an asymptotic error expansion in powers of h,

$$T(h) = \tau_0 + \tau_1 h^{\gamma_1} + \tau_2 h^{\gamma_2} + \cdots, \qquad 0 < \gamma_1 < \gamma_2 < \cdots \qquad (5.1\text{-}9)$$

where τ_0 is the exact solution of the continuous problem. For most common numerical methods, $\gamma_i = i\gamma$, $i = 1, 2, \ldots$, Referring to (1.11-2) and $p = 1$ (Euler's method), $\gamma_i = i\gamma$ with $\gamma = 1$. τ_1 is related to $\varepsilon(x_n)$, as in (5.1-4).

Richardson [12] proposed that a discrete approximation $T(h)$ with an asymptotic error expansion of the form (5.1-9) with $\gamma_i = i\gamma$ could be improved in accuracy by forming linear combinations of $T(h)$ at different values of h in such a way that the largest error term is successively eliminated. Richardson called this procedure "the deferred approach to the limit" because after a large number of linear combinations one has a very close approximation to the true value τ_0. Obviously (5.1-7) represents the first step in an extrapolation to the limit for the case where $\gamma_i = i$, $i = 1, 2, \ldots$.

Consider for a moment the evaluation of the integral $\int_a^b f(x)\, dx$ by means of the trapezoidal rule,

$$T(h) = h\left[\tfrac{1}{2}f(a) + f(a + h) + \cdots + \tfrac{1}{2}f(b)\right] \qquad (5.1\text{-}10)$$

which has an asymptotic error expansion of the form (5.1-9) with $O(h^2)$, i.e., $\gamma_i = i\gamma$, $\gamma = 2$. Following (5.1-4) and (5.1-5), we can use (5.1-10) with h and $h/2$. If we form the linear combination

$$T_1^{(0)} = \frac{4T(h/2) - T(h)}{3} \qquad (5.1\text{-}11)$$

where the subscript 1 denotes the first linear combination and the reason for using the superscript 0 will become clear shortly, then $T_1^{(0)}$ has an asymptotic error expansion of $O(h^4)$,

$$T_1^{(0)} = \tau_0 - \frac{\tau_2}{4} h^4 - \cdots \qquad (5.1\text{-}12)$$

Proceeding, we form the combination

$$T_2^{(0)} = \frac{16 T_1^{(0)}(h/2) - T_1^{(0)}(h)}{15} \qquad (5.1\text{-}13)$$

which is accurate to $O(h^6)$,

$$T_2^{(0)} = \tau_0 + \frac{\tau_3}{64} h^6 + \cdots \qquad (5.1\text{-}14)$$

We now can generalize the extrapolation procedure. Let us denote by $T_0^{(k)}$ the approximation obtained by the numerical method with 2^k subintervals. For example, the values $y_n(h)$ obtained by (1.11-2) would be denoted $T_0^{(k)}$ with $h_k = (x - a)/2^k$, and the values $T(h)$ obtained by (5.1-10) would also be denoted $T_0^{(k)}$ with $h_k = (b - a)/2^k$. Let m denote the number of linear combinations performed. The two important cases are $\gamma_i = i$ and $\gamma_i = 2i$, for which the above procedures can be generalized to $T_m^{(k)}$ which is computed by the recursion relations

$$T_m^{(k)} = \frac{2^m T_{m-1}^{(k+1)} - T_{m-1}^{(k)}}{2^m - 1}; \qquad \gamma = 1 \qquad (5.1\text{-}15)$$

$$T_m^{(k)} = \frac{4^m T_{m-1}^{(k+1)} - T_{m-1}^{(k)}}{4^m - 1}; \qquad \gamma = 2 \qquad (5.1\text{-}16)$$

It is convenient to arrange these values in a table, called the T table

$$T_0^{(0)}$$

$$T_0^{(1)} \quad T_1^{(0)}$$

$$T_0^{(0)} \quad T_1^{(1)} \quad T_2^{(0)}$$

$$T_0^{(3)} \quad T_1^{(2)} \quad T_2^{(1)} \quad T_3^{(0)}$$

$$\vdots \qquad \vdots \qquad \vdots \qquad \vdots$$

The entries in the table, other than the first column are computed by either (5.1-15) or (5.1-16) which is represented by (with the arrows showing the calculation sequence)

$$T_{m-1}^{(k)} \searrow$$
$$T_{m-1}^{(k+1)} \longrightarrow T_m^{(k)}$$

It is not necessary that the first column of the T table be generated on the basis of halving the previous value of h. A general sequence of subinterval lengths $h_0 > h_1 > \cdots > h_m$ can be used. Then the general recursion relation for $\gamma_i = i\gamma$ can be written

$$T_m^{(k)} = \frac{(h_k/h_{m+k})^\gamma T_{m-1}^{(k+1)} - T_{m-1}^{(k)}}{(h_k/h_{m+k})^\gamma - 1} \qquad (5.1\text{-}17)$$

If $h_k = h_0/2^k$, (5.1-17) reduces to (5.1-15) and (5.1-16) for $\gamma = 1, 2$ respectively.

The application of Richardson's idea to the evaluation of integrals by the trapezoidal rule is due to Romberg [13], the procedure normally referred to as Romberg integration. A complete account of Romberg integration is given by Davis and Rabinowitz [7].

The following two theorems state the conditions for convergence of the extrapolation procedure (5.1-17) with $\gamma = 2$. Similar theorems exist for $\gamma = 1$. They are presented without proof and the reader may consult the references cited.

Theorem 5.1.2 [8].   A necessary and sufficient condition that

$$\lim_{m \to \infty} T_m^{(0)} = \tau_0$$

for all functions $T(h)$ continuous from the right at $h = 0$ for which $\gamma = 2$ in (5.1-17) is that

$$\sup_{k \geq 0} \frac{h_{k+1}}{h_k} < 1 \qquad (5.1\text{-}18)$$

where sup represents least upper bound.

Theorem 5.1.3 [8].   If $T(h)$ has an error expansion of the form (5.1-9) with $\gamma_i = 2i$ and (5.1-18) is satisfied, then as $k \to \infty$,

$$T_m^{(k)} - \tau_0 = (-1)^m h_k^2 \cdots h_{k+m}^2 (\tau_{m+1} + O(h_k^2)) \qquad (5.1\text{-}19)$$

If, in addition,

$$\inf_{k \geq 0} \frac{h_{k+1}}{h_k} > 0 \qquad (5.1\text{-}20)$$

where inf represents greatest lower bound, then there exist constants θ_m such that, for each $m > 0$,

$$|T_m^{(0)} - \tau_0| \leq \theta_{m+1} h_k^2 \cdots h_{k+m}^2 \qquad (5.1\text{-}21)$$

In short, (5.1-19) states that each column of the T table converges to τ_0 faster than the preceding one, and (5.1-21) states that the principal diagonal $T_m^{(0)}$ converges to τ_0 faster than any column. Theorem 5.1.3 provides an estimate of the truncation error at any location in the T table. It can also be shown that if the τ_i obey mild restrictions on rate of growth, $T_m^{(0)}$ converges superlinearly to τ_0 in that $|T_m^{(0)} - \tau_0| \leq K_m$ and

$$\lim_{m \to \infty} \frac{K_{m+1}}{K_m} = 0$$

as is the case when $T(h)$ can be extended to a function which is analytic at $h = 0$ [8].

We may extend the foregoing treatment slightly by requiring that $T_m^{(k)}$ be a rational function

$$T_m^{(k)} = \frac{p_0^{(k)} + p_1^{(k)} h^2 + \cdots + p_{m/2}^{(k)} h^m}{q_0^{(k)} + q_1^{(k)} h^2 + \cdots + q_{m/2}^{(k)} h^m} \qquad (5.1\text{-}22)$$

defined by the requirement that

$$T_m^{(k)}(h_i) = T(h_i), \qquad i = k, \ldots, k+m \tag{5.1-23}$$

If we do, Bulirsch and Stoer [2] have shown that one gets the following recursion relation

$$T_m^{(k)} = T_{m-1}^{(k+1)} + \frac{T_{m-1}^{(k+1)} - T_{m-1}^{(k)}}{(h_k/h_{m+k})^2[1 - (T_{m-1}^{(k+1)} - T_{m-1}^{(k)})/(T_{m-1}^{(k+1)} - T_{m-2}^{(k+1)})] - 1} \tag{5.1-24}$$

with the computational diagram

$$T_{m-2}^{(k)}$$

$$T_{m-2}^{(k+1)} \textemdash T_{m-1}^{(k)} \searrow$$

$$T_{m-2}^{(k+2)} \qquad T_{m-1}^{(k+1)} \longrightarrow T_m^{(k)}$$

We will not be concerned with the technicalities of obtaining (5.1-24), but only observe that extrapolation based on (5.1-24) rather than (5.1-17) usually gives faster convergence.

5.2. EXTRAPOLATION ALGORITHMS FOR ODE

The same principle as applied to the numerical evaluation of an integral can also be used for the numerical integration of ODEs. This follows from the fact that most numerical integration methods for ODEs have an error expansion of the form (5.1-9). In this section we will present three algorithms based on extrapolation to the limit coupled with Euler's method, the trapezoidal rule, and the modified midpoint rule.

5.2.1. The Euler–Romberg Method

We desire to generate approximations y_n to the true solution $y(x_n)$ at mesh points x_n of the scalar initial value problem $y' = f(x, y)$; $y(x_0) = y_0$. If we use Euler's method, the approximation y_n to $y(x_n)$ has a truncation error in powers of h,

$$y_n = y(x_n) + \tau_1(x_n)h + \tau_2(x_n)h^2 + \tau_3(x_n)h^3 + \cdots \tag{5.2-1}$$

where the coefficients of the error expansion depend on x. Values of y_n can be generated using Euler's method (1.6-7) based on a sequence of h_k, $k = 0, 1, 2, \ldots$ and the $y_n(h_k)$ obtained can be denoted as $Y_0^{(k)}$. The initial step length h_0 is for the moment the total interval length, $x_n - x_0$.

Obviously, since Euler's method has the error expansion (5.2-1) which is of the form (5.1-9) with $\gamma_i = i$, $i = 1, 2, \ldots$, extrapolation to the limit can be applied to (5.2-1). Since (5.2-1) contains odd as well as even powers of h, the extrapolation recursion relations must be modified slightly. If successive interval halving is used, i.e., $h_k = h_0/2^k$, the recurrence relation analgous to (5.1-15) is

$$Y_m^{(k)} = \frac{2^m Y_{m-1}^{(k+1)} - Y_{m-1}^{(k)}}{2^m - 1} \tag{5.2-2}$$

For an arbitrary sequence h_k, $k = 0, 1, 2, \ldots$, analogous to (5.1-17),

$$Y_m^{(k)} = \frac{(h_k/h_{k+m}) Y_{m-1}^{(k+1)} - Y_{m-1}^{(k)}}{(h_k/h_{k+m}) - 1} \tag{5.2-3}$$

This method is termed the Euler–Romberg algorithm. Convergence of (5.2-3) is guaranteed if $f(x, y)$ satisfies a Lipschitz condition [1].

In the discussion so far we have considered the extrapolation procedure as applied over the interval x_0 to x_n, where the initial step length h_0 in the sequence h_k, $k = 0, 1, 2, \ldots$, was equal to $x_n - x_0$. Actually what we desire to do in the Euler–Romberg algorithm is to use extrapolation at *each* step in the integration, $x_1, x_2, \ldots$, where $x_{i+1} - x_i = h_0$. The use of extrapolation at each step in the integration is termed *local* extrapolation to the limit. The procedure is self-starting and the choice of h_0 is fairly arbitrary since the step length is automatically reduced until the required accuracy is achieved at each step of the solution.

Let us assume we are at (x_n, y_n) and we want to compute y_{n+1} at x_{n+1}, where $x_{n+1} - x_n = h_0$, by the Euler–Romberg algorithm. The procedure is started by computing an initial y_{n+1}, denoted by $Y_0^{(0)}$, by the normal Euler alogrithm using the full step length h_0,

$$Y_0^{(0)} = y_n + h_0 f_n \tag{5.2-4}$$

Then the first column $Y_0^{(k)}$ of the Euler–Romberg table

$$Y_0^{(0)}$$
$$Y_0^{(1)} \quad Y_1^{(0)}$$
$$Y_0^{(2)} \quad Y_1^{(1)} \quad Y_2^{(0)}$$
$$\vdots \qquad \vdots \qquad \vdots$$

is generated by Euler's method with successively smaller step lengths, $h_1, h_2, \ldots$, In the case of successive halving, $h_k = h_0/2^k$, the procedure can be illustrated by Figure 5.1 due to McCalla [11]. Thus, when $h_k = h_0/2^k$ for $Y_0^{(1)}$ use $h_1 = h_0/2$ and apply Euler's method twice to get from x_n to x_{n+1}. For $Y_0^{(2)}$ use $h_1 = h_0/4$ and apply Euler's method four times. Obviously, the same procedure holds when $h_1, h_2, \ldots$ is a prescribed sequence in which h_0/h_k, $k = 1, 2, \ldots$, is an integer value. The elements of the Y table are computed according to either (5.2-2) or (5.2-3), depending on the step-length

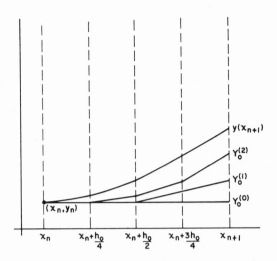

Figure 5.1. Successive halving in Euler's method. [Adapted and reprinted from T. R. McCalla, "Introduction to Numerical Methods and Fortran Programming," p. 342; copyright 1962, John Wiley & Sons.]

sequence used. Since the table is easily generated column by column, a convenient convergence test is that adjacent elements in a column agree within some prespecified fractional tolerance ε_Y,

$$\left| \frac{Y_m^{(k)} - Y_m^{(k-1)}}{Y_m^{(k-1)}} \right| \leq \varepsilon_Y$$

When convergence is obtained, $Y_m^{(k)}$ is used as the value y_{n+1} and the procedure is repeated to obtain y_{n+2}.

The algorithm may be summarized as follows:

1. $Y_0^{(0)} = y_n + h_0 f_n$.
2. Compute $Y_m^{(k)}$ from (5.2-2) or (5.2-3). Storage can be reduced by using the array notation

$$Y_1$$
$$Y_2 \quad Y_1$$
$$Y_3 \quad Y_2 \quad Y_1$$
$$Y_4 \quad Y_3 \quad Y_2 \quad Y_1$$
$$\vdots \quad \vdots \quad \vdots \quad \vdots$$

where

$$Y_k = \frac{(h_k/h_{k+m})Y_{k+1} - Y_k}{(h_k/h_{k+m}) - 1} \tag{5.2.-5}$$

and m is the column number.

3. If $k > 1$, go to step 4. If $k = 1$, return to step 2.
4. Test convergence by

$$\left| \frac{Y_k - Y_{k-1}}{Y_{k-1}} \right| \leq \varepsilon_Y$$

If convergence is obtained, set $y_{n+1} = Y_k$ and return to step 1.
If convergence is not obtained set $k = k + 1$ and if $k < K_{\max}$ return
to step 2. $K_{\max}$ is the largest number in the k set.

We can mention several important points in connection with the Euler–
Romberg algorithm. This algorithm possesses two important advantages,
namely its simple recursion relation, and its automatic control of accuracy.
One disadvantage is that because (5.2-1) contains odd powers of h, the convergence of the columns and main diagonal of the Y table is slower than for
an integration method with error terms in powers of h^2 only. Thus, it is of
interest to see if other methods with such error expansions can be combined
with extrapolation to give faster convergence for the same level of accuracy.
Finally we point out that the Euler–Romberg method is based on *local*
extrapolation where the extrapolation is carried out at *each* step in the integration. The other alternative is *global* extrapolation wherein the particular
numerical method is first applied over the *entire* range of integration from
$x_0 = a$ to $x = b$ for a decreasing sequence of step lengths, and then extrapolation is applied to the values obtained at each original mesh point,
$x_n = x_0 + nh_0$.

5.2.2. The Trapezoidal Rule

The trapezoidal rule has two distinct advantages over Euler's method;
it is more accurate and it is A-stable. Thus, an obvious choice for a method
with an h^2 error expansion to combine with extrapolation is the trapezoidal
rule. If we let $T(x, h) = y_n(h)$ and

$$y_{n+1} = y_n + (h/2)(f_{n+1} + f_n) \tag{5.2-6}$$

then

$$T(x, h) = y(x_n) + \tau_1(x)h^2 + \tau_2(x)h^4 + \cdots \tag{5.2-7}$$

where $\tau_1(x), \tau_2(x), \ldots$ are not the same functions as in (5.2-1). A very important point is that in order to preserve the A-stability of the trapezoidal rule
it is necessary to use *global* extrapolation rather than local extrapolation.
This is proved in Section 5.3.2. Another point of importance which somewhat
diminishes the general utility of (5.2-6) combined with extrapolation is that
in order for (5.2-7) to hold it is necessary that (5.2-6) be solved *exactly* at each

step [8]. Thus, the set of nonlinear algebraic equations at each step must be solved to an order of accuracy smaller than the level of the truncation error. However, the Newton–Raphson procedure described in Section 4.3.3 usually gives a highly accurate solution in a few iterations.

The algorithm is used in the following way:

1. Integrate the ODE system from $x_0 = a$ to $x = b$ using (5.2-6) with the basic step length h_0. Denote these values as $T_{0,n}^{(0)}$, $n = 1, 2, \ldots, N$, $N = (b - a)/h_0$. The additional subscript n is required because there will be a separate T table corresponding to each mesh point $x_n = x_0 + nh_0$.

2. Integrate the ODE system from $x_0 = a$ to $x = b$ using the sequence of step lengths $h_k = h_0/2^k$, $k = 1, 2, \ldots$. At each of the original mesh points, $x_n = x_0 + nh_0$, let the computed value be set equal to $T_{0,n}^k$, $n = 1, 2, \ldots, N$; $k = 1, 2, \ldots, K$. The series of values $T_{0,n}^k$ now form the first column of a T table at each mesh point x_n.

3. Perform extrapolation with (5.1-16) at each x_n, $x_n = x_0 + nh_0$, until the final diagonal element, i.e., compute the complete table.

4. If adjacent diagonal elements at a particular mesh point differ by more than a preset criterion, set $K = K + 1$ and repeat step 1 for the new K. If adjacent diagonal elements at all mesh points satisfy the convergence criterion, set $y_n = T_{K,n}^{(0)}$ as the final result.

Because of its A-stability, this algorithm is particularly suited to the integration of ODE systems with widely separated eigenvalues. For this reason, computational results based on this method will be presented in Chapter 6 where we consider such systems.

5.2.3. The Modified Midpoint Method

The principal drawback of the trapezoidal rule and global extrapolation is that implicit equations must be solved at each step. However, the h^2 error expansion is advantageous because higher accuracy than with an h expansion can be obtained in the same number of operations. An explicit method with an h expansion based on the combination of the midpoint method (1.6-31) and local rational extrapolation has been formulated by Gragg [8] and Bulirsch and Stoer [2].

The system $y' = f(x, y)$ is integrated from x_0 to $x_0 + h_0$ by the modified midpoint rule using first a step length h_0, then h_1, etc. Two possible step-size sequences are

$$h_k = h_0/2^k, \qquad k = 0, 1, 2, \ldots$$

$$h_k = \{h_0, h_0/2, h_0/3, h_0/4, h_0/5, h_0/6\}$$

The result is a sequence of approximations $T(h_k, x)$ to $y(x)$ such that

$$T(0, x) = \lim_{h_k \to 0} T(h_k, x) = y(x) \tag{5.2-8}$$

The modified midpoint rule is defined as follows. Let $y(\xi_i, h)$ be the approximate value obtained at ξ_i, where $x_0 < \xi_i < x_0 + h_0$. Thus, we are considering the method as applied to the interval $[x_0, x_0 + h_0]$. The method is then applied in an identical fashion from x_n to $x_n + h_0$, $n = 1, 2, \ldots$.

Since the midpoint rule is a two step method, we start the integration with Euler's method

$$y(\xi_1, k) = y_0 + hf(x_0, y_0) \tag{5.2-9}$$

and then proceed by means of the midpoint rule,

$$y(\xi_{i+1}, h) = y(\xi_{i-1}, h) + 2hf\{\xi_i, y(\xi_i, h)\}, \qquad i = 1, 2, \ldots, l \tag{5.2-10}$$

At $\xi_l = x_0 + h_0 = x_1$ we form

$$S(h, x_1) = \tfrac{1}{2}\left[y(\xi_l, h) + y(\xi_{l-1}, h) + hf\{\xi_l, y(\xi_l, h)\} \right] \tag{5.2-11}$$

and then $T(h, x_1)$ the approximation to $y(x_1)$ is related to $S(h, x_1)$ by

$$T(h, x_1) = S(h/2, x_1) \tag{5.2-12}$$

It is shown by Gragg and Bulrisch and Stoer that the asymptotic error expansions of both $y(x, h)$ and $T(h, x)$ contain only even powers of h.

The algorithm (5.2-9) to (5.2-12) can then be combined with extrapolation. Either polynomial or rational extrapolation can be used. If rational extrapolation is used, the extrapolated values $T_m^{(k)}$ can be computed from $T(h_k, x)$ by

$$T_{-1}^{(k)} = 0, \qquad T_0^{(k)} = T(h_k, x) \tag{5.2-13}$$

combined with (5.1-24).

The algorithm is used in the following way:

1. Select a basic integration step length h_0 which determines the grid points for the integration, $x_n = x_0 + nh_0$, $n = 1, 2, \ldots, N$.
2. Use (5.2-9)–(5.2-11) to compute $S(h_k, x_1)$ based on h_k, $k = 0, 1, 2, \ldots$, K where h_0/h_k is an integer. These values are identified as $T(2h_k, x_1)$ from (5.2-12) and are denoted by $T_0^{(k)}$, the first column of the T table.
3. Compute the remaining entries of the T table by (5.1-17) or (5.1-24), depending on the type of extrapolation. A convergence test based on the difference between adjacent column elements, $T_m^{(k)}$ and $T_m^{(k-1)}$, or adjacent main diagonal elements, $T_{m+1}^{(0)}$ and $T_m^{(0)}$, can be used. The converged value is taken as y_1.
4. Repeat steps 2 and 3 for $x_n = x_0 + nh_0$, $n = 2, 3, \ldots, N$.

The algorithm is thus based on local extrapolation.

In this section three methods based on extrapolation have been presented. These are summarized below:

Method	Application	Error expansion	Core algorithm
Euler–Romberg	Local	h	Euler method
Trapezoidal rule	Global	h^2	Trapezoidal rule
Modified midpoint	Local	h^2	Midpoint rule

5.3. STABILITY AND ERROR ANALYSIS OF EXTRAPOLATION METHODS

In this section we wish to indicate the elements of numerical stability analysis as applied to the extrapolation algorithms in Section 5.2. Each of these methods is of single-step type, although a fundamental difference exists in whether the extrapolation is executed locally (Euler–Romberg method and the modified midpoint method) or globally (the trapezoidal rule). We will first consider the stability of the Euler–Romberg method. Then we will show that in order to preserve A-stability, the trapezoidal rule must be used in conjunction with global extrapolation rather than local extrapolation. The complexity of the modified midpoint rule prohibits a stability analysis as in the first two methods. However, extensive computational results reported at the end of this and the next chapter serve to indicate roughly the stability properties of the modified midpoint rule.

Any single-step method can be expressed in the form

$$y_{n+1} = \mu_1(h_0 \lambda) y_n \tag{5.3-1}$$

where $x_{n+1} - x_n = h_0$ and $\mu_1(h_0 \lambda)$ is the characteristic root of the method. An extrapolation algorithm applied *locally* over the interval from x_n to x_{n+1} can be expressed as

$$y_{n+1} = \beta(h_0 \lambda, m, k) y_n \tag{5.3-2}$$

where k is the number of step lengths for which the core algorithm is used over the overall step length h_0, and m is the number of times that extrapolation is carried out (e.g., the number of columns in the Y table). To simplify the analysis we will assume that each time extrapolation is carried out the extrapolation is stopped at $m = M$, $k = K$, so that

$$y_{n+1} = \beta(h_0 \lambda, M, K) y_n \tag{5.3-3}$$

We are also assuming that the sequence of step lengths h_k, $k = 1, 2, \ldots, K$, is related to h_0, e.g., $h_k = h_0/2^k$, $k = 1, 2, \ldots, K$.

At n steps we may write

$$y_n = [\beta(h_0 \lambda, M, K)]^n y_0 \tag{5.3-4}$$

so that when applied to $y' = \lambda y$ with $\mathrm{Re}(\lambda) < 0$, stability requires that

$$|\beta(h_0 \lambda, M, K)| \leq 1 \qquad (5.3\text{-}5)$$

In order to determine the stability bounds of the algorithm it is necessary to determine for fixed M and K the values of $h_0 \lambda$ for which the equality in (5.3-5) holds. We will now consider local extrapolation by means of the Euler–Romberg method and the trapezoidal rule, the latter case to illustrate that A-stability is lost if local extrapolation is used.

5.3.1. The Euler–Romberg Method

The first column of the Euler–Romberg Y table is generated by

$$Y_0^{(k)} = [1 + (h_0 \lambda/2^k)]^{2^k} y_n$$
$$= [\mu_1(h_0 \lambda/2^k)]^{2^k} y_n, \qquad k = 0, 1, 2, \dots \qquad (5.3\text{-}6)$$

The T table recurrence relation is (5.2-2),

$$Y_m^{(k)} = \frac{2^m Y_{m-1}^{(k+1)} - Y_{m-1}^{(k)}}{2^m - 1}$$

which can be expressed as a linear combination of the first column elements,

$$Y_m^{(k)} = \sum_{i=0}^{m} c_{m,m-i} Y_0^{(k+1)} \qquad (5.3\text{-}7)$$

The coefficients obey the recursion relation

$$c_{m,m-i} = \frac{2^m c_{m-1,m-i} - c_{m-1,m-1-i}}{2^m - 1}, \qquad c_{m-1,m} = c_{m-1,-1} = 0 \qquad (5.3\text{-}8)$$

Combining (5.3-6) and (5.3-7) we obtain

$$Y_m^{(k)} = \sum_{i=0}^{m} c_{m,m-i} [\mu_1(h_0 \lambda/2^{k+i})]^{2^{k+i}} y_n \qquad (5.3\text{-}9)$$

If the extrapolated value $Y_M^{(K)}$ is taken as the value y_{n+1}, then comparing (5.3-3) and (5.3-9) we have the characteristic root,

$$\beta(h_0 \lambda, M, K) = \sum_{i=0}^{M} c_{M,M-i} [\mu_1(h_0 \lambda/2^{K+i})]^{2^{K+i}} \qquad (5.3\text{-}10)$$

Consider, for example, the $m = 1$ column for which

$$Y_1^{(k)} = \{2[\mu_1(h_0 \lambda/2^{k+1})]^{2^{k+1}} - [\mu_1(h_0 \lambda/2^k)]^{2^k}\} y_n \qquad (5.3\text{-}11)$$

For $k = 0$,

$$Y_1^{(0)} = \{1 + h_0 \lambda + h_0^2 \lambda^2/2\} y_n \qquad (5.3\text{-}12)$$

which, if the extrapolation is stopped at this point, is identical to the explicit second-order Runge–Kutta methods. In continuing, consider the $m = 2$ column for which $c_{2,2} = \frac{1}{3}$, $c_{2,1} = -2$, and $c_{2,0} = \frac{8}{3}$. Thus,

$$Y_2^{(k)} = \frac{8}{3} Y_0^{(k+2)} - 2 Y_0^{(k+1)} + \frac{1}{3} Y_0^{(k)} \qquad (5.3\text{-}13)$$

For $k = 0$, (5.3-13) becomes

$$Y_2^{(0)} = \{\frac{8}{3}(1 + \frac{1}{4} h_0 \lambda)^4 - 2(1 + \frac{1}{2} h_0 \lambda)^2 + \frac{1}{3}(1 + h_0 \lambda)\} y_n \qquad (5.3\text{-}14)$$

The complexity for larger values of m soon becomes apparent.

In order to determine the stability region corresponding to various values of K and M, it is necessary to determine the values of $h_0 \lambda$ corresponding to K and M that the equality in (5.3-5) holds. This has been carried out for $K = 0$ (the diagonal elements in the Y table) and various values of M for real, negative values of $h_0 \lambda$. Figure 5.2 shows $\beta(h_0 \lambda, M, 0)$ for $M = 1$

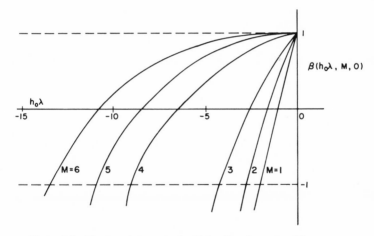

Figure 5.2. Characteristic roots for Euler–Romberg algorithm.

through 6 as a function of $h_0 \lambda$. The points at which $\beta(h_0 \lambda, M, 0)$ becomes ± 1 determine the real stability bounds for the particular value of M. These bounds are summarized below:

M	1	2	3	4	5	6
$(h\lambda_0)_{max}$	-2	-2.785	-4.23	-9.06	-10.88	-13.5

As the number of extrapolations is increased, the overall algorithm becomes more stable. In the limit of an infinite number of extrapolations the method approaches A-stability.

5.3.2. Trapezoidal Rule

It is the purpose of this subsection to illustrate that if the trapezoidal rule were used with *local* extrapolation the A-stability of the method is lost. Obviously, *global* extrapolation will have no effect on the A-stability, since extrapolation is used only after the integration has been carried out over the total interval $[a, b]$.

The development used in the previous subsection can be used here. If the trapezoidal rule were used with local extrapolation the first column of the T table is given by

$$T_0^{(k)} = \left[\frac{1 + (h_0\,\lambda/2^{k+1})}{1 - (h_0\,\lambda/2^{k+1})}\right]^{2^k} y_n$$

$$= \left[\mu_1(h_0\,\lambda/2^{k+1})\right]^{2^k} y_n \qquad (5.3\text{-}15)$$

The T table recurrence relation (5.1-15) results in an equation of the form of (5.3-7). By reasoning similar to that used in obtaining (5.3-10) we obtain the characteristic root

$$\beta(h_0\,\lambda, M, K) = \sum_{i=0}^{M} c_{M,M-i}\left[\frac{1 + (h_0\,\lambda/2^{K+1+i})}{1 - (h_0\,\lambda/2^{K+1+i})}\right]^{2^{K+i}} \qquad (5.3\text{-}16)$$

Let us consider $M = 1$, $K = 0$, the first element in column 2. In this case (5.3-16) becomes

$$\beta(h_0\,\lambda, 1, 0) = \frac{4}{3}\left[\frac{1 + (h_0\,\lambda/4)}{1 - (h_0\,\lambda/4)}\right]^2 - \frac{1}{3}\left[\frac{1 + (h_0\,\lambda/2)}{1 - (h_0\,\lambda/2)}\right] \qquad (5.3\text{-}17)$$

which gives

$$\beta(h_0\,\lambda, 1, 0) = \frac{96 - 18(h_0\,\lambda)^2 - 5(h_0\,\lambda)^3}{96 - 96(h_0\,\lambda) + 30(h_0\,\lambda)^2 - 3(h_0\,\lambda)^3} \qquad (5.3\text{-}18)$$

Taking only real, negative $h_0\,\lambda$ values, we wish to show that $\beta(h_0\,\lambda, 1, 0)$ as given by (5.3-18) is not an A-stable characteristic root. If this can be done, we can reason by induction that the characteristic roots corresponding other M and K are not necessarily A-stable either.

It is straightforward to show that the regions in which $|\beta(h_0\,\lambda, 1, 0)| < 1$ are $-25.8 < h_0\,\lambda < 0$ and $h_0\,\lambda > 1.85$. Thus, in local extrapolation the trapezoidal rule loses A-stability. This is further verified by the observation that in the limit of $h_0\,\lambda \to \infty$,

$$\lim_{h_0\lambda \to \infty} \frac{1}{3}\left\{4\left[\frac{1 + (h_0\,\lambda/4)}{1 - (h_0\,\lambda/4)}\right]^2 - \left[\frac{1 + (h_0\,\lambda/2)}{1 - (h_0\,\lambda/2)}\right]\right\} = \frac{5}{3} \qquad (5.3\text{-}19)$$

5.3.3. The Modified Midpoint Rule

The modified midpoint rule with rational extrapolation is so involved as to preclude a stability analysis of the type in Sections 5.3.1 and 5.3.2. However, we can make some general statements about the stability of the method. The core algorithm, the midpoint rule, was shown to be unstable for all real, negative $h_0\lambda$ values in Section 3.3. Thus, even though the process of extrapolation can be expected to improve the stability characteristics of the core algorithm because the core algorithm is unstable, we cannot expect a significantly stable overall algorithm to result. We will pursue this point further in the examples, particularly in Chapter 6.

5.4. PUBLISHED NUMERICAL RESULTS

An extensive comparison of the computational aspects of Runge–Kutta, predictor–corrector, and extrapolation methods has been made by Clark [4]. The methods used by Clark on several test ODEs were:

1.   Fourth-order Runge–Kutta method (2.3-13).
2.   Fourth-order Adams P–C method, (4.7-3) and (4.7-4), in the PECE mode with starting values from fourth-order Runge–Kutta.
3.   The modified midpoint rule with extrapolation: (a) Bulirsch–Stoer rational function extrapolation (as described in Section 5.2.3), (b) polynomial extrapolation (5.1-16).

The following problems were solved numerically with each of the four algorithms:

1. $y' = -y$, $y(0) = 1$. Integrations from $x = 0$ to $x = 20$ were performed for several step sizes for each method. At $x = 20$ the relative error, $(y_n - e^{-20})/e^{-20}$ was computed and plotted versus the number of derivative evaluations required to $x = 20$.
2.   Euler's equation of motion for rigid bodies.
3.   Bessel's equation of order 16.
4.   The restricted three-body problem.

The following conclusions were reached in the study:

1.   For $y' = -y$ the Bulirsch–Stoer method required fewer derivative evaluations than the polynomial extrapolation for the same relative error. Both extrapolation methods ($M = 6$) required fewer derivative evaluations than the other two methods for relative errors less than 10^{-4}. For larger relative errors, the Adams method was preferable to the Runge–Kutta method. The speeds (in derivative evaluations per second) of the Runge–Kutta method and both extrapolation methods with $M = 6$ were nearly the

same. The Adams method took about five-thirds as long as the other three methods per function evaluation.

2. For Euler's equation the results obtained by using the Bulirsch–Stoer method and polynomial extrapolation ($M = 6$) were nearly the same, both requiring fewer derivative evaluations than the other two methods for relative errors less than 10^{-5}. For Bessel's equation, the Bulirsch–Stoer method with $M = 6$ is preferable for relative errors less than 10^{-3}.

3. A double precision study of the order M of the extrapolation was made on $y' = -y$ and the restricted three-body problem. The Bulirsch–Stoer method proved more efficient than the polynomial extrapolation. The order M of the optimum method increased with decreasing relative error.

In general, the following recommendations were made. The choice of which of the four algorithms to use depends on the level of desired accuracy. For example, to obtain a relative error of 10^{-16} for $y' = -y$, polynomial extrapolation with $M = 10$ takes about two and one-half times as long as $M = 4$. But the Runge–Kutta and Adams methods would require about one hundred and sixty times as long for relative errors of 10^{-16}. Even at errors of 10^{-9} these methods would take about nine times as long as extrapolation with $M = 6$. On the other hand, for relative errors of 10^{-2}, extrapolation with $M = 10$ would take about seven times as long as the fourth-order Adams method.

5.5. NUMERICAL EXPERIMENTS

We now present the results of computations on some of the systems of Chapter 2 using the modified midpoint method of Bulirsch and Stoer. Further results using this alogrithm and the trapezoidal rule with extrapolation will be given in Chapter 6.

The extrapolation algorithm used here is a FORTRAN IV version of DIFSYS as given by Bulirsch and Stoer. The program has been modified to include more intermediate details.

Two important input variables are required by the program. These are h_0, the initial step size to use in the calculation, and EPS, the allowable tolerance in two successive extrapolations. Actually the choice of h_0 is not too important since the program has an automatic step size adjustment. Thus if m is the number of extrapolations the program defines

$$m < 7, \quad h_{0,\,new} = 1.5 h_{0,\,old}$$
$$m \geq 7, \quad h_{0,\,new} = 0.9 h_{0,\,old}$$

to increase or decrease the step size. The program also uses the sequence

$$h_0, \quad h_0/2, \quad h_0/4, \quad h_0/8, \quad \ldots$$

to generate the extrapolation values. All calculations are carried out in single precision.

To illustrate the explicit features of this algorithm we have constructed Table 5.1, in which the results of integration of System I, $y' = -y$, are shown.

TABLE 5.1

System I. Single Precision Modified Midpoint Alogorithm, $h_0 = 0.05$, eps $= 1_{10}{}^{-3}$

Counter	x	h_0	y (calculated)	ε (error)	System evaluations (v)
0	0		1.0	0	1
5	0.05	0.05	0.9506108	$-6.2_{10}{}^{-4}$	49
7	0.125	0.075	0.8840196	$1.52_{10}{}^{-3}$	105
7	0.1925	0.0675	0.8279507	$3.06_{10}{}^{-3}$	105
7	0.25325	0.06075	0.7804320	$4.16_{10}{}^{-3}$	105
5	0.307925	0.054675	0.7383381	$3.37_{10}{}^{-3}$	49
$\vdots$	$\vdots$	$\vdots$	$\vdots$	$\vdots$	$\vdots$

Progress in Calculation, First Series of Extrapolations

x values	Extrapolated y	Counter
0.0250	0.0	0
0.0125		
0.0250		
0.0375	0.5908082	1
0.0083		
0.0167		
0.0250		
0.0333		
0.0417	0.9606364	2
0.0062		
0.0187		
0.0250		
0.0312		
0.0375		
0.0437	0.9231223	3
0.0042		
0.0083		
0.0125		
0.0167		
0.0208		
0.0250		
0.0292		
0.0333		
0.0375		
0.0417		
0.0458	0.9513791	4

TABLE 5.1 (continued)

Progress in calculation, First Series of Extrapolations

x values	Extrapolated y	Counter
0.0031		
0.0062		
0.0094		
0.0125		
0.0156		
0.0187		
0.0219		
0.0250		
0.0281		
0.0313		
0.0344		
0.0375		
0.0406		
0.0437		
0.0469	0.9506108	5

End step 1, $|0.9513791 - 0.9506108| = |7.7_{10^{-4}}| < 1.0_{10^{-3}}$

Table 5.1 is divided into two parts, the first showing the final output and the second illustrating the calculations involved in the first step to $x = 0.05$. The heading "counter" refers to the number of extrapolation steps m used in the calculation, and the heading "system evaluations" refers to the number v. To determine v a counter was inserted in the subroutine counting the number of times the RHS of the ODE is called in the integration routine.

The extrapolation was started with $h_0 = 0.05$ and EPS $= 1.0_{10^{-3}}$. In the first series of extrapolations in the calculation the interval $\{0, 0.05\}$ is successively cut into smaller and smaller parts and an extrapolation carried out to yield a final estimated value of $y_{0.05}$; each estimated value of $y_{0.05}$ is then compared to the previous one until, after five such calculations, the difference is $7.7_{10^{-4}}$. Since this is smaller than EPS $= 1.0_{10^{-3}}$, the program generates the estimate $y_{0.05} = 0.9506108$ and proceeds further into x space. Since only five extrapolations were used, the value of h_0 is then increased to 0.0750 for the next series of calculations. As seen in Table 5.1 this next series follows as before, but with seven extrapolations required to yield $y_{0.125} = 0.8840196$. Thus, in the next set of calculations h_0 is decreased to 0.0675 yielding $y_{0.1925} = 0.8279507$. Obviously this is continued in the same manner for the remainder of the calculation.

System I was also integrated using EPS $= 0.1$ and h $= 0.5$. Some of the results are shown in Table 5.2. It is obvious by comparison of Tables 5.1

TABLE 5.2

SYSTEM I. SINGLE PRECISION, MODIFIED MIDPOINT METHOD

x^a	ε (error)	Counter	v
	$h_0 = 0.05$, EPS $= 0.1$		
0	0	0	1
0.0500	$-2.8_{10}-2$	3	21
0.1250	$-6.1_{10}-2$	3	21
0.2375	$-9.5_{10}-2$	3	21
0.4062	$-1.2_{10}-1$	3	21
0.6594	$-1.3_{10}-1$	3	21
1.039	$-1.1_{10}-1$	3	21
1.608	$-7.4_{10}-2$	3	21
2.462	$-4.7_{10}-2$	3	21
3.744	$4.2_{10}-2$	3	21
4.224	$3.5_{10}-2$	23	357
4.945	$1.4_{10}-2$	3	21
6.026	$-1.3_{10}-2$	3	21
7.698	$-5.0_{10}-3$	3	21
10.08	$-8.2_{10}-3$	3	21
	$h_0 = 0.5$, EPS $= 0.1$		
0	0	0	1
0.5000	$7.5_{10}-3$	5	49
0.8750	$2.3_{10}-2$	15	223
1.437	$-7.2_{10}-3$	3	21
2.281	$-3.0_{10}-2$	3	21
2.914	$-3.4_{10}-3$	13	221
3.863	$-1.4_{10}-2$	3	21
5.287	$1.4_{10}-3$	3	21
7.422	$1.7_{10}-2$	3	21
10.62	$2.5_{10}-2$	3	21
	$h_0 = 0.05$, EPS $= 1.0_{10}-3$		
0	0	0	1
0.1925	$3.0_{10}-3$	7	105
0.4637	$6.1_{10}-3$	7	105
0.7603	$7.8_{10}-3$	7	105
1.084	$8.4_{10}-3$	7	105
1.439	$8.2_{10}-3$	7	105
1.827	$7.4_{10}-3$	7	105
2.308	$6.4_{10}-3$	6	73
2.987	$5.5_{10}-3$	7	105
3.946	$3.3_{10}-3$	5	49
5.228	$8.2_{10}-4$	3	21
7.656	$-4.0_{10}-5$	13	21

TABLE 5.2 (continued)

SYSTEM I. SINGLE PRECISION, MODIFIED MIDPOINT METHOD

x^a	ε (error)	Counter	v
$h_0 = 0.5$, EPS $= 1.0_{10^{-3}}$			
0	0	0	1
0.3387	$1.0_{10^{-2}}$	7	105
0.6521	$1.3_{10^{-2}}$	7	105
0.9549	$1.2_{10^{-2}}$	7	105
1.286	$1.1_{10^{-2}}$	7	105
1.648	$1.0_{10^{-2}}$	6	73
2.096	$8.1_{10^{-3}}$	5	49
2.655	$6.2_{10^{-3}}$	7	105
3.444	$4.2_{10^{-3}}$	5	49
4.712	$1.7_{10^{-3}}$	5	49
7.073	$-1.1_{10^{-4}}$	5	49

[a] Every fourth value shown.

and 5.2 that EPS $= 0.1$ does not lead to very accurate answers. In most cases the counter values were less than seven, indicating that larger h_0 values could be used. Other data not reported here with $h_0 = 0.05$ and EPS $= 1.0_{10^{-5}}$ show that errors of the order of $4.0_{10^{-4}}$ can easily be achieved. On this basis one may postulate that the algorithm does achieve accurate answers. The disturbing feature, however, concerns the number of function evaluations and thus computer time. If one used a third-order Runge–Kutta with $h = 0.05$ and covered the interval $\{0, 7\}$, approximately 420 function evaluations would be required. For the extrapolation algorithm with $h_0 = 0.05$ and EPS $= 1.0_{10^{-3}}$ approximately four times as many function evaluations would be required. On this basis it is difficult to suggest that the extrapolation is better than the standard Runge–Kutta approach. Results for Systems V and VII are shown in Table 5.3. The results for System V, at least in terms of accuracy, are quite good; System VII is not integrated as accurately, but the answers are acceptable. The use of $h_0 = 0.05$ (not shown) tends to decrease the error especially in the System VII case. In the case of System V, there seems to be further room for increasing h_0 since all of the counter values are less than 7. However, both systems require an excessive number of function evaluations.

In all cases for the values of h_0 used, the algorithm is stable. Thus, the use of extrapolation seems to convert the unstable midpoint method into a stable algorithm. An additional study of the stability properties of the modified midpoint method will be presented in Chapter 6.

TABLE 5.3

SYSTEMS V AND VII. SINGLE PRECISION, MODIFIED MIDPOINT
METHOD

$h_0 = 0.5$, EPS $= 1.0_{10} - 5$

System V

x^a	ε_2 (error)	Counter	v
0	0	0	1
0.1015	$-3.7_{10} - 5$	5	49
0.3190	$-1.5_{10} - 4$	7	105
0.4966	$-2.1_{10} - 4$	6	73
0.6409	$-2.3_{10} - 4$	3	21
0.8091	$-2.7_{10} - 4$	6	73
0.9048	$-2.6_{10} - 4$	5	49
1.091	$-2.3_{10} - 4$	6	73
1.368	$-1.2_{10} - 4$	6	73
1.564	$-1.1_{10} - 5$	5	49
1.763	$1.4_{10} - 4$	4	33
1.987	$3.1_{10} - 4$	4	33
2.152	$5.0_{10} - 4$	5	49
2.332	$7.0_{10} - 4$	5	49

System VII

x	ε (error)	Counter	v
0	0	0	1
0.2739	$-3.7_{10} - 3$	9	217
0.3209	$-3.7_{10} - 3$	6	73
0.3607	$-3.7_{10} - 3$	7	105
0.3990	$-3.8_{10} - 3$	8	153
0.4317	$-3.8_{10} - 3$	8	153
0.4595	$-3.7_{10} - 3$	6	73
0.4859	$-3.7_{10} - 3$	6	73
0.5109	$-3.7_{10} - 3$	6	73
0.5354	$-3.7_{10} - 3$	6	73
0.5622	$-3.6_{10} - 3$	7	105
0.5845	$-3.6_{10} - 3$	8	153
0.6065	$-3.6_{10} - 3$	8	153
0.6253	$-3.5_{10} - 3$	6	73

[a] Every fourth value shown.

5.5.1. A Nonlinear System

The extrapolation algorithm was also used on a two variable nonlinear system representing the concentration and temperature behavior in a chemical reactor [10],

$$y_1' = \mathcal{H}_1 - \mathcal{H}_2\, y_1 - \mathcal{H}_3\, y_1 \exp(-\mathcal{H}_4/y_2)$$
$$y_2' = \Omega_1 - \Omega_2\, y_2 - \Omega_3\, y_1 \exp(-\mathcal{H}_4/y_2) \qquad (5.5\text{-}1)$$
$$y_1(0) = 5.0_{10-3}, \qquad y_2(0) = 300$$

where $\mathcal{H}_1 = 2.5_{10-5}$, $\mathcal{H}_2 = 5.0_{10-3}$, $\mathcal{H}_3 = 7.86_{10+12}$, $\mathcal{H}_4 = 11363$, $\Omega_1 = 1.735$, $\Omega_2 = 5.678_{10-3}$, $\Omega_3 = 7.86_{10+16}$. During the integration, y_1 and y_2 will approach steady state values of 4.41_{10-3} and 306 respectively.

In the extrapolation calculations values of h_0 from 0.05 to 90.0 and EPS from 0.1 to 1.0_{10-5} were used. In addition the fourth-order classic Runge–Kutta procedure was used for comparison purposes.

Table 5.4 shows certain selected results from these calculations. Unfortunately there seems to be little difference in the cases $h_0 = 0.05$ and 20.0; the actual changes in y_1 and y_2 are all about the same order of magnitude although the intermediate values are quite different. In terms of computing time the $h_0 = 0.05$ case took approximately 2.7 sec, whereas the $h_0 = 20.0$ case took about 1.4 sec. For a constant EPS the computation time decreases to a minimum asymptotic value of about 1.2 sec. While not shown, the computation time also decreases to an asymptotic minimum with increasing EPS at a fixed h_0. Because the program being used calculates an optimum step size (roughly) there is little difference in the results obtained for various h_0.

Table 5.4 also shows the fixed step size fourth-order Runge–Kutta results with $h = 20.0$. The computation time here was 0.55 sec. as compared to the extrapolation time with $h_0 = 20.0$ of 1.4 sec. Once again the Runge–Kutta procedure seems preferable.

It is important to realize that since the extrapolation algorithm determines its own step size, the values of x at which one obtains the numerical results are not simple multiples of h_0. This makes a comparison with other algorithms difficult. Obviously this feature could be removed by changing the parameters which relate to decreasing and increasing the step size or by using an auxiliary single-step method to generate the desired values. Both of these approaches work against the accuracy of the algorithm however.

5.5.2. Summary

1. Extrapolation converts the midpoint rule, an unstable algorithm, into a stable one for the examples studied in this chapter. (We will see, however,

TABLE 5.4

NONLINEAR SYSTEM. SINGLE PRECISION[a]

$h_0 = 0.05$, EPS $= 1.0_{10}^{-5}$			$h_0 = 20.0$, EPS $= 1.0_{10}^{-5}$			Runge–Kutta, $h = 20.0$		
x	y_1	y_2	x	y_1	y_2	x	y_1	y_2
0	5.0_{10}^{-3}	300.0	0	5.0_{10}^{-3}	300.0	0	5.0_{10}^{-3}	300.0
0.05	4.9999_{10}^{-3}	300.001	20.0	4.9685_{10}^{-3}	300.63	100	4.8517_{10}^{-3}	302.62
0.125	4.9998_{10}^{-3}	300.004	50.0	4.9227_{10}^{-3}	301.47	200	4.7261_{10}^{-3}	305.09
0.2375	4.9996_{10}^{-3}	300.008	95.0	4.8576_{10}^{-3}	302.48	300	4.6281_{10}^{-3}	304.92
0.4062	4.9993_{10}^{-3}	300.013	162.5	4.7694_{10}^{-3}	303.59	400	4.5564_{10}^{-3}	305.38
0.6593	4.9989_{10}^{-3}	300.022	263.75	4.6604_{10}^{-3}	304.63	500	4.5064_{10}^{-3}	305.64
1.039	4.9983_{10}^{-3}	300.035	350.87	4.5862_{10}^{-3}	305.17	600	4.4727_{10}^{-3}	305.79
1.608	4.9974_{10}^{-3}	300.054	404.08	4.5546_{10}^{-3}	305.36	700	4.4506_{10}^{-3}	305.87
2.462	4.9961_{10}^{-3}	300.082	477.89	4.5167_{10}^{-3}	305.57	800	4.4364_{10}^{-3}	305.91
3.744	4.9940_{10}^{-3}	300.125	588.61	4.4768_{10}^{-3}	305.76	900	4.4274_{10}^{-3}	305.94
5.666	4.9910_{10}^{-3}	200.188	754.68	4.4428_{10}^{-3}	305.89	1000	4.4218_{10}^{-3}	305.95
8.549	4.9865_{10}^{-3}	300.282	844.36	4.4326_{10}^{-3}	305.92	1040	4.4202_{10}^{-3}	305.96
12.87	4.9796_{10}^{-3}	300.420	978.88	4.4233_{10}^{-3}	305.95			
19.36	4.9695_{10}^{-3}	300.621	1180.66	4.4168_{10}^{-3}	305.97			
29.09	4.9544_{10}^{-3}	300.908						
43.68	4.9322_{10}^{-3}	301.310						
65.58	4.8997_{10}^{-3}	301.853						
...	...	...						
1163.	4.4171_{10}^{-3}	305.967						

[a] Every fifth value used

in Chapter 6 that the stability region is quite small from computational results on stiff ODEs.)

2. From the examples tested, the variable step Bulirsch–Stoer extrapolation algorithm results in more function evaluations than Runge–Kutta methods for moderate levels of accuracy. For extremely high levels of accuracy, extrapolation will probably prove to be more efficient, as was exhibited by Clark [4].

REFERENCES

1. Bauer, F. L., Rutishauser, H. and E. Stiefel, New aspects in numerical quadrature, *Proc. of Symposia in Applied Mathematics*, **15**, 199 (1963).
2. Bulirsch, R. and Stoer, J. Numerical treatment of ordinary differential equations by extrapolation methods, *Numer. Math.* **8**, 1 (1966).
3. Bulirsch, R. and Stoer, J. Asymptotic upper and lower bounds for results of extrapolation methods, *Numer. Math.* **8**, 93 (1966).
4. Clark, N. W., A study of some numerical methods for the integration of systems of first-order ordinary differential equations, *Argonne National Laboratory Report*, ANL-7428 (March, 1968).
5. Dahlquist, G., A special stability problem for linear multistep methods, *BIT*, **3**, 27 (1963).
6. Dahlquist, G., "A Numerical Method for some Ordinary Differential Equations with Large Lipschitz Constants," *Proc. IFIP Congress, Supplement*, Booklet I, 32 (1968).
7. Davis, P. J. and Rabinowitz, P., "Numerical Integration," Random House (Blaisdell), New York, 1967.
8. Gragg, W. B., On extrapolation algorithms for ordinary initial value problems, *SIAM J. Numer. Anal.* **2**, 384 (1965).
9. Henrici, P., "Elements of Numerical Analysis," Chapter 14, Wiley, New York, 1964
10. Lapidus, L., "Digital Computation For Chemical Engineers," McGraw-Hill, New York, 1962.
11. McCalla, T. R., "Introduction to Numerical Methods and FORTAN Programming," Wiley, New York, 1967.
12. Richardson, C. and J. Gaunt, The deferred approach to the limit, *Trans. Roy. Soc. London*, **226**, 300 (1927).
13. Romberg, W., Vereinfachte Numerische Integration, *Norske Vid. Selsk. Forh. (Trondheim)*, **28**, 30 (1955).

6

Numerical Integration of Stiff Ordinary Differential Equations

Many physical systems give rise to ordinary differential equations the magnitudes of the eigenvalues of which vary greatly. Such situations arise in the study of the flow of a chemically reacting gas [4], exothermic chemical reaction in a tubular reactor [1], process dynamics and control [8], and circuit theory [2]. It is common to refer to such systems as *stiff*.

6.1. DEFINITION OF THE PROBLEM

Let us examine the particular problems associated with the numerical integration of stiff equations. To do this, consider the linear ODE of System VI, (2.15-6), $\mathbf{y}' = \mathbf{A}\mathbf{y}$, where $\mathbf{y} = [y_1, y_2, y_3]$ and

$$\mathbf{A} = \begin{bmatrix} -0.1 & -49.9 & 0 \\ 0 & -50 & 0 \\ 0 & 70 & -120 \end{bmatrix}$$

With $\mathbf{y}(0) = [2, 1, 2]$, the solution is

$$y_1(x) = e^{-0.1x} + e^{-50x}$$
$$y_2(x) = e^{-50x}$$
$$y_3(x) = e^{-50x} + e^{-120x} \qquad (6.1\text{-}1)$$

where the eigenvalues of $\mathbf{A}$ are $\lambda_1 = -120$, $\lambda_2 = -50$, and $\lambda_3 = -0.1$. $y_1, y_2,$ and y_3 each has a rapidly decaying component, corresponding to λ_1 or λ_2, which soon becomes insignificant. After a brief initial phase of the solution, in which the λ_1 and λ_2 components are not negligible, we would like to proceed in a numerical solution with a step length h which is determined only by the component of the solution corresponding to λ_3.

However, for a stable numerical solution, most methods require that $|h\lambda_i|$, $i = 1$, 2 be bounded by a small number, of the order 1 to 10. Since λ_1 is the largest negative eigenvalue of $\mathbf{A}$, the stability of a method will be governed by $|120h|$. For example, for Euler's method it is necessary that $|120h| < 2$, giving the maximum $h = \frac{1}{60}$. Although the component of the solution corresponding to λ_1 may be of no practical interest, the criterion of absolute stability forces us to use an extremely small value of h over the entire range of integration. As a result, the computation time necessary to integrate a highly stiff system can become excessive.

Usually stiff problems are most critical for systems of ODE $(m > 1)$, although we will see in Section 6.7.1 that a single ODE with a relatively constant component and a rapidly decaying solution component presents the same numerical stability problems. A stiff system of ODEs will be characterized by the property that the ratio of the largest to the smallest eigenvalue is much greater than one. We will be concerned with cases in which the ODE system is inherently stable, i.e., $\mathrm{Re}(\lambda_i) < 0$, $i = 1$, 2, ..., m, because such systems present the most serious stability problems if the eigenvalues are widely separated.

If we have a nonlinear system of ODEs, the eigenvalues of the ODEs are those of the Jacobian matrix, $[\partial \mathbf{f}/\partial \mathbf{y}]$. Of course, if $\mathbf{A} = \mathbf{A}(x)$ in the linear system and in general in the nonlinear system, the eigenvalues will vary in magnitude over the range of the integration. This point can be seen by considering the following system of ODEs [15] representing a set of chemicals reaction rate equations,

$$y_1' = -0.04y_1 + 10^4 y_2 y_3$$
$$y_2' = 0.04y_1 - 10^4 y_2 y_3 - 3_{10^7} y_2{}^2$$
$$y_3' = 3_{10^7} y_2{}^2 \tag{6.1-2}$$
$$y_1(0) = 1; \qquad y_2(0) = 0; \qquad y_3(0) = 0$$

The Jacobian matrix of (6.1-2) is

$$\mathbf{A} = \begin{bmatrix} -0.04 & 10^4 y_3 & 10^4 y_2 \\ 0.04 & -10^4 y_3 - 6_{10^7} y_2 & -10^4 y_2 \\ 0 & 6_{10^7} y_2 & 0 \end{bmatrix}$$

which can easily be shown to be singular. The three eigenvalues of $\mathbf{A}$ are given by

$$\lambda_1 = 0 \qquad\qquad (6.1\text{-}3)$$

and

$$\lambda^2 + (0.04 + 10^4 y_3 + 6_{10^7} y_2)\lambda + (0.24_{10^7} y_2 + 6_{10^{11}} y_2{}^2) = 0 \qquad (6.1\text{-}4)$$

At $x = 0$, the three eigenvalues are $\lambda_1 = 0$, $\lambda_2 = 0$, $\lambda_3 = -0.04$. The asymptotic behavior of (6.1-2) is to $y_1 = 0$, $y_3 = 0$, $y_3 = 1$, since $\sum y_i = 1$ always. Thus, for large values of x, the three eigenvalues approach $\lambda_1 = 0$, $\lambda_2 = 0$, $\lambda_3 = -10^4$. Even though the eigenvalues are varying considerably, it is necessary to choose h in accordance with the largest value achieved by an eigenvalue. Therefore, integration of (6.1-2) by Euler's method would require $h < 2_{10^{-4}}$. The total interval of interest for (6.1-2) is $0 \leq x \leq 40$, so that roughly 2_{10^5} steps would be necessary, resulting in an excessive amount of computing time.

Let us examine the nature of the problem in more detail. Consider the numerical solution of $\mathbf{y}' = \mathbf{A}\mathbf{y}$. The exact solution at x_{n+1} is related to that at x_n by (3.4-13), whereas a single-step method, for example, is governed by a relationship of the form (3.4-14). For a single-step method applied to a system of m ODEs, the characteristic difference equation will have m roots, all of which are identical in form, but which depend individually on the m eigenvalues of the ODE. For example, for Euler's method, $\mathbf{M}(h\mathbf{A}) = \mathbf{I} + h\mathbf{A}$, and the characteristic roots are $\mu_{1i} = 1 + h\lambda_i$, $i = 1, 2, \ldots, m$. For a single-step method, each of the roots represents an approximation to $\exp(\lambda_i h)$, whereas in a multistep method only the principal root is an approximation to $\exp(\lambda_i h)$, the others being extraneous. The necessity to use a small h for absolute stability when large λ_i are present is evident.

It is natural to require close approximations to $\exp(\lambda_i h)$ in the neighborhood of the origin, and this is normally the consideration in determining the principal root. At points where the λ_i have large negative real parts, the exact solution components corresponding to these eigenvalues are negligible when compared to the other solution components. It is only necessary then that the principal root also be negligible for the stiff eigenvalues. In other words, we would like the principal root to approximate accurately both the small and the large eigenvalues.

The problem associated with stiff systems is twofold: *stability* and *accuracy*. If a method with a finite absolute stability boundary is used, large negative real parts of some λ_i will force the step length used to be excessively small. On the other hand, if an A-stable method is used, e.g., the trapezoidal rule, the stability problem is avoided, but for a reasonable step length h the solution component corresponding to the largest eigenvalue will be approximated very inaccurately. In addition, the convergence requirements for the iterative solution of the nonlinear algebraic equations arising at each step in an implicit method place restrictions on the largest value of h that can be used. These restrictions vary considerably depending on the particular iterative technique used.

In this chapter we present numerical integration routines which can be used for stiff ODEs. We consider explicit and implicit single-step methods and predictor–corrector methods. The ultimate basis for comparison of stable methods is accuracy and computation time. Detailed numerical results of the use of many of the methods on example stiff problems will appear in Section 6.7.

6.2 EXPLICIT SINGLE-STEP METHODS

Explicit methods have finite regions of stability, so that the principal object of explicit methods for stiff systems will be to extend the stability boundary (usually the real stability boundary) as much as possible for a given order of accuracy.

6.2.1. Treanor's Method

A modified Runge–Kutta method for the integration of stiff ODEs has been proposed by Treanor [16]. The method is motivated by the observation that many stiff systems are of the form $y' = -P(y - \tilde{y})$ where P is a large number and $\tilde{y}$ is a function of x which is slowly varying. $\tilde{y}$ can be approximated by a power series in x containing unknown parameters which are determined in the course of the integration. In particular, the method assumes that $y_i' = f_i(x, \mathbf{y})$ can be approximated in an interval by

$$y_i' = -(P_i)_n y_i + (A_i)_n + (B_i)_n x + (C_i)_n x^2 \qquad (6.2\text{-}1)$$

The four constants A_i, B_i, C_i, and P_i can be evaluated by determining the value of $f_i(x, \mathbf{y})$ at four points in the interval $[x_n, x_n + h]$ and solving for the constants from evaluation of (6.2-1) at the four points. The four points are x_n, $x_n + \frac{1}{2}h$, $x_n + \frac{1}{2}h$ and x_{n+1}, as in the fourth-order Runge–Kutta method.

The following algorithm is obtained for each component of the vector $\mathbf{y}$,

$$y^{(1)}_{n+1/2} = y_n + (h/2)f_n$$
$$y^{(2)}_{n+1/2} = y_n + (h/2)f^{(1)}_{n+1/2}$$
$$y^{(3)}_{n+1} = y_n + h[2f^{(2)}_{n+1/2} F_2 + f^{(1)}_{n+1/2} PhF_2 + f_n(F_1 - 2F_2)] \qquad (6.2\text{-}2)$$
$$y_{n+1} = y_n + hf_n F_1 + hv_3(Py_n + f_n) + hv_2(Py^{(1)}_{n+1/2} + f^{(1)}_{n+1/2})$$
$$+ hv_2(Py^{(2)}_{n+1/2} + f^{(2)}_{n+1/2}) + hv_1(Py^{(3)}_{n+1} + f^{(3)}_{n+1})$$

where

$$F_1 = \frac{e^{-Ph} - 1}{-Ph}, \qquad F_2 = \frac{e^{-Ph} - 1 + Ph}{(Ph)^2}$$

$$F_3 = \frac{e^{-Ph} - 1 + Ph - \frac{1}{2}(Ph)^2}{-(Ph)^3} \qquad (6.2\text{-}3)$$

and

$$v_1 = -F_2 + 4F_3, \qquad v_2 = 2(F_2 - 2F_3)$$

$$v_3 = 4F_3 - 3F_2 \qquad (6.2\text{-}4)$$

The only undetermined parameter is P. In (6.2-1) we have used P_i whereas in (6.2-2) and (6.2-3) a scalar P has been used. The relation is as follows. The values P_i are determined as the ratio of the terms in the first two steps of (6.2-2)

$$P_i = -\frac{f^{(2)}_{i,n+1/2} - f^{(1)}_{i,n+1/2}}{y^{(2)}_{i,n+1/2} - y^{(1)}_{i,n+1/2}} \qquad (6.2\text{-}5)$$

where the division is defined to mean that an element in the vector in the numerator is divided by the corresponding element in the vector in the denominator. The value of P used in (6.2-2) and (6.2-3) is taken to be the largest value of P_i computed from (6.2-5). If this value is negative, P is set equal to zero. Using a scalar P ensures that each of the m ODEs is differenced with the same step length.

If the ODEs were uncoupled, the P_i would represent the local eigenvalues of the individual equations. Taking the single value of P equal to the largest P_i makes the algorithm approximate the corresponding solution component. The algorithm is used as follows:

1. Select an initial step length h.
2. Compute $y^{(1)}_{n+1/2}$ and $y^{(2)}_{n+1/2}$ from the first two equations of (6.2-2).
3. Compute P as the largest P_i from (6.2-5). If all $P_i < 0$, set $P = 0$.
4. Compute y_{n+1} from the last equation of (6.2-2).
5. If $|y_{n+1} - y_n|/|y_{n+1}| > \varepsilon_{\max}$, set $h = h/2$ and return to step 2. If $|y_{n+1} - y_n|/|y_{n+1}| < \varepsilon_{\min}$, set $h = 2h$ and return to step 2.

The stability region for real negative $h\lambda$ has been computed by Lomax and Bailey [11] and is shown in Figure 6.1. If h is made very small, the method is identical to the fourth-order Runge–Kutta method. For $-2 < h\lambda_i < 0$, the method is stable for any value of P. If eigenvalues exist for which $h\lambda_i < -2$, the method is conditionally stable with the region shown in Figure 6.1. If $Ph = 8$, the real stability boundary is -10, compared to the fourth-order Runge–Kutta value of -2.785.

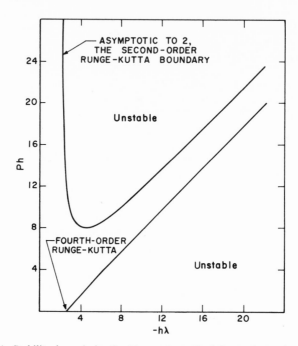

Figure 6.1. Stability boundaries for Treanor's method for real negative $h\lambda$ [11].

Treanor's method has the advantage of being explicit, which makes it easy to program. In addition, for nonstiff ODEs it reduces identically to the fourth-order Runge–Kutta method, while improving on the stability properties of the fourth-order Runge–Kutta method for stiff ODEs. The method still has the disadvantage of being only conditionally stable. If all the eigenvalues are such that $-2 < h\lambda_i < 0$, the method is stable for any P, but when $h\lambda_i < -2$, the stability corridor shown in Figure 6.1 starts to form and the choice of P becomes critical. A more serious disadvantage is that because of the form of the approximation (6.2-1), the method can only be used when the Jacobian matrix of the original ODE has large main diagonal elements, not large off-diagonal elements.

6.2.2. Lawson's Method

Lawson [9] has considered the problem of designing A-stable methods of Runge–Kutta type for stiff ODEs. Similar in many respects to the semi-implicit Runge–Kutta methods of Section 2.5, Lawson's method has the additional feature of a transformation of the stiff system into a nonstiff system.

In integrating the ODE (1.1-5), if we define the new variable

$$\mathbf{z}(x) = \exp(-x\mathbf{A})\mathbf{y}(x) \qquad (6.2\text{-}6)$$

then (1.1-5) becomes

$$\mathbf{z}' = \exp(-x\mathbf{A})[\mathbf{f}(x, \exp(x\mathbf{A})\mathbf{z}) - \mathbf{A}\exp(x\mathbf{A})\mathbf{z}], \qquad \mathbf{z}(0) = \mathbf{y}_0 \quad (6.2\text{-}7)$$

At this point $\mathbf{A}$ is an undetermined $m \times m$ matrix. If we write (6.2-7) as $\mathbf{z}' = \mathbf{g}(x, \mathbf{z})$, then the Jacobian matrix of the transformed system is

$$\mathbf{g}_z = \exp(-x\mathbf{A})[\mathbf{f}_y - \mathbf{A}]\exp(x\mathbf{A}) \qquad (6.2\text{-}8)$$

Thus, the eigenvalues of $\mathbf{g}_z$ are those of $\mathbf{f}_y - \mathbf{A}$. Requiring that all these eigenvalues are small will ensure that (6.2-7) is nonstiff.

If we apply the general Runge–Kutta method (2.1-1) and (2.1-2) to (6.2-7) we obtain

$$\mathbf{k}_1 = \exp(-x_n\mathbf{A})[(\mathbf{f}(x_n, \exp(x_n\mathbf{A})\mathbf{z}_n) - \mathbf{A}\exp(x_n\mathbf{A})\mathbf{z}_n]$$

$$\mathbf{p}_i = \exp[(x_n + c_i h)\mathbf{A}]\left[\mathbf{z}_n + h\sum_{j=1}^{i-1} a_{ij}\mathbf{k}_j\right] \qquad (6.2\text{-}9)$$

$$\mathbf{k}_i = \exp[-(x_n + c_i h)\mathbf{A}][(\mathbf{f}x_n + c_i h, \mathbf{p}_i) - \mathbf{A}\mathbf{p}_i], \qquad i = 2, 3, \ldots, v$$

$$\mathbf{z}_{n+1} = \mathbf{z}_n + \sum_{i=1}^{v} w_i\mathbf{k}_i$$

The algorithm can be rewritten in the more convenient form

$$\mathbf{k}_1{}^* = \mathbf{f}(x_n, \mathbf{y}_n) - \mathbf{A}\mathbf{y}_n$$

$$\mathbf{p}_i{}^* = \exp(c_i h\mathbf{A})\mathbf{y}_n + h\sum_{j=1}^{i-1} a_{ij}\exp[(c_i - c_j)h\mathbf{A}]\mathbf{k}_j{}^*$$

$$\mathbf{k}_i{}^* = \mathbf{f}(x_n + c_i h, \mathbf{p}_i{}^*) - \mathbf{A}\mathbf{p}_i{}^* \qquad (6.2\text{-}10)$$

$$\mathbf{y}_{n+1} = \exp(h\mathbf{A})\mathbf{y}_n + h\sum_{i=1}^{v} w_i\exp[(1 - c_i)h\mathbf{A}]\mathbf{k}_i{}^*$$

$$0 = c_1 \leq c_2 \leq \cdots \leq c_v \leq 1$$

The most logical choice for $\mathbf{A}$ is the Jacobian matrix $\mathbf{f}_y$, which need only be computed periodically. However, an algorithm for computing $\exp(c_i h\mathbf{A})$ is required. Lawson proves that if the approximation to $\exp(c_i h\mathbf{A})$ has a spectral radius less than one for all h, then the algorithm (6.2-10) is A-stable. As we know the exponential approximations which possess this property are the diagonal Padé approximants.

Since this method involves the Padé approximation and necessitates the periodic computation of the Jacobian matrix, it has similar properties to the semi-implicit Runge–Kutta methods. Although it involves more computation per step than most of the semi-implicit Runge–Kutta methods, the order of accuracy is higher.

6.2.3. Osborne's Method

In general, an object of stiff methods is to have a characteristic root (single-step methods) which has the property that $\mu_1(h\lambda) \to 0$ as $h\lambda \to -\infty$. In other words, we want the root behavior to be the same as $\exp(h\lambda)$ as $h\lambda \to -\infty$. A convenient form available for the charactristic root is certain of the Padé approximations, $P(h\lambda)/Q(h\lambda)$.

Osborne [12] has considered this problem for linear stiff ODEs. The object of Osborne's work is to find single-step formulas the characteristic root of which has the property that $\mu_1(h\lambda) = O(1/h\lambda)$ as $h\lambda \to -\infty$. We will sketch the technique here to illustrate Osborne's approach. Assume for the scalar, linear ODE $y' = \lambda y$ that y can be represented by

$$y_0 + \sum_{i=1}^{m} a_i x^i \tag{6.2-11}$$

in the range $-h < x < 0$. If $-h < \xi_1 < \xi_2 < \cdots < \xi_m < 0$, then a relation of the form

$$y_0 = \mu_1(\lambda)y_{-1} \tag{6.2-12}$$

can be obtained by eliminating $a_1, \ldots, a_m$ between the equations

$$y_{-1} = y_0 + \sum_{i=1}^{m} a_i(-h)^i \tag{6.2-13}$$

and

$$\lambda y_0 + \sum_{i=1}^{m} a_i \xi_j^{i-1}(\lambda \xi_j + i) = 0, \qquad j = 1, 2, \ldots, m \tag{6.2-14}$$

We can rewrite (6.2-14) as

$$\mathbf{Ca} = -\lambda y_0 \mathbf{e} \tag{6.2-15}$$

where $\mathbf{e}$ is the unit vector and

$$\mathbf{C} = \begin{bmatrix} \lambda\xi_1 + 1 & (\lambda\xi_1 + 2)\xi_1 & \cdots & (\lambda\xi_1 + m)\xi_1^{m-1} \\ \vdots & \vdots & & \vdots \\ \lambda\xi_m + 1 & (\lambda\xi_m + 2)\xi_m & \cdots & (\lambda\xi_m + m)\xi_m^{m-1} \end{bmatrix} \tag{6.2-16}$$

Let us define

$$p(\lambda) = \det \mathbf{C} \tag{6.2-17}$$

and

$$p_i(\lambda) = \det[\mathbf{C} + (\mathbf{e} - \mathbf{\kappa}_i(\mathbf{C}))\mathbf{e}_i^T] \tag{6.2-18}$$

where $\mathbf{\kappa}_i(\mathbf{C}) = \mathbf{Ce}_i$, the ith column of $\mathbf{C}$. Then

$$a_i = -\lambda y_0 p_i(\lambda)/p(\lambda) \tag{6.2-19}$$

where $p_i(\lambda)$ is a polynomial of degree $m - 1$ in λ and $p(\lambda)$ is a polynomial of degree m in λ unless $\xi_m = 0$ when it is of degree $m - 1$.
Returning to (6.2-12) we see that

$$\mu_1(\lambda) = p(\lambda) \Big/ \left[p(\lambda) - \lambda \sum_{i=1}^{m} (-h)^i p_i(\lambda) \right] \qquad (6.2\text{-}20)$$

If $\xi_m = 0$, $p(\lambda)$ and $p_i(\lambda)$ are of degree $m - 1$, so that $\lambda \sum_1^m (-h)^i p_i(\lambda)$ is of degree m. Thus, $\mu_1(\lambda) = O(1/\lambda)$ as $\lambda \to -\infty$. If $m = 2$, $\xi_1 = -h$, and $\xi_2 = 0$, $\mu_1(\lambda)$ becomes the characteristic root of the trapezoidal rule.

To determine $\mu_1(\lambda)$ from (6.2-17), (6.2-18), and (6.2-20) it is necessary to specify the points $\xi_1, \xi_2, \ldots, \xi_{m-1}$ ($\xi_m = 0$). These are chosen so that the denominator is of degree m in λ and $\mu_1(\lambda) \to 0$ as $\lambda \to -\infty$. Also they should be chosen so that the truncation error is minimized. Osborne shows that for $m = 2$ and $\xi_1 > -h$, choosing ξ_1 to minimize the truncation error yields

$$\mu_1(\lambda) = \frac{1 + h\lambda/3}{1 - \frac{2}{3}h\lambda + \frac{1}{6}(h\lambda)^2} \qquad (6.2\text{-}21)$$

which is the $n = 2$, $m = 1$ Padé approximant to exp $(h\lambda)$. Recall that this is the same characteristic root (3.7-10) which was obtained for a method employing second derivatives.

The algorithm would be used in the direct single-step format, $y_{n+1} = \mu_1(h\lambda)y_n$. Although the method was devised for linear stiff systems, Osborne reported success in integrating nonlinear problems using local linearization about each step in the integration. The value of Osborne's approach is that the general procedure of designing a characteristic root to have certain specified properties offers promise for special situations like stiff equations.

6.3. IMPLICIT SINGLE-STEP METHODS

The main advantage of implicit methods over explicit methods is that it is possible to achieve A-stability. The main disadvantage is that nonlinear algebraic equations must be solved at each step. Standard methods of solving these nonlinear algebraic equations (Gauss–Seidel, Newton–Raphson) either become unstable or else are quite slow. Nevertheless, the condition of A-stability is quite important when dealing with stiff systems, and the increased h often will more than compensate for the required iterations in terms of computation time.

In this section we consider three main classes of implicit methods: implicit and semi-implicit Runge–Kutta methods, the trapezoidal rule with global extrapolation, and methods employing second derivatives of y. Each of these has been treated in detail previously, however, in the third

class of methods free parameters, which can be chosen so that the characteristic root is matched to $\exp(h\lambda_1)$, will be introduced.

6.3.1. Implicit and Semi-Implicit Runge–Kutta Methods

Because of the possibility of both A-stability and high accuracy, implicit
Runge–Kutta methods make good prospects for stiff ODEs. The main
disadvantage is that nonlinear algebraic equations, involving the computation
of the Jacobian matrix, must be solved at each step in the integration. Of
particular interest are the semi-implicit Runge–Kutta methods of Section
2.5.4. Only one matrix inversion, at most, is required for the calculation of
each k_i. Actually these methods correspond to using one Newton–Raphson
iteration in the calculation of each k_i with $\mathbf{A}$ being the Jacobian matrix
appropriately evaluated.

6.3.2. Trapezoidal Rule with Extrapolation

The second class of single-step methods which are candidates for stiff
ODEs are those used in conjunction with extrapolation. Of these, the trapezoidal rule combined with global extrapolation is the logical choice because
it is A-stable.

6.3.3. The Method of Liniger and Willoughby

Theorem 3.3.3 limits the order of accuracy of A-stable linear multistep
methods of the form (3.1-1) to $p \leq 2$. However, we have seen that higher
order A-stable methods can be achieved that are not of the linear multistep
type, e.g., the implicit Runge–Kutta methods of Chapter 2. Even though
these higher-order A-stable methods exist, the problem is that the characteristic root, e.g., a Padé approximant, may still be a poor approximation to
$\exp(\lambda_1 h)$, where λ_1 is the largest eigenvalue. An important contribution to
this problem was made by Liniger and Willoughby [10] who introduced the
concept of *exponential fitting*. In particular, if a general single-step method
can be written

$$y_{n+1} = \mu_1(h\lambda_i)y_n \qquad (6.3-1)$$

then we not only require A-stability, i.e., $|\mu_1(h\lambda_1)| < 1$ for $\mathrm{Re}(\lambda_i) < 0$, but
also that $\mu_1(h\lambda_i)$ approximate the stiff solution component, $\exp(h\lambda_1)$.

The method of Liniger and Willoughby is based on the general class of
single-step methods employing second derivatives (3.7-1)

$$y_{n+1} = \alpha_1 y_n + h[\beta_0 y'_{n+1} + \beta_1 y_n'] + h^2[\gamma_0 y''_{n+1} + \gamma_1 y''_n] \qquad (6.3-2)$$

Three specific algorithms based on (6.3-2) have been proposed by Liniger and Willoughby:

(1) $\quad y_{n+1} = y_n + h[(1 - \beta_1)y'_{n+1} + \beta_1 y'_n]$ (6.3-3)

(2) $\quad y_{n+1} = y_n + (h/2)[(1 + a)y'_{n+1} + (1 - a)y_n']$
$\qquad\quad - (h^2/4)[(b + a)y''_{n+1} - (b - a)y''_n]$ (6.3-4)

(3) $\quad y_{n+1} = y_n + (h/2)[(1 + a)y'_{n+1} + (1 - a)y_n']$
$\qquad\quad - (h^2/12)[(1 + 3a)y''_{n+1} - (1 - 3a)y''_n]$ (6.3-5)

For certain specific parameter values these formulas revert to earlier methods:

(1) $\beta_1 = 1$ Euler's method
$\quad\ \beta_1 = 0$ Backward Euler method
$\quad\ \beta_1 = \frac{1}{2}$ Trapezoidal rule
(2) $a = b = 0$ Trapezoidal rule

Method 3, (6.3-5), is seen to be simply method 2, (6.3-4), with $b = \frac{1}{3}$.
The truncation errors for the three methods are

(1) $\quad -h^2 \int_0^1 (\zeta - \beta_1)y''(x + \zeta h)\, d\zeta$ (6.3-6)

(2) $\quad (h^3/4) \int_0^1 [2\zeta^2 - 2(1 - a)\zeta + (b - a)]y^{[3]}(x + \zeta h)\, d\zeta$ (6.3-7)

(3) $\quad -(h^4/12) \int_0^1 \zeta[2\zeta^2 - 3(1 - a)\zeta + (1 - 3a)]y^{[4]}(x + \zeta h)\, d\zeta$ (6.3-8)

The characteristic equation of (6.3-2) has the single root (3.7-3)

$$\mu_1 = \frac{1 + \beta_1 h\lambda + \gamma_1 h^2 \lambda^2}{1 - \beta_0 h\lambda - \gamma_0 h^2 \lambda^2}$$ (6.3-9)

For method 1, for example,

$$\mu_1^{(1)} = \frac{1 + \beta_1 h\lambda}{1 - (1 - \beta_1)h\lambda}$$ (6.3-10)

where the superscript indicates the method number. We will require each of the methods 1–3 to be A-stable. The condition that $|\mu_1| < 1$ for all $\mathrm{Re}(\lambda) < 0$ for (6.3-10) results in the condition $\beta_1 < \frac{1}{2}$. In the same way the following two conditions are obtained for the A-stability of (6.3-4) and (6.3-5),

(2) $\quad a > 0, \quad b > 0$
(3) $\quad a > 0$

Each of the three methods can be placed in the general single-step form (6.3-1). The characteristic root of (6.3-3) has already been given by (6.3-10).

For methods 2 and 3 the roots are

$$\mu_1^{(2)} = \frac{4 + 2(1 - a)h\lambda + (b - a)h^2\lambda^2}{4 - 2(1 + a)h\lambda + (b - a)h^2\lambda^2} \qquad (6.3\text{-}11)$$

$$\mu_1^{(3)} = \frac{12 + 6(1 - a)h\lambda + (1 - 3a)h^2\lambda^2}{12 - 6(1 + a)h\lambda + (1 + 3a)h^2\lambda^2} \qquad (6.3\text{-}12)$$

Having obtained conditions for A-stability, the next consideration is the accuracy with which the characteristic roots approximate $\exp(h\lambda_i)$, $i = 1, 2,$ $\ldots, m$, consistent with these conditions. As we outlined in detail in Section 3.7, the conventional procedure is to equate (6.3-9) to the Taylor series expansion of $\exp(h\lambda)$ about $h\lambda = 0$ and match powers of $h\lambda$ up to the desired order. This procedure was illustrated in Section 3.7 for $p = 3$. For the nonstiff eigenvalues, this is highly effective since those $h\lambda_i$ are relatively small. However, the stiff eigenvalues frequently are very large, e.g., $|h\lambda_i| \gg 1$, for which a Taylor series expansion of $\exp(h\lambda_1)$ about $h\lambda_1 = 0$ converges very poorly. Thus, conventional methods are inaccurate for stiff systems where it is necessary to require accuracy and stability for *both* large $|h\lambda_i|$ and $h\lambda_i \to 0$.

In methods 1–3 the free parameters β_1, a, and b can be adjusted to provide accurate representation of the stiff components within the constraint of maintaining A-stability. In particular, we desire for a certain value $h\lambda = q_0$ that

$$\mu_1(q_0) = e^{q_0} \qquad (6.3\text{-}13)$$

We say that the method is *exponentially fitted* at q_0, if (6.3-13) is used to determine the free parameters. Requiring that method 1 be fitted at a given q_0 yields the following relation for β_1,

$$\beta_1(q_0) = -q_0^{-1} - (e^{-q_0} - 1)^{-1} \qquad (6.3\text{-}14)$$

Note that the trapezoidal rule ($\beta_1 = \frac{1}{2}$) and the backward Euler ($\beta_1 = 0$) correspond to exponential fitting at $q_0 = 0$ and $q_0 = \infty$ respectively.

Method 3 contains the free parameter a the value of which from exponential fitting is

$$a(q_0) = \tfrac{1}{3}[q_0^2 + 6q_0 + 12 - e^{q_0}(q_0^2 - 6q_0 + 12)]$$
$$\times [e^{q_0}(q_0^2 - 2q_0) + q_0^2 + 2q_0]^{-1} \qquad (6.3\text{-}15)$$

Method 2 contains two free parameters, a and b, and can be exponentially fitted to two values of $h\lambda$, q_0 and q_1,

$$a(q_0, q_1) = 2[r(q_1) - r(q_0)]/[q_1 r(q_0) - q_0 r(q_1)] \qquad (6.3\text{-}16)$$

$$b(q_0, q_1) = 2(q_0 - q_1)/[q_0 r(q_1) - q_1 r(q_0)] \qquad (6.3\text{-}17)$$

where

$$r(q) = q^2(1 - e^{-q})[-(2 + q) + (2 + q)e^{-q}]^{-1} \qquad (6.3\text{-}18)$$

If q_0 and q_1 are real or complex conjugates, real parameter values a and b are obtained from (6.3-16)–(6.3-18). For the case in which q_0 are q_1 are complex conjugates the reader is referred to the original work for the appropriate relations.

The authors prove that the procedure of exponential fitting is compatible with both A-stability and accuracy in the limit $h\lambda \to 0$ for $0 > q_0 > -\infty$ and $0 > q_0$, $q_1 > -\infty$. If q_0 and q_1 are complex conjugates, the region of A-stability is somewhat restricted. We will use methods 1 and 3 for the examples at the end of the chapter.

6.3.4. General Observations on Implicit Methods

The general problem with nonexponentially fitted A-stable implicit methods is that the characteristic root is often not asymptotic to zero as $h\lambda \to -\infty$, e.g., for the trapezoidal rule, $\mu_1 \to -1$ as $h\lambda \to -\infty$. The components corresponding to the largest eigenvalues are simulated inaccurately initially when they are nonnegligible and then remain in the solution as slowly decaying oscillations. Thus, if too large an h is chosen, even though the method may be A-stable, it becomes inaccurate. The reader will recall that this behavior was encountered in Section 3.10.3 for nonstiff systems in which h was chosen too large.

Two strategies have been proposed to deal with this problem, i.e., A-stable methods not exponentially fitted to large $h\lambda$. Liniger and Willoughby [10] have suggested that small step sizes be used during the initial phase of the solution when the stiff components have not yet died out. However, once the stiff components have become negligible, large steps are used based on the rate of change of the nonstiff solution components. Even though the stiff components will be poorly approximated from then on, these components will not injure the overall solution as long as their magnitude remains small. If the stiff components never reappear, A-stability will maintain small the numerical errors associated with the stiff components and preserve the accuracy of the overall solution.

The second procedure was suggested by Dahlquist [3] particularly for use with the trapezoidal rule and global extrapolation. The trapezoidal rule is given by (5.2-6) and the initial filtering procedure suggested by Dahlquist is:

1. Use (5.2-6) to compute y_1 and y_2.
2. Replace y_1 by $(y_0 + 2y_1 + y_2)/4$.
3. Use (5.2-6) to compute y_2 and y_3.
4. Replace y_2 by $(y_1 + 2y_2 + y_3)/4$.
5. Use (5.2-6) to compute $y_3, y_4, y_5, \ldots$.

The necessity for an initial filtering procedure or an initial small step length is avoided by the procedure of exponential fitting. The same large step length can be used throughout the entire integration. We note, however, that Liniger and Willoughby's method 1 is only of order $p = 1$. Since it is a linear multistep method, we could anticipate this result from Theorem 3.3.3. Thus, once the integration is past the initial phase we would be at a disadvantage with respect to the trapezoidal rule for which $p = 2$. One possible course of action is to change β_1 from the exponentially fitted value to $\frac{1}{2}$, the value corresponding to the trapezoidal rule. Methods 2 and 3 give increased degrees of freedom which can be used to fit exponentially at more than one $h\lambda$ value and/or increase the order of accuracy.

The characteristic root behavior for real negative $h\lambda$ of many of the common A-stable implicit methods is shown in Figure 6.2. As we see, μ_1 for

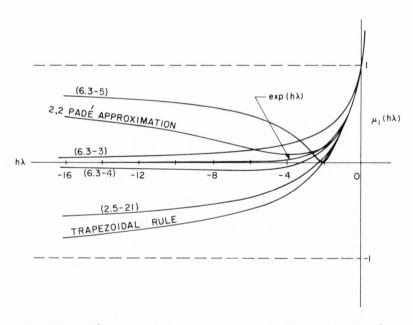

Figure 6.2. μ_1 ($h\lambda$) for several A-stable numerical methods for real negative $h\lambda$.

the trapezoidal rule is asymptotic to -1, and μ_1 for Calahan's method (2.5-21) is asymptotic to -0.735. Liniger and Willoughby's methods 1, (6.3-3), and 3, (6.3-5), are best from the standpoint of the asymptotic value of μ_1.

6.4. PREDICTOR–CORRECTOR METHODS

6.4.1. Gear's Method

Gear [7] has considered the problem of devising predictor–corrector methods for stiff systems. He considers methods of the linear multistep type (3.1-1) in which the corrector is iterated to convergence. The stability and accuracy requirements of the methods are shown in general form in Figure 6.3. In the complex $h\lambda$ plane it is required that the methods be stable for

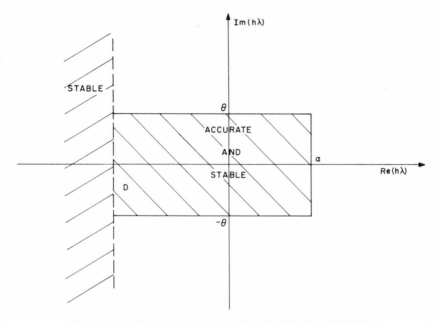

Figure 6.3. Stability and accuracy regions for Gear's method [7].

$h\lambda < D$ and that the methods be both accurate and stable in the rectangle bounded by D, α and $\pm\theta$. In this way the restrictions imposed by Theorem 3.3.3 are relaxed. Using the requirements shown in Figure 6.3, methods of order as high as 6 have been obtained for reasonable values of the parameters D, α, and θ.

As $h\lambda \to -\infty$, the roots of (3.1-1) become the roots of $\sigma(\xi) = 0$. $\sigma(\xi)$ must have stable roots and degree k. This condition leads to the stability regions shown for different k in Figure 6.4. It is seen that up to $k = 6$ methods are obtained for which $D < -6.1$, $\theta < 0.5$ and α approximately equal to zero.

The corrector equation is iterated using the Newton–Raphson method. The Jacobian matrix $\mathbf{A}_n$ is not evaluated for every iteration and not necessarily at every step.

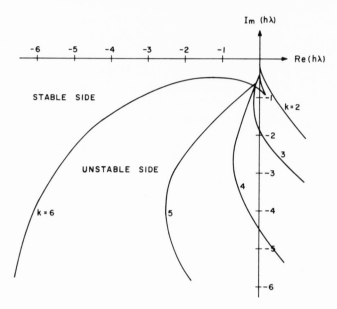

Figure 6.4. Stability regions for different values of k in Gear's method [7].

Gear expresses the predictor corrector method in the following form,

$$\mathbf{a}_n^{(0)} = \mathbf{B}\mathbf{a}_{n-1} \qquad (6.4\text{-}1)$$

$$\mathbf{a}_n^{(m+1)} = \mathbf{a}_n^{(m)} - \mathbf{l}\mathbf{F}(\mathbf{a}_n^{(m)}) \qquad (6.4\text{-}2)$$

where

$$\mathbf{F} = [\mathbf{I} + \mathbf{A}h\beta_k/\alpha_k']^{-1}[\mathbf{a}_1 - h\mathbf{f}(x, \mathbf{a}_0)] \qquad (6.4\text{-}3)$$

and $\mathbf{a} = [\mathbf{y}, h\mathbf{y}', \ldots, h^k\mathbf{y}^{[k]}/k!]$. $\mathbf{l}$ is given in Table 6.1. β_k/α_k is equal to $-l_0$ and $\mathbf{B}$ is the Pascal triangle matrix, $B_{ij} = (\tbinom{j}{i})$, (0 if $i > j$).

TABLE 6.1

COEFFICIENTS FOR GEAR'S METHOD

k	1	2	3	4	5	6
l_0	1	2/3	6/11	24/50	120/274	720/1764
l_1	1	1	1	1	1	1
l_2	0	1/3	6/11	35/50	225/274	1624/1764
l_3	0	0	1/11	1/5	85/274	735/1764
l_4	0	0	0	1/50	15/274	175/1764
l_5	0	0	0	0	1/274	21/1764
l_6	0	0	0	0	0	1/1764

Gear has used this algorithm to construct a program for the automatic integration of ODEs. The step length h and the order k are determined automatically by the program. The basis for the choice of h and k is minimization of computation to achieve a given error bound. Since it is not possible to calculate the accumulated truncation error [this would require solution of (5.1-2)], local truncation errors are controlled under the assumption that the equations are stable. For each order method, the local truncation error is known in the form $C_k h^k y^{[k]}$. A value δ is selected such that the local truncation error be kept less than $h\delta$. h is then determined from

$$h = \left[\frac{\delta}{C_k y^{[k]}}\right]^{1/(k-1)} \tag{6.4-4}$$

where $y^{[k]}$ is estimated by the program.

Periodically the maximum power of 2 by which h can be increased is calculated for orders $k - 1$, k and $k + 1$, where k is the current order. After h or k is changed, at least k steps must be taken before h can be changed again. The Jacobian matrix $\mathbf{A}$ must be reevaluated periodically. This is done when the corrector fails to converge. The corrector is iterated up to M times. The iteration is stopped if changes are less than a prescribed amount, but if this criterion is not met in M iterations, $\mathbf{A}$ is reevaluated. If the iteration again fails, h is halved and the iteration is repeated. The maximum number M of iterations is determined on the basis of two factors. If M is too small, the reevaluation of $\mathbf{A}$ may have to be performed too frequently. If M is too large, too many function evaluations may be performed before the iteration is abandoned. Gear found from many applications that $M = 3$ is an optimal value.

Because of the complexity of Gear's program, we have not attempted to reproduce it for use in the examples section. However, a large amount of practical computation with this algorithm on stiff systems has been reported by Gear [7] and Ratliff [13]. These results indicate superiority of the algorithm when compared to classical predictor–corrector methods of comparable order.

6.5. OTHER STIFF METHODS

There is much current interest in the problem of devising numerical methods of stiff ODEs. In this section we will outline or simply reference some of the most recent techniques proposed to deal with this problem. In each case we have, at most, tried to outline the approach taken. The reader can refer to the original reference for details of the methods.

Dahlquist [3] has introduced an interesting idea based on local polynomial approximations. He considers cases in which $\mathbf{y}' = \mathbf{f}(x, \mathbf{y})$ can be separated into

$$\mathbf{u}' = \mathbf{A}\mathbf{u} + \mathbf{h}(\mathbf{u}, \mathbf{v}, x) \qquad (6.5\text{-}1)$$

$$\mathbf{v}' = \mathbf{g}(\mathbf{u}, \mathbf{v}, x) \qquad (6.5\text{-}2)$$

$$\mathbf{u}(0) = \mathbf{u}_0; \qquad \mathbf{v}(0) = \mathbf{v}_0 \qquad (6.5\text{-}3)$$

where the stiffness in the system is a result of widely separated eigenvalues of $\mathbf{A}$. If $\mathbf{r}(x)$ is a polynomial of degree two, then

$$\mathbf{p}' = -\mathbf{A}\mathbf{p} + \mathbf{r}(x) \qquad (6.5\text{-}4)$$

has the solution

$$\mathbf{p} = \mathbf{A}^{-1}\mathbf{r}(x) + \mathbf{A}^{-2}\mathbf{r}'(x) + \mathbf{A}^{-3}\mathbf{r}''(x) \qquad (6.5\text{-}5)$$

The idea is to construct an approximate particular solution of (6.5-1) sectionally consisting of polynomials of the form (6.5-5). Let $\mathbf{p}_0(t)$ be a first approximation to this particular solution in $[x_n, x_{n+1}]$. This can be obtained by linearization from the previous step. Let $\mathbf{p}_1(x)$ be the polynomial solution of (6.5-4) when $\mathbf{r}(x)$ is the quadratic interpolation polynomial defined by

$$\mathbf{r}(x_i) = \mathbf{h}(\mathbf{p}_0(x_i), \mathbf{v}(x_i), x_i) \qquad (6.5\text{-}6)$$

Then $\mathbf{p}_2(x)$ is the polynomial solution of (6.5-5) when $\mathbf{r}(x)$ is defined by

$$\mathbf{r}(x_i) = \mathbf{h}(\mathbf{p}(x_i), \mathbf{v}(x_i), x_i) \qquad (6.5\text{-}7)$$

The iterative scheme will converge rapidly to $\mathbf{p}(t)$ when $\|\mathbf{A}^{-1}\| \, \|\mathbf{h}_u\|$ is small. The approximate particular solution technique is applied to (6.5-1) and a conventional technique is applied to (6.5-2). In addition, Dahlquist discusses the redefinition of variables so that the coupling between $\mathbf{u}$ and $\mathbf{v}$ can be reduced.

Richards, *et al.* [14] considered the problem of numerical integration of large stiff systems. By using a qualitative picture of the solution trajectories of stiff systems, they proposed a method based on a modification of Euler's method. Using Euler's method as the basic algorithm, we continually monitor y'_{n+1} and y_n'. If y'_{n+1} undergoes a sudden reversal from y_n', indicating the onset of instability, then the initial computed value of y_{n+1} is abandoned and replaced by an interpolated value at $x_{n+1} = x_n + wh$ computed by Euler's method from y_n. w is computed by

$$w = \frac{\mathbf{f}_n^{\mathrm{T}}(\mathbf{f}_n - \mathbf{f}_{n+1})}{\|\mathbf{f}_{n+1} - \mathbf{f}_n\|^2} \qquad (6.5\text{-}8)$$

The method is extremely simple and explicit though not highly accurate.

Success was reported by the authors on a fourth-order system when compared to simple Euler integration.

Fowler and Warten [6] have derived an explicit second-order method for large systems of stiff ODEs. We will not outline the method here because of the involved step-size control in the use of the algorithm. The authors reported very favorable comparisons with the fourth-order Runge–Kutta method on several systems.

6.6 PUBLISHED NUMERICAL RESULTS

Extensive calculations with the fourth-order Runge–Kutta method, Treanor's method, and the trapezoidal rule on a problem involving the flow of a gas in chemical nonequilibrium behind a normal shock wave have been carried out by Lomax and Bailey [11]. The system involved 13 nonlinear ODEs, the eigenvalues of which varied over three orders of magnitude. While they found that Treanor's method gave a considerable reduction in computing time when compared to the standard Runge–Kutta method, the authors concluded that if more than one stiff eigenvalue is present, the ODE should be locally linearized and integrated implicitly using the simplest implicit method compatible with the desired accuracy.

Lawson [9] tested his method on two nonlinear second-order systems. The second diagonal Padé approximant was used for $\exp(h\mathbf{A})$ in each case and the method was compared to the fourth-order Runge–Kutta method. The amount of computation per step was comparable for the two methods, since the matrix–vector multiplications required in the evaluation of $\mathbf{p}_i^*$ and $\mathbf{y}_{n+1}$ are exactly compensated by one less such operation in each derivative evaluation. In each problem Lawson's method permitted a considerable increase in h for comparable error.

Detailed computations with various forms of Gear's method have been presented by Ratliff [13]. The report examined the following questions:

1. The efficiency of the number of corrector iterations for various order methods.

2. The comparison of stiff and nonstiff methods (classical P–C methods) on both stiff and nonstiff ODEs in terms of accuracy and computing time.

3. The comparison of fixed-order methods and automatic order selection in terms of accuracy and computing time.

The following conclusions were drawn on the basis of the numerical solution of four ODE systems:

1. Two corrector iterations are more efficient for low-order methods and three iterations for high-order methods.

2. It is preferable to use low-order methods when a large error can be tolerated, and high-order methods when more accuracy is desired. Nonstiff methods used on stiff ODEs were in all cases inferior to the stiff methods. For nonstiff ODEs there was no appreciable difference between the stiff and nonstiff methods with respect to accuracy and computing time, indicating that a good strategy is to use a stiff method when integrating an arbitrary system of ODEs which may be stiff or nonstiff.

3. When using stiff methods it is preferable to vary the order of the method during the course of the integration.

6.7. NUMERICAL EXPERIMENTS

6.7.1. ODE Examples

The following problems will be studied:

System	Exact solution
1. $y' = -200(y - F(x)) + F'(x)$ $y(0) = 10$ $F(x) = 10 - (10 + x)e^{-x}$	$y(x) = F(x) + 10e^{-200x}$
2. System VI	(6.1-1)
3. $y_1' = 0.2(y_2 - y_1)$ $y_2' = 10y_1 - (60 + 0.125x)y_2 + 0.124x$ $y_1(0) = 0 \qquad y_2(0) = 0$	Not obtained
4. (6.1-2)	Nonlinear

6.7.2. Methods Used

Method	Designation	Comments
1. Fourth-order Runge–Kutta (2.3-13)	RK4	
2. Adams fourth-order P–C (4.7-3) and (4.7-4)	DEQ	One corrector evaluation, RK4 to start
3. Treanor's method (6.2-2)–(6.2-5)	TM	Automatic step length control
4. Modified midpoint rule (5.2-9)–(5.2-12)	DIFSYS	
5. Trapezoidal rule (1.6-20)	TR	Initial filtering procedure in Section 6.3.4 used
6. Trapezoidal rule with extrapolation	TR–EX	Initial filtering procedure in Section 6.3.4 used Extrapolation to $M = 3$

7. Calahan's method   CAL   Step-length adjustment in
   (2.5-21)   Section 6.3.4 used
8. Liniger–Willoughby   LW1
   Method 1 (6.3-3)
9. Liniger–Willoughby   LW3
   Method 3 (6.3-5)

6.7.3. Computational Results

Example 1 is a single ODE with a solution containing a rapidly decaying component and a slowly decaying component. The eigenvalue is -200, and the solution is desired from $x = 0$ to $x = 15$. Thus, the solution component $\exp(-200x)$ becomes negligible almost immediately compared to the $\exp(-x)$ component. The results of the numerical integration of example 1 are presented in Table 6.2. The R_n columns show $[y_n - y(x_n)]/y(x_n)$ at two values of

TABLE 6.2

COMPUTATIONAL RESULTS FOR EXAMPLE 1

Method	h	R_n at $x = 0.4$	$x = 1.0$	Time (sec)
RK4	0.01	$1.0_{10}-5$	$2.0_{10}-9$	11
DEQ	0.005	$3.0_{10}-9$	$2.0_{10}-9$	18
TM	0.2^a	$6.7_{10}-8$	$1.0_{10}-9$	16.5
DIFSYS	d	b	b	b
TR	0.2	$1.85_{10}-2$	$4.3_{10}-5$	2
TR–EX	0.2	$1.4_{10}-4$	$1.0_{10}-8$	36
CAL	$0.01/0.2^c$	$1.7_{10}-2$	$4.0_{10}-8$	1
LW1	0.2	$1.1_{10}-3$	$5.0_{10}-8$	3
LW3	0.2	$1.8_{10}-3$	$9.0_{10}-8$	4

^{a} Automatic step-size control.
^{b} Unstable.
^{c} h changed from 0.01 to 0.2 at $x = 0.1$.
^{d} Initial step size 0.1, extrapolations performed until error $< 10^{-5}$.

x, 0.4 and 10. The time column indicates the total computation time in seconds on an IBM 7094. The two points $x = 0.4$ and $x = 10$ were chosen as representative of the errors early and late in the integration. For a stable method we expect that the solution will be quite accurate at $x = 10$, where the extraneous solutions have become negligible. All computations were performed in single precision and an entry of zero in the columns indicates eight-place accuracy.

RK4 and DEQ are both highly accurate but quite time-consuming. The automatic step-size selection routine in TM in this case determined an h comparable to RK4 and actually required more time. DIFSYS was unstable for h values comparable to RK4 and DEQ. TR, CAL, LW1, and LW3 were roughly comparable. TR–EX was more accurate, as expected, but required an excessive amount of time.

Example 2 is System VI, the exact solution of which is given by (6.1-1). The eigenvalues are -120, -50, and -0.1. This system is interesting because it contains two stiff eigenvalues so that three different characteristic times appear. Results of the integration of System VI are shown in Table 6.3.

TABLE 6.3

COMPUTATIONAL RESULTS FOR EXAMPLE 2

Method	h	R_{1n}		R_{2n}	R_{3n}	Time (sec)
		$x = 0.4$	$x = 10$	$x = 0.4$	$x = 0.4$	
RK4	0.01	$2.0_{10^{-7}}$	$5.4_{10^{-7}}$	$3.0_{10^{-1}}$	$3.0_{10^{-1}}$	20
DEQ	0.01	$2.0_{10^{-8}}$	$8.1_{10^{-7}}$	$9.5_{10^{-1}}$	7.4_{10^5}	23
TM	0.2^a	$4.0_{10^{-4}}$	$1.35_{10^{-4}}$	1.1_{10^5}	1.2_{10^5}	1
DIFSYS	d	$5.0_{10^{-4}}$	$2.16_{10^{-4}}$	$9.4_{10^{-1}}$	8.3_{10^2}	22
TR	0.2	$1.0_{10^{-3}}$	$2.7_{10^{-4}}$	6.5_{10^7}	1.3_{10^5}	1.3
TR–EX	0.2	$4.0_{10^{-5}}$	$8.1_{10^{-7}}$	5.7_{10^1}	8.0_{10^1}	30
CAL	$0.01/0.2^c$	$2.0_{10^{-3}}$	$2.7_{10^{-6}}$	2.5_{10^5}	1.6_{10^5}	1
LW1	0.2	$4.0_{10^{-3}}$	$1.1_{10^{-2}}$	5.0_{10^5}	5.0_{10^5}	3

[a] Automatic step-size control.
[b] Unstable.
[c] h changed from 0.01 to 0.2 at $x = 0.1$.
[d] Initial step size 0.1, extrapolations performed until error $< 10^{-5}$.

Again it was desired to integrate from $x = 0$ to $x = 15$ and errors are tabulated at $x = 0.4$ and $x = 10$. TM was somewhat less accurate than RK4, but required only one second compared to 20 seconds for RK4. DIFSYS was again unstable, as evidenced at $x = 10$. CAL and LW1 were roughly comparable and somewhat more accurate than TR.

Example 3 was not solved analytically and the results from RK4 with $h = 0.005$ were taken as the exact solution. The eigenvalues are approximately -60 and -0.17, changing somewhat over the course of the integration from $x = 0$ to $x = 15$. As seen in Table 6.4, TM again was considerably less time consuming for accuracies comparable to RK4 and DEQ. Because the largest eigenvalue, -60, was smaller than in the previous two examples, DIFSYS was stable. CAL and LW1 were both very fast, CAL being particularly attractive.

TABLE 6.4

COMPUTATIONAL RESULTS FOR EXAMPLE 3

Method	h	R_{1n}		R_{2n}		Time (sec)
		$x = 0.4$	$x = 10$	$x = 0.4$	$x = 10$	
RK4	0.01	0	$8.5_{10^{-6}}$	0	$1.3_{10^{-7}}$	16
DEQ	0.005	0	0	0	$4.3_{10^{-7}}$	41
TM	0.2^a	$2.9_{10^{-2}}$	$3.9_{10^{-5}}$	$1.6_{10^{-4}}$	$2.1_{10^{-6}}$	5
DIFSYS	d	$4.3_{10^{-1}}$	$3.1_{10^{-4}}$	$4.0_{10^{-2}}$	$8.5_{10^{-4}}$	12
TR	0.2	$5.7_{10^{-1}}$	$2.3_{10^{-5}}$	$8.1_{10^{-4}}$	$8.5_{10^{-6}}$	2
TR–EX	0.2	0	$1.5_{10^{-6}}$	0	$8.5_{10^{-7}}$	44
CAL	$0.01/0.2^c$	0	$7.7_{10^{-5}}$	$1.5_{10^{-4}}$	$2.1_{10^{-3}}$	1
LW1	0.2	$2.9_{10^{-1}}$	$5.4_{10^{-3}}$	$5.7_{10^{-5}}$	$8.5_{10^{-4}}$	4

[a] Automatic step-size control.
[b] Unstable.
[c] h changed from 0.01 to 0.2 at $x = 0.1$.
[d] Initial step size 0.1, extrapolations performed until error $<10^{-5}$.

Example 4 is (6.1-2), a very stiff set of nonlinear ODEs. As noted, the eigenvalues change from 0, 0, -0.04 to 0, 0, -10^4 over the range $x = 0$ to $x = 40$. In fact, most of this change occurs in the first few instants, e.g., $|\lambda_{max}|$ changes from 0.04 to 2405 within $x = 0$ to $x = 0.02$. Thus, this example represents the severest test of the methods of all the examples. The results of the methods on (6.1-2) are shown in Table 6.5. All of the explicit methods eventually became unstable. With $h = 0.001$, RK4 became unstable after $x \geq 16$, where $|\lambda_{max}| \cong 2.78_{10^3}$. On the basis of the time to compute to $x = 10$, RK4 would require 138 sec to get to $x = 40$ (if it were stable). DEQ with $h = 0.001$ was unstable after $x = 0.012$. DIFSYS with $h = 0.001$ was unstable after $x = 0.358$. TM was strongly influenced by the off-diagonal elements and was completely unstable.

The semi-implicit method CAL was stable with $h = 0.005$ up to $x = 1$ and $h = 0.02$ for $x > 1$. However, slowly occurring oscillations could not be avoided because of the asymptotic root behavior of the method. For $h = 0.05$, for $x > 1$, the numerical solution converged to the wrong values without exhibiting oscillatory behavior.

The full implicit methods, TR, TR–EX, LW1 and LW3, were most applicable. Even though these methods are all A-stable, the necessity to solve

TABLE 6.5

COMPUTATIONAL RESULTS FOR EXAMPLE 4

Method	h	R_{1n}		R_{2n}		R_{3n}		Time (sec)
		$x = 0.4$	$x = 10$	$x = 0.4$	$x = 10$	$x = 0.4$	$x = 10$	
RK4	0.001	0	0	0	0	0	0	b
DEQ	0.001	b	b	b	b	b	b	b
TM	0.01^a	b	b	b	b	b	b	b
DIFSYS	0.001^c	b	b	b	b	b	b	b
TR	0.2	1.35_{10}^{-3}	1.05_{10}^{-3}	2.12_{10}^{-1}	2.4_{10}^{-1}	9.0_{10}^{-2}	1.5_{10}^{-2}	9.3
TR–EX	0.2	1.72_{10}^{-5}	3.6_{10}^{-4}	3.5_{10}^{-2}	4.3_{10}^{-4}	6.8_{10}^{-4}	1.2_{10}^{-3}	34
CAL	0.005/0.02 at $R=1.0$	2.4_{10}^{-3}	1.01_{10}^{-1}	2.5_{10}^{0}	6.0_{10}^{-1}	1.62_{10}^{-1}	5.4_{10}^{-1}	10
LW1	0.02	1.6_{10}^{-4}	4.9_{10}^{-4}	2.4_{10}^{-4}	1.3_{10}^{-4}	3.2_{10}^{-3}	4.4_{10}^{-4}	20
LW3	0.02	5.9_{10}^{-4}	7.1_{10}^{-5}	2.9_{10}^{-3}	1.1_{10}^{-3}	4.0_{10}^{-2}	1.9_{10}^{-3}	23.3

[a] Automatic step size control.
[b] Unstable.
[c] Initial step size 0.1, extrapolations performed until error $< 10^{-5}$.

nonlinear algebraic equations (see Section 4.3) presented limitations on the size of h (as well, of course, as accuracy considerations). For example, convergence of the Newton–Raphson method used with TR required $h \leq 0.25$. The initial filtering procedure was effective in eliminating oscillations due to inaccuracy for $h \leq 0.1$. The total time for $x = 0$ to $x = 40$ was 9.3 sec. For the increased accuracy of TR–EX, 34 sec were required. Because of the size of the stiff eigenvalues, the exponential fitting procedure in LW1 and LW3 caused a computer overflow since the largest exponential argument that can be handled is 174.673. Thus, the allowable maximum h is 0.02 in each method. Computing times were 20 and 23.3 sec respectively. TR with initial filtering turned out to be the best method for (6.1-2).

From the examples the following conclusion can be drawn:

1. RK4 and DEQ are both highly accurate, DEQ requiring more computation time because of the small absolute stability bound. Because of the small finite stability bound, it is not recommended that either of these methods be used for stiff ODEs.

2. DIFSYS has even less desirable stability properties than RK4 and DEQ, confirming our earlier observation on the poor stability properties of the midpoint rule. Of all the methods used DIFSYS is the least desirable for stiff ODEs.

3. TM with automatic control of h is generally not effective as an all-purpose stiff routine. TM usually decreases the time from that required by RK4 for comparable accuracy, however, in some cases, the automatic step-size control may decrease h to values comparable to RK4. In addition, its utility is limited to those ODE with only large diagonal elements in the Jacobian.

4. The four implicit methods studied, TR, CAL, LW1, and LW3, were roughly comparable in terms of accuracy and computing time, and each resulted in significant savings of time over the explicit methods. TR–EX resulted in the highest accuracy in each example but at the expense of considerable computing time. For systems that are only moderately stiff, CAL is slightly more accurate than the other three. However, for highly stiff systems, LW1, with proper scaling to avoid computer overflows, appears to be the best even though it is the least accurate. If exponential fitting is not used, either the initial filtering procedure or the step length adjustment is necessary to prevent oscillations from inaccurate simulation of the stiff eigenvalues.

5. In general, an A-stable method is necessary for stiff ODEs. In addition to A-stability, the characteristics that a general method for nonlinear stiff ODE would possess are accurate simulation of the stiff eigenvalues (exponential fitting) and explicit or semi-implicit form. The importance of the

latter requirement can be seen if we consider large ($m > 15$) systems of stiff ODEs.

For small systems, we have seen that all the current methods are effective in reducing the computing time for a given level of accuracy over that for conventional methods. For large systems, computing time becomes an acute consideration. Current methods based on analytical solution of the locally linearized equations or a conventional method of solving implicit equations require evaluation of the Jacobian matrix or its eigenvalues. For example, computing times for standard matrix inversion and eigenvalue determination of an $m \times m$ matrix on an IBM 7094 are:

	10	20	30	40	50
inversion, seconds	0.083	0.533	1.684	3.917	7.567
eigenvalues, seconds	1.5	11.2	60	—	—

It is obvious that a method suitable for large systems must not require eigenvalue determination. Implicit methods usually require at least one matrix inversion per step if the Jacobian matrix is reevaluated at each step. Only a few elements of the matrix are likely to change significantly from step to step so it may be possible to update the matrix at each step using only a small number of derivative evaluations.

REFERENCES

1. Amundson, N. R. and Luss, D., Qualitative and quantitative observations on the tubular reactor, *Canad. J. Chem. Eng.* **46**, 425 (1968).
2. Brayton, R. K., Gustavson, R. G., and Liniger, W., A numerical analysis of the transient behavior of a transistor circuit, *IBM J. Res. Develop.* **10**, 292 (1966).
3. Dahlquist, G., A numerical method for some ordinary differential equations with large Lipschitz constants, *Proc. IFIP Congress*, Supplement, Booklet I: 32–36, 1968.
4. Emanuel, G., Problems underlying the numerical integration of the chemical and vibrational rate equations in a near-equilibrium flow, AEDC-TDR-63-82, Aerospace Corp., El Segundo, Calif. (1963).
5. Emanuel, G., Numerical analysis of stiff equations, SSD-TDR-63-380, Aerospace Corp., El Segundo, Calif. (1964).
6. Fowler, M. E. and Warten, R. M., A numerical integration technique for ordinary differential equations with widely separated eigenvalues, *IBM J. Res. Dev.*, **11**, 537 (1967).
7. Gear, C. W., The automatic integration of stiff ordinary differential equations, *Proc. IFIP Congress*, Supplement, Booklet A: 81–85 (1968).
8. Kalman, R. E., Toward a thoery of difficulty of computation in optimal control, *Proc. IBM Scientific Computing Symposium on Control Theory and Application*, 1966.
9. Lawson, J. D., Generalized Runge–Kutta processes for stable systems with large Lipschitz constants, *J. SIAM Numer. Anal.* **4**, 372 (1967).

10. Liniger, W. and Willoughby, R. A., Efficient numerical integration of stiff systems of ordinary differential equations,, *IBM Research Report* RL-1970 (Dec., 1967).

11. Lomax, H. and Bailey, H. E., A critical analysis of various numerical integration methods for computing the flow of a gas in chemical nonequilibrium, *NASA Technical Note*, NASA TN D-4109 (1967).

12. Osborne, M. R., A new method for the integration of stiff systems of ordinary differential equations, *Proc. IFIP Congress*, Math. Booklet A: 86–90, 1968.

13. Ratliff, K., A comparison of techniques for the numerical integration of ordinary differential equations, *Univ. of Illinois Dept. of Computer Science Report*, No. 274, (July, 1968).

14. Richards, R. I., W. D. Lanning and M. D. Torrey, Numerical integration of large highly-damped, nonlinear systems, *SIAM Rev.* **1**, 3 (1965).

15. Robertson, H. H., The solution of a set of reaction rate equations, in " Numerical Analysis," (J. Walsh, ed.) Thompson Book Co., Washington (1967).

16. Treanor, C. E., A method for numerical integration of coupled first order differential equations with greatly different time constants, *Math. Comp.* **20**, 39 (1966).

Index

Mathematics in Science and Engineering

A Series of Monographs and Textbooks

Edited by RICHARD BELLMAN, *University of Southern California*

1. T. Y. Thomas. Concepts from Tensor Analysis and Differential Geometry. Second Edition. 1965

2. T. Y. Thomas. Plastic Flow and Fracture in Solids. 1961

3. R. Aris. The Optimal Design of Chemical Reactors: A Study in Dynamic Programming. 1961

4. J. LaSalle and S. Lefschetz. Stability by by Liapunov's Direct Method with Applications. 1961

5. G. Leitmann (ed.). Optimization Techniques: With Applications to Aerospace Systems. 1962

6. R. Bellman and K. L. Cooke. Differential-Difference Equations. 1963

7. F. A. Haight. Mathematical Theories of Traffic Flow. 1963

8. F. V. Atkinson. Discrete and Continuous Boundary Problems. 1964

9. A. Jeffrey and T. Taniuti. Non-Linear Wave Propagation: With Applications to Physics and Magnetohydrodynamics. 1964

10. J. T. Tou. Optimum Design of Digital Control Systems. 1963.

11. H. Flanders. Differential Forms: With Applications to the Physical Sciences. 1963

12. S. M. Roberts. Dynamic Programming in Chemical Engineering and Process Control. 1964

13. S. Lefschetz. Stability of Nonlinear Control Systems. 1965

14. D. N. Chorafas. Systems and Simulation. 1965

15. A. A. Pervozvanskii. Random Processes in Nonlinear Control Systems. 1965

16. M. C. Pease, III. Methods of Matrix Algebra. 1965

17. V. E. Benes. Mathematical Theory of Connecting Networks and Telephone Traffic. 1965

18. W. F. Ames. Nonlinear Partial Differential Equations in Engineering. 1965

19. J. Aczel. Lectures on Functional Equations and Their Applications. 1966

20. R. E. Murphy. Adaptive Processes in Economic Systems. 1965

21. S. E. Dreyfus. Dynamic Programming and the Calculus of Variations. 1965

22. A. A. Fel'dbaum. Optimal Control Systems. 1965

23. A. Halanay. Differential Equations: Stability, Oscillations, Time Lags. 1966

24. M. N. Oguztoreli. Time-Lag Control Systems. 1966

25. D. Sworder. Optimal Adaptive Control Systems. 1966

26. M. Ash. Optimal Shutdown Control of Nuclear Reactors. 1966

27. D. N. Chorafas. Control System Functions and Programming Approaches (In Two Volumes). 1966

28. N. P. Erugin. Linear Systems of Ordinary Differential Equations. 1966

29. S. Marcus. Algebraic Linguistics; Analytical Models. 1967

30. A. M. Liapunov. Stability of Motion. 1966

31. G. Leitmann (ed.). Topics in Optimization. 1967

32. M. Aoki. Optimization of Stochastic Systems. 1967

33. H. J. Kushner. Stochastic Stability and control. 1967

34. M. Urabe. Nonlinear Autonomous Oscillations. 1967

35. F. Calogero. Variable Phase Approach to Potential Scattering. 1967

36. A. Kaufmann. Graphs, Dynamic Programming, and Finite Games. 1967

37. A. Kaufmann and R. Cruon. Dynamic Programming: Sequential Scientific Management. 1967

38. J. H. Ahlberg, E. N. Nilson, and J. L. Walsh. The Theory of Splines and Their Applications. 1967

39. Y. Sawaragi, Y. Sunahara, and T. Nakamizo. Statistical Decision Theory in Adaptive Control Systems. 1967

40. R. Bellman. Introduction to the Mathematical Theory of Control Processes Volume I. 1967 (Volumes II and III in preparation)

41. E. S. Lee. Quasilinearization and Invariant Imbedding. 1968

42. W. Ames. Nonlinear Ordinary Differential Equations in Transport Processes. 1968

43. W. Miller, Jr. Lie Theory and Special Functions. 1968

44. P. B. Bailey, L. F. Shampine, and P. E. Waltman. Nonlinear Two Point Boundary Value Problems. 1968

45. Iu. P. Petrov. Variational Methods in Optimum Control Theory. 1968

46. O. A. Ladyzhenskaya and N. N. Ural'tseva. Linear and Quasilinear Elliptic Equations. 1968

47. A. Kaufmann and R. Faure. Introduction to Operations Research. 1968

48. C. A. Swanson. Comparison and Oscillation Theory of Linear Differential Equations. 1968

49. R. Hermann. Differential Geometry and the Calculus of Variations. 1968

50. N. K. Jaiswal. Priority Queues. 1968

51. H. Nikaido. Convex Structures and Economic Theory. 1968

52. K. S. Fu. Sequential Methods in Pattern Recognition and Machine Learning. 1968

53. Y. L. Luke. The Special Functions and Their Approximations (In Two Volumes). 1969

54. R. P. Gilbert. Function Theoretic Methods in Partial Differential Equations. 1969

55. V. Lakshmikantham and S. Leela. Differential and Integral Inequalities (In Two Volumes). 1969

56. S. H. Hermes and J. P. LaSalle. Functional Analysis and Time Optimal Control. 1969

57. M. Iri. Network Flow, Transportation, and Scheduling: Theory and Algorithms. 1969

58. A. Blaquiere, F. Gerard, and G. Leitmann. Quantitative and Qualitative Games. 1969

59. P. L. Falb and J. L. de Jong. Successive Approximation Methods in Control and Oscillation Theory. 1969

60. G. Rosen. Formulations of Classical and Quantum Dynamical Theory. 1969

61. R. Bellman. Methods of Nonlinear Analysis, Volume I. 1970

62. R. Bellman, K. L. Cooke, and J. A. Lockett. Algorithms, Graphs, and Computers. 1970

63. E. J. Beltrami. An Algorithmic Approach to Nonlinear Analysis and Optimization. 1970

64. A. H. Jazwinski. Stochastic Processes and Filtering Theory. 1970

65. P. Dyer and S. R. McReynolds. The Computation and Theory of Optimal Control. 1970

66. J. M. Mendel and K. S. Fu (eds.). Adaptive, Learning, and Pattern Recognition Systems: Theory and Applications. 1970

67. C. Derman. Finite State Markovian Decision Processes. 1970

68. M. Mesarovic, D. Macko, and Y. Takahara. Theory of Hierarchial Multilevel Systems. 1970

69. H. H. Happ. Diakoptics and Networks. 1971

70. Karl Astrom. Introduction to Stochastic Control Theory. 1970

71. G. A. Baker, Jr. and J. L. Gammel (eds.). The Padé Approximant in Theoretical Physics. 1970

72. C. Berge. Principles of Combinatorics. 1971

73. Ya. Z. Tsypkin. Adaptation and Learning in Automatic Systems. 1971

74. Leon Lapidus and John H. Seinfeld. Numerical Solution of Ordinary Differential Equations. 1971

In preparation

Harold Greenberg. Integer Programming

E. Polak. Computational Methods in Optimization: A Unified Approach

Leon Mirsky. Transversal Theory

Thomas G. Windeknecht. A Mathematical Introduction to General Dynamical Processes

Andrew P. Sage and James L. Melsa. System Identification

R. Boudarel, J. Delmas, and P. Guichet. Dynamic Programming and Its Application to Optimal Control

William Stenger and Alexander Weinstein. Methods of Intermediate Problems for Eigenvalues Theory and Ramifications